Encyclopaedia of Mathematical Sciences

Volume 32

Editor-in-Chief: R.V. Gamkrelidze

Yu.V. Egorov M.A. Shubin (Eds.)

Partial Differential Equations III

The Cauchy Problem. Qualitative Theory of Partial Differential Equations

Springer-Verlag Berlin Heidelberg GmbH

Scientific Editors of the Series:
A.A. Agrachev, E.F. Mishchenko, N.M. Ostianu, L.S. Pontryagin
Editors: V.P. Sakharova, Z.A. Izmailova

Title of the Russian edition:
Itogi nauki i tekhniki, Sovremennye problemy matematiki,
Fundamental'nye napravleniya, Vol. 32,
Differentsial'nye uravneniya s chastnymi proizvodnymi 3
Publisher VINITI, Moscow 1988

Mathematics Subject Classification (1991):
35A05, 35Jxx, 35Kxx

ISBN 978-3-540-52003-0

Library of Congress Cataloging-in-Publication Data
Differentsial'nye uravneniia s chastnymi proizvodnymi 3. English.
Partial differential equations III / Yu.V. Egorov, M.A. Shubin (eds.).
p. cm.—(Encyclopaedia of mathematical sciences; v. 32)
Translation of: Differentsial'nye uravneniia s chastnymi proizvodnymi 3.
Translated from the Russian by M. Grinfeld.
Includes bibliographical references and indexes.
Contents: The Cauchy problem / S.G. Gindikin, L.R. Volevich—Qualitative theory of second order linear partial differential equations / V.A. Kondrat'ev. E.M. Landis.
ISBN 978-3-540-52003-0 ISBN 978-3-642-58173-1 (eBook)
DOI 10.1007/978-3-642-58173-1
1. Differential equations, Partial. I. Egorov, IU.V. (IUriĭ, Vladimirovich)
II. Shubin, M.A. (Mikhail Aleksandrovich), 1944- .
III. Grinfeld, M. IV. Title. V. Title: Partial differential equations three. VI. Series.
QA377.D6413 1991 515'.353—dc20 90-43176

Originally published by Springer-Verlag Berlin Heidelberg New York in 1991

Typesetting: Asco Trade Typesetting Ltd., Hong Kong
41/3140-543210—Printed on acid-free paper

List of Editors, Authors and Translators

Editor-in-Chief

R.V. Gamkrelidze, Academy of Sciences of the USSR, Steklov Mathematical Institute, ul. Vavilova 42, 117966 Moscow, Institute for Scientific Information (VINITI), ul. Usievicha 20 a, 125219 Moscow, USSR

Consulting Editors

Yu.V. Egorov, Department of Mathematics, Moscow State University, Leninskie Gory, 119899 Moscow, USSR

M.A. Shubin, Department of Mathematics, Moscow State University, Leninskie Gory, 119899 Moscow, USSR

Authors

S.G. Gindikin, A.N. Belozersky, Laboratory of Molecular Biology and Bioorganic Chemistry, Building "A", Moscow State University, Moscow, GSP-234, 119899, USSR

V.A. Kondrat'ev, Moscow State University, Leninskie Gory, 119899 Moscow, USSR

E.M. Landis, Moscow State University, Leninskie Gory, 119899 Moscow, USSR

L.R. Volevich, Keldysh Institute of Applied Mathematics of the Academy of Sciences of the USSR, Miusskaya Sq., 125047 Moscow, USSR

Translator

M. Grinfeld, Department of Mathematics, Bath University, Claverton Down, Bath BA2 7AY, UK

Contents

I. The Cauchy Problem

S.G. Gindikin, L.R. Volevich

Translated from the Russian
by M. Grinfeld

Contents

Introduction

The present paper is a survey of results on the correctness[1] of the Cauchy problem for operators both with constant and with variable coefficients. We mainly concentrate on results that apply to as general operators as possible, in particular, without distinguishing the principal part. Statements that apply for specific classes of operators, such as hyperbolic, parabolic, etc., are discussed only insofar as they illustrate general constructions. Such specific statements may be found in more detail in papers devoted to the particular type of equation. In our selection of results dealing with equations with constant coefficients, we chose the ones at least a significant part of which also covers the case of variable coefficients.

The *Cauchy problem* is a classical problem in partial differential equations: it arises naturally in physical problems. The first examples of partial differential equations appeared in the middle of the XVIIIth century in the framework of mathematical physics. D'Alembert, Euler, and Bernoulli studied the equation of a vibrating string from different points of view. The main goal, as they saw it, was a derivation of a general solution (an integral) of the equation. The primary observation was that while the general solution of an ordinary differential equation depends on some arbitrary constants, a general solution of a partial differential equation should depend on some arbitrary functions. D'Alembert's solution of the string equation contained two arbitrary functions of one (space) variable. The works of the classics of the XVIIIth century contained in an inchoate form many basic ideas of the future theory: characteristics, separation of variables, expansion of a solution in a basis (harmonics).

Having noticed that the arbitrariness in the solution can be removed by fixing the initial position and initial velocity of the string, Euler, in principle, considered the Cauchy problem for the string equation. The famous dispute of Euler and D'Alembert concerning the class of functions in which the string equation should be solved, not only aided in making precise the concept of function in analysis, but also foresaw the general problem of choosing a function space in which an equation should be solved. Contrary to D'Alembert's opinion, who thought that the discussion should be restricted to analytic solutions, Euler maintained that the string can assume any shape, which can be depicted by "a free hand movement" (see below for a discussion of Hadamard's ideas). A significant number of equations in mathematical physics, in particular Laplace's equation and the two- and three-dimensional wave equations, appear in the works of Euler and D'Alembert. In cases other than the string equation, only special solutions were constructed; moreover, Euler considered three-dimensional problems to lie beyond the scope of the analysis of his time.

[1] The term "well-posed" is frequently used instead of "correct" in this context; we use "correct" and "correctness" throughout both for brevity and consistency.

In the XIXth century, mathematical physics continued to be the main source of new partial differential equations and of problems involving them. The study of Laplace's equation and of the wave equation had assumed a more systematic nature. In the beginning of the century, Fourier added the heat equation to the aforementioned two. Marvellous progress in obtaining precise solution representation formulas is connected with Poisson, who obtained formulas for the solution of the Dirichlet problem in a disc, for the solution of the Cauchy problems for the heat equation, and for the three-dimensional wave equation. The physical setting of the problem led to the gradual replacement of the search for a general solution by the study of boundary value problems, which arose naturally from the physics of the problem. Among these, the Cauchy problem was of utmost importance. Only in the context of first order equations, the original quest for general integrals justified itself. Here again the first steps are connected with the names of D'Alembert and Euler; the theory was being intensively developed all through the XIXth century, and was brought to an astounding completeness through the efforts of Hamilton, Jacobi, Frobenius, and E. Cartan.

In terms of concrete equations, the studies in general rarely concerned equations of higher than second order, and at most in three variables. Classification of second order equations was undertaken in the second half of the XIXth century (by Du Bois-Raymond). An increase in the number of variables was not sanctioned by applications, and led to the little understood ultra-hyperbolic case. Increase in the order of the equation required the removal of a significant psychological block: instead of solving the classification problem, the problem of distinguishing different classes of equations had to be attempted.

As far as the general theory is concerned, the first important results in this direction are connected with the name of Cauchy, who, in general was more interested in general problem settings in a century when concrete problems were valued higher in analysis, his results were not widely disseminated, and aroused interest only when they were rediscovered thirty years later by S. Kovalevskaya. The Cauchy-Kovalevskaya theorem states that for a non-characteristic Cauchy problem for a system with analytic coefficients and with analytic initial data, there is (local) existence and uniqueness in the class of analytic functions.

The study of holomorphic solutions to the characteristic Cauchy problem was carried on by Riquier, Janet, and others. In the process, analogs of the Cauchy problem with initial data prescribed on manifolds of different dimension, appeared. A complete theory of such problems (couched in terms of Pfaffian equations) was constructed by E. Cartan and Kähler.

The idea to renounce analyticity in the study of the Cauchy problem (or its generalizations) appeared comparatively early: it was tempting to substitute smoothness for analyticity in the Cauchy-Kovalevskaya theorem. Moreover, because of Weierstrass' work, it was very much in vogue to approximate continuous functions by analytic ones, especially by polynomials. An additional argument in favour of this approach was provided by the Holmgren theorem, which under the conditions of the Cauchy-Kovalevskaya theorem gives uniqueness in the class of functions with a finite number of continuous derivatives. Hadamard showed that a naïve strategy of passing from analytic functions to smooth ones cannot in general be successful, as nearby initial conditions can correspond to substantially differing solutions. Using the Cauchy problem for Laplace's equation as an example, Hadamard showed that an arbitrarily small perturbation of initial data can lead to an arbitrarily large perturbation of the solution. To this example, Hadamard added the remark that no physical problem known at that time led to the Cauchy problem for elliptic equations. In this context, Hadamard [1932] introduced the concept of a *correct* (*well posed*) *initial* or *boundary value problem* for partial differential equations. A problem is called correct if it has unique solution that depends continuously on the initial data of the problems, in a class of functions of finite smoothness.

Hadamard [1932] proved correctness of the Cauchy problem for general linear hyperbolic second order equations. In the case of constant coefficients the corresponding results followed from the work of Volterra and his students. The case of variable coefficients and two variables was considered before by Riemann, but his results did not become widely known. Hadamard suggested a generalisation of Riemann's method to the case of many variables, which allowed him to construct the fundamental solution for the Cauchy problem for a second order hyperbolic equation (the Riemann function), and, using it, to express the solution of a linear second order hyperbolic equation in terms of the right hand side and the Cauchy data. As the existence of a Riemann function was proved for small times, Hadamard obtained formulas for the solution in a small (but independent of the initial data and of the right hand side) time interval. Additional applications of this procedure made it possible to construct the solution on any finite time interval. However, at each step "loss of smoothness" of the solution, which is characteristic of hyperbolic equaitrons, occurred. Thus, in order to ensure existence of a classical $C^{(2)}$ solution on a time interval of length T, a high degree of smoothness (depending on T) had to be assumed.

The use of energy estimates enabled Schauder and Sobolev [1950] to get rid of the exaggerated smoothness assumptions. Schauder established solubility of the Cauchy problem by approximating the coefficients by analytic functions and using the Cauchy-Kovalevskaya theorem. Sobolev used methods similar to the ones used by Hadamard. Considering generalized solutions in classes of functions that are square summable together with their derivatives (Sobolev spaces), Sobolev obtained sharp regularity results for solutions of hyperbolic equations.

Another approach to the Cauchy problem, based on direct methods (finite differences) and estimates of the energy integral, was suggested by Courant, Friedrichs, and Lewy [1928].

Some other methods in the general theory were being developed from the start of the century. The connection between equations with constant coefficients and the algebraic properties of their symbols, in particular using operational calculus, was being mooted. This led to some explicit formulas for solutions, viz. by Fredholm, Zeilon, Herglotz. On the other hand, general methods for equations with variable coefficients were being invented, notably the parametrix method of Picard, Hilbert, Hendrick, and Levi (see Levi [1907]).

The work of Petrovskij in the thirties proved to be of great significance. As far as the Cauchy problem is concerned, Petrovskij tried to describe systems for which the Cauchy problem is correct in the sense of Hadamard. In the case of constant coefficients, he succeeded in obtaining quite effective final results.

Petrovskij [1938] considered the Cauchy problem for a general system with coefficients that depend only on time,

$$\frac{\partial^{n_i} u_i}{\partial t^{n_i}} = \sum_{j=1}^{N} \sum A_{ij}^{(k_0,\ldots,k_n)}(t) \frac{\partial^{k_0+\cdots+k_n} u_j}{\partial t^{k_0} \ldots \partial_{x_n}^{k_n}}, \tag{1}$$

$$k_0 < n_j, \quad i = 1, \ldots, N,$$

$$\left.\frac{\partial^l u_i}{\partial t^l}\right|_{t=0} = \varphi_{il}(x), \quad i = 1, \ldots, N, l = 0, \ldots, n_i - 1. \tag{2}$$

The main result is as follows: problem (1), (2), is correct in $C^{(k)}$ on the interval $[0, T]$ if and only if condition (A) is satisfied: the system of ordinary differential equations

$$\frac{d^{n_i} v_i}{dt^{n_i}} = \sum_{j=1}^{N} \sum A_{ij}^{(k_0,\ldots,k_n}(t)(i\xi_1)^{k_1} \ldots (i\xi_n)^{k_n} \frac{\alpha^{k_0} v_j}{dt^{k_0}}, \tag{3}$$

obtained from (1) by a formal Fourier transform with respect to the space variables, has a fundamental matrix of solutions which grows in ξ not faster than some power of $|\xi|$.

In the case of constant coefficients, condition (A) is equivalent to requiring that imaginary parts of the roots of the algebraic equation

$$\det \left\| (i\lambda)^{n_i} \delta_{ij} - \sum_{k_0,\ldots,k_n} A_{ij}^{(k_0,\ldots,k_n}(i\lambda)^{k_0}(i\xi_1)^{k_1} \ldots (i\xi_n)^{k_n} \right\| = 0$$

go to $-\infty$ at least as fast as $\text{const} \ln(1 + |\xi|)$. As we shall see below, this is equivalent to asking that the imaginary parts of the roots are uniformly (in ξ) bounded from below by a constant. Such systems became known as correct in the sense of Petrovskij.

Let us explicitly state these conditions for a scalar equation. For the operator $P\left(\frac{1}{i}\frac{\partial}{\partial t}, \frac{1}{i}\frac{\partial}{\partial x_1}, \ldots, \frac{\partial}{\partial x_n}\right)$ the *correctness conditions in the sense of Petrovskij* reduce to the following conditions on the symbol $P(\tau, \xi)$: (i) P can be solved with respect to the highest power of τ, that is $P = \tau^k + Q$, $\deg_\tau Q \leqslant k - 1$; (ii) there exists γ such that $P(\tau, \xi) \neq 0$ for $\operatorname{Im} \tau < \gamma$, $\xi \in \mathbb{R}^n$.

In the case of t-dependent coefficients, the fundamental matrix of (3) is found by integrating the system of ordinary differential equations, and condition (A) in general is not formulated in terms of algebraic conditions on the coefficients. Petrovskij singled out two classes of systems: strictly hyperbolic systems (which generalize the wave equation), and $2b$-parabolic ones (which generalize the heat equation), for which condition (A) necessarily holds for t-dependent coefficients, so that we have correctness of the Cauchy problem.

At the same time, Petrovskij [1937] studied the Cauchy problem for strictly hyperbolic equations with variable coefficients. As regards $2b$-parabolic equations, it turned out later that a theory of the Cauchy problem for variable coefficients can also be constructed (see Ladyzhenskaya [1950], Ejdel'man [1964], Ladyzhenskaya, Solonnikov, and Uraltseva [1967]).

At the end of the forties, Schwarz [1950/51] reformulated Petrovskij's results in the language of the then nascent distribution theory. He showed that Petrovskij's correctness condition is a necessary and sufficient correctness condition for the Cauchy problem in the space of distributions of slow growth that depend smoothly on time. In addition Schwarz noted that differential operators with respect to space variables in Petrovskij's results could be replaced by convolution operators.

The development of distribution (generalized functions) theory was accompanied by a study of Cauchy problems for systems with constant coefficients in very general settings. This refers first of all to works dealing with exponential uniqueness classes and correctness of the Cauchy problem. Starting with the work of Holmgren, Tikhonov, Täcklind, and Ladyzhenskaya on parabolic equations, function spaces with exponential growth in which the Cauchy problem is soluble (correctness classes) or has a unique solution (uniqueness classes) were investigated for different classes of solutions and systems. Gel'fand and Shilov described exponential uniqueness classes for general systems with constant coefficients, while Shilov and his students considered correctness questions with the same degree of generality. In this survey, we do not discuss questions that are specific to systems with constant coefficients. Their detailed exposition can be found in the monograph of Gel'fand and Shilov [1958]. The result of Hörmander [1983–1985] (Ch. XII) on the correctness of an equation in D', is also beyond the scope of this survey.

The theory of generalized functions (distributions), the method of *a-priori* estimates, and the theory of pseudodifferential operators, have all found their reflection in the present state of the general theory of the Cauchy problem. In the framework of a general theory for the constant coefficient case, it is now natural to work in terms of convolution equations, while in the variable coefficient case it is often natural to conduct the discussion in the general framework of pseudodifferential operators. Both these circumstances have been taken into account in the present survey.

The survey consists of three parts. Chapter 1 and 2 are devoted to the Cauchy problem for differential equations with constant coefficients and for more general convolution equations. In the first chapter we have collected questions relating to the Cauchy problem in spaces of slowly growing and rapidly decreasing functions and distributions, while in Chapter 2 analogous questions for spaces of exponentially growing or decaying functions and distributions (exponential correctness classes) are discussed. Chapter 3 is devoted to extending a number of results obtained for the constant coefficient case, to the case of variable coefficients.

We use consecutive numeration of formulas in each chapter. Referring to a formula from inside the same chapter, we give only its number, while if reference is made to a formula from a different chapter, we also give the number of the chapter. Thus (1.20) means formula (20) from Chapter 1.

In each section we, as a rule, have one central claim (a theorem, proposition, etc.) Then, referring to, say, the theorem from Section 2.3 we write Theorem 2.3. If there is more than one theorem (proposition) in a section, we shall indicate the number of the theorem (proposition) and of the section.

Chapter 1
The Cauchy Problem in Slowly Increasing and Rapidly Decreasing Functions and Distributions (Constant Coefficients)

Schwarz's [1950] book on the theory of distributions appeared in 1950. The language of the theory of tempered distributions proved to be exceptionally well-suited both for the formulation of Petrovskij's theory itself, and for its extension (due to Schwarz) to the case of initial data with polynomial growth. In essence, the present chapter is a modernized exposition of Petrovskij's theory, using the language of distribution theory. We also discuss some natural generalizations of the theory and related topics. Proofs of the principal proposition of this chapter can be found in Volevich and Gindikin [1972]. As it turns out, most of the results of Petrovskij's theory do not rely on specific features of differential operators. In view of that, we conduct our discussion in the framework of a wider class of equations, namely, convolution equations.

All the results of this and the following chapter are obtained with the help of the same construction. This construction is best explained using a simple example not directly connected with the Cauchy problem, namely, the question of solubility of convolution (and in particular, differential) equations in Schwarz spaces on all of $\mathbb{R}^n$.

§ 1. Convolution Equations in Schwarz Spaces on $\mathbb{R}^n$

1.1. Schwarz Spaces of Test Functions. The space $\mathscr{S}(\mathbb{R}^n)$ consists of infinitely differentiable functions which decrease faster than any power of $|x|$ as $|x| \to \infty$ together with all their derivatives.

The precise meaning of this statement is that $\phi \in \mathscr{S}$ has finite Hölder norms

$$|\varphi|_{(l)}^{(s)} = \sup_{x \in \mathbb{R}^n, |\alpha| \leqslant s} (1 + x^2)^{l/2} |D^\alpha \varphi(x)|. \tag{1}$$

Here $s \geqslant 0$ is an integer, l is a real number, $\mathscr{D}^\alpha = \left(\frac{1}{i}\frac{\partial}{\partial x_1}\right)^{\alpha_1} \cdots \left(\frac{1}{i}\frac{\partial}{\partial x_n}\right)^{\alpha_n}$. Let us denote by $C_{(l)}^{(s)}$ the Banach space of s-times continuously differentiable functions with norm (1). Then we have the continuous embeddings

$$C_{(l)}^{(s)} \subset C_{(l')}^{(s')}, \quad s \geqslant s', \quad l \geqslant l',$$

that is, $C_{(l)}^{(s)}$ form a scale of Banach spaces which we denote by $\{C_{(l)}^{(s)}\}$. The space $\mathscr{S}$ is the projective limit of this scale:

$$\mathscr{S} = C_{(\infty)}^{(\infty)} \overset{\text{def}}{=} \bigcap_{s,l} C_{(l)}^{(s)}(\mathbb{R}^n).$$

$\mathscr{S}$ is a countably normed space, and therefore a Fréchet space.

By definition, continuity of an operator $A\colon \mathscr{S} \to \mathscr{S}$ means that for any sufficiently large s, l there exist s', l', such that estimate

$$|A\varphi|_{(l)}^{(s)} \leqslant \text{const}\, |\varphi|_{(l')}^{(s')}. \tag{2}$$

holds. Let us remark that though $\mathscr{S}$ is not dense in $C_{(l')}^{(s')}$, the closure of $\mathscr{S}$ in the $|\ |_{(l')}^{(s')}$ norm will contain $C_{(l'')}^{(s')}$ for any $l'' \geqslant l'$. Therefore the operator A can be extended as a continuous operator from $C_{(l'')}^{(s')}$ to $C_{(l)}^{(s)}$. This means (again, by definition) that A acts continuously on the scale $\{C_{(l)}^{(s)}\}$. Thus, a continuous operator A on $\mathscr{S}$ is automatically extended to the scale $\{C_{(l)}^{(s)}\}$.

Let us note that the theory of Schwarz spaces can be formulated equivalently either in terms of projective limits of Banach spaces or in terms of Banach scales. It is frequently convenient to pass from one viewpoint to the other.

By taking inductive and projective limits, the space $C_{(l)}^{(s)}$ can be used to construct further spaces, which are important in what follows:

$$\mathscr{O} = \bigcup_l C_{(l)}^{(\infty)} = \bigcup_l \bigcap_s C_{(l)}^{(s)}, \tag{3}$$

$$\mathscr{M} = \bigcap_s C_{(-\infty)}^{(s)} = \bigcap_s \bigcup_l C_{(l)}^{(s)}. \tag{3'}$$

The space $\mathscr{O}$ consists of C^∞-functions, all the derivatives of which grow as $|x| \to \infty$ not faster than some power of $|x|$ which depends only on the function. The space $\mathscr{M}$ consists of C^∞-functions, each derivative of which has its own polynomial growth rate. Clearly, $\mathscr{O} \subset \mathscr{M}$. The example of the function $\exp(i|x|^2) \in \mathscr{M}$ shows that the inclusion is proper.

The space $\mathscr{M}$ is a ring with respect to multiplication. If $a \in \mathscr{M}$ and $\varphi \in \mathscr{S}$, then $a(x)\varphi(x) \in \mathscr{S}$, that is, elements of $\mathscr{M}$ are *multipliers* in $\mathscr{S}$.

In any of the spaces $C_{(l)}^{(s)}$, $l > n$, we can define the classical *Fourier transform*

$$\mathscr{F}\colon \varphi(x) \mapsto \tilde{\varphi}(\xi) = (2\pi)^{-n/2} \int \exp(-ix\cdot\xi)\varphi(x)\, dx,$$

which maps this space into $C_{(s)}^{(l')}$, $l' + n < l$. Moreover, we have the estimate

$$|\tilde{\varphi}|_{(s)}^{(l')} \leqslant \text{const}\, |\varphi|_{(l)}^{(s)}, \quad l' + n < l, \tag{4}$$

which it is natural to call "*Parseval's inequality*" for Hölder norms. Since similar estimates can be obtained for the inverse Fourier transform $\mathscr{F}^{-1}$, we deduce from this that the Fourier transform is an automorphism of $\mathscr{S}$, that is

$$\mathscr{F}\mathscr{S} = \mathscr{S}. \tag{5}$$

In view of the above, there corresponds to every symbol $a(\xi) \in \mathscr{M}$ a *pseudo-differential operator* (PDO) $a(D)$ which maps the space $\mathscr{S}$ into itself:

$$\begin{aligned}(a(D)\varphi)(x) &= \mathscr{F}^{-1}_{\xi\to x}a(\xi)\mathscr{F}_{x\to\xi}\varphi \\ &= (2\pi)^{-n/2}\int \exp(ix\cdot\xi)a(\xi)\hat{u}(\xi)\,d\xi.\end{aligned}$$

1.2. $\mathscr{S}$ Considered as a Countably Hilbert Space. A most important property of $\mathscr{S}$ is that it can be regarded not only as a countably normed, but also as a countably Hilbert space.

As $(1 + |\xi|^2)^{s/2} \in \mathscr{M}$, the PDO $(1 + |\mathscr{D}|^2)^{s/2}$ is defined on $\mathscr{S}$, and we can consider on $\mathscr{S}$ the family of Hilbert norms

$$\|\varphi\|^{(s)}_{(l)} = \|(1 + |x|^2)^{l/2}(1 + |D|^2)^{s/2}\varphi\|, \tag{6}$$

for any l, s, where $\|\cdot\|$ is the usual L_2 norm on $\mathbb{R}^n$. Let us denote by $H^{(s)}_{(l)}$ the Hilbert spaces obtained as completions of $\mathscr{S}$ in the norm (6). We have the continuous embeddings:

$$H^{(s+k)}_{(l)} \subset C^{(s)}_{(l)} \subset H^{(s)}_{(l-m)}, \quad k, m > \frac{n}{2}, \quad s \geqslant 0$$

where the first result follows from the classical Sobolev embedding theorem, and the second one is trivially true. Hence it follows that

$$\begin{gathered}\mathscr{S} = H^{(\infty)}_{(\infty)} \overset{\text{def}}{=} \bigcap_{s,l} H^{(s)}_{(l)}, \quad \mathscr{O} = \bigcup_{l} H^{(\infty)}_{(l)} = \bigcup_{l}\bigcap_{s} H^{(s)}_{(l)}, \\ \mathscr{M} = \bigcap_{s} H^{(s)}_{(-\infty)} = \bigcap_{s}\bigcup_{l} H^{(s)}_{(l)}.\end{gathered} \tag{7}$$

The scale $\{H^{(s)}_{(l)}\}$ is self-conjugate, that is

$$(H^{(s)}_{(l)})' = H^{(-s)}_{(-l)}. \tag{8}$$

This property is an important advantage of Hilbert norms as compared to the Hölder norms.

In view of (7) and (8), the conjugate spaces $\mathscr{S}'$ and $\mathscr{O}'$ can be defined in terms of inductive and projective limits:

$$\mathscr{S}' = H^{(-\infty)}_{(-\infty)} \overset{\text{def}}{=} \bigcup_{s,l} H^{(s)}_{(l)}, \quad \mathscr{O}' = \bigcap_{l} H^{(-\infty)}_{(l)} = \bigcap_{l}\bigcup_{s} H^{(s)}_{(l)}. \tag{7'}$$

This equality is to be understood as an algebraic isomorphism of linear spaces. As far as topological isomorphism is concerned, it can be proved for the spaces $\mathscr{S}$ and $\mathscr{S}'$, $\mathscr{S}$ being a projective limit of Hilbert spaces. At the same time, the question of topology on $\mathcal{O}$ and $\mathcal{O}'$ is quite complicated, and we do not consider it here. In this context we call an operator $A\colon \mathcal{O} \to \mathcal{O}$ with domain and range in $\mathcal{O}$, continuous, if it admits the following estimates: $\forall l\ \exists\ l'$, and $\forall s\ \exists\ s'$ such that

$$\|A\varphi\|_{(l')}^{(s)} \leqslant \text{const}\ \|\varphi\|_{(l)}^{(s')} \quad \forall \varphi \in \mathcal{S}.$$

Let us remark that the continuity condition above is stronger than the continuity condition for an operator A on the inductive limit of countably normed spaces $H_{(l)}^{(\infty)}$, $l \in \mathbb{R}$.

If the operator A maps $\mathcal{O}'$ into itself, and for the choice of s, l, s', l' above

$$\|A\varphi\|_{(l)}^{(s')} \leqslant \text{const}\ \|\varphi\|_{(l')}^{(s)} \quad \forall \varphi \in \mathcal{S},$$

we shall call A a continuous operator from $\mathcal{O}'$ into $\mathcal{O}'$.

For comparison, we remind the reader that continuity of operator $A\colon \mathcal{S} \to \mathcal{S}$ is determined by estimates (2), or by analogous estimates in the $\|\ \|_{(l)}^{(s)}$ norms, while continuity of an operator $A\colon \mathcal{S}' \to \mathcal{S}'$ is determined by availability of the following estimates: $\forall s, l\ \exists\ s', l'$ such that

$$\|A\varphi\|_{(l')}^{(s')} \leqslant \text{const}\ \|\varphi\|_{(l)}^{(s)}.$$

Elements of $\mathcal{S}'$ are called distributions (generalized functions) of temperate (polynomial or power law) growth, or just tempered distributions, while the elements of $\mathcal{O}'$ are called rapidly decreasing distributions.

On the space $\mathcal{S}'$ we can define the Fourier transform by conjugation; moreover in view of (5)

$$\mathcal{F}\mathcal{S}' = \mathcal{S}'. \tag{5'}$$

It can be shown that the norm $\|\ \|_{(l)}^{(s)}$ of (6) is equivalent to the norm $'\|\ \|_{(l)}^{(s)}$, in which PDO $(1 + |\mathcal{D}|^2)^{s/2}$ and the operator of multiplication by $(1 + |x|^2)^{l/2}$ exchange places. For this pair of norms we have the "*Parseval equality*" (compare with the Parseval inequality from the previous section):

$$'\|\hat{\varphi}\|_{(s)}^{(l)} = \|\varphi\|_{(l)}^{(s)}.$$

Hence it follows that

$$\mathcal{F} H_{(l)}^{(s)} = H_{(s)}^{(l)}.$$

From (7), (8) and these equalities we deduce that

$$\mathcal{F}\mathcal{O}' = \mathcal{M}. \tag{9}$$

In conclusion, let us emphasize again that the availability of equivalent families of Hölder ($|\ |_l^s$) and Hilbert ($\|\ \|_{(l)}^{(s)}$) norms is the main advantage of working with the spaces $\mathcal{S}$ and $\mathcal{S}'$ and with the scales corresponding to these spaces.

1.3. Convolution Operators

Definition. *A convolution operator* on the space $\mathcal{S}$ (or $\mathcal{O}$, $\mathcal{S}'$, $\mathcal{O}'$) is any continuous operator that commutes with shifts.

Proposition. *Let $\Phi = \mathscr{S}, \mathscr{O}$. For every convolution operator A on Φ there exists a distribution $f \in \Phi'$ such that*

$$(A\varphi)(x) = (f, IT_x\varphi), \tag{10}$$

where $(I\phi)(x) = \phi(-x)$, $(T_x\phi)(y) = \phi(x + y)$, and (f, ψ) is the value of the functional $f \in \Phi'$ acting on the test function $\psi \in \Phi$.

Conversely, if for each $\phi \in \Phi$ the right-hand side of (10) *is a function in Φ for $f \in \Phi'$,* (10) *defines a convolution operator.*

In the case of a functional defined by a function, (10) *has the form of the usual convolution:*

$$(f * \varphi)(x) = \int_{\mathbb{R}^n} f(y)\varphi(x - y)\, dy.$$

Let us sketch the proof of the proposition. We set a correspondence between any continuous operator A on Φ and a family of distributions $f_x \in \Phi'$ where $(f_x, IT_x\varphi) = (A\varphi)(x)$. (Here we use the fact that $\Phi = \mathscr{S}, \mathscr{O}$ are invariant with respect to the shifts T_x and reflections I.) It can be directly verified that commutativity of A with shifts is equivalent to the distributions f_x being independent of x.

As regards the converse statement, note that the right-hand side of (10) defines a closed operator on $\mathscr{S}, \mathscr{O}'$. Then continuity of this operator follows from the closed graph theorem, which holds in Fréchet spaces (and thus, in $\mathscr{S}$) and in their inductive limits (and thus in $\mathscr{O}$).

Under the conditions of the proposition, let us call A the convolution operator corresponding to the *convolutor* f, using the notation

$$A\varphi = \mathrm{con}_f\, \varphi = f * \varphi. \tag{10'}$$

We denote the set of convolutors by $\mathfrak{L}(\Phi)$.

Obviously, convolution operators on $\mathscr{S}'$, $\mathscr{O}'$ are the transpose operators of convolution operators on spaces of test functions $\mathscr{S}, \mathscr{O}$. Let us denote the operator on $\Phi' = \mathscr{S}', \mathscr{O}'$ which is the transpose of $\mathrm{con}_f\, f \in \mathfrak{L}(\mathscr{S})$ by con_{If}. This is consistent notation, since on $\Phi \cap \Phi'$ this operator coincides with the operator $\mathrm{con}_{If}\colon \Phi \to \Phi$. Since $\mathscr{S}$ and $\mathscr{O}$ are invariant with respect to reflections I,

$$\mathfrak{L}(\Phi) = \mathfrak{L}(\Phi'), \quad \Phi = \mathscr{S}, \mathscr{O}. \tag{11}$$

It is clear that composition of convolution operators con_f and con_g, $f, g \in \mathfrak{L}(\Phi)$, $\Phi = \mathscr{S}, \mathscr{O}$, commutes with shifts and therefore, in view of the proposition, is a convolution operator con_h for some $h \in \mathfrak{L}(\Phi)$. On the other hand, as we have just pointed out, for $f \in \mathfrak{L}(\Phi)$ and $g \in \Phi'$ the convolution $f * g \in \Phi'$can be defined. It is immediately verified that $f * g$ and h coincide; that is, we have

Lemma. *Let $f, g \in \mathfrak{L}(\Phi)$, where $\Phi = \mathscr{S}, \mathscr{O}$. Then $\mathrm{con}_f\, \mathrm{con}_g = \mathrm{con}_{f*g}$.*

1.4. Description of Convolutors

Theorem.

$$\mathfrak{L}(\mathscr{S}) = \mathfrak{L}(\mathscr{S}') = \mathfrak{L}(\mathscr{O}) = \mathfrak{L}(\mathscr{O}') = \mathscr{O}'. \tag{12}$$

Sketch of the proof. To prove all the equalities of (12) it is enough to demonstrate the inclusions

$$\mathcal{O}' \subset \mathfrak{L}(\mathcal{S}), \quad \mathcal{O}' \subset \mathfrak{L}(\mathcal{O}'), \tag{12'}$$

$$\mathfrak{L}(\mathcal{S}) \subset \mathcal{O}'. \tag{12''}$$

In fact, then $\mathfrak{L}(\mathcal{S}) = \mathcal{O}'$, and by (11) $\mathfrak{L}(\mathcal{S}') = \mathcal{O}'$. By definition, $\mathfrak{L}(\mathcal{O}) \subset \mathcal{O}'$, and then $\mathfrak{L}(\mathcal{O}') \subset \mathcal{O}'$. The two final equalities of (12) follow then by (12′) and (11). Let us outline the proof of (12′). Let $f \in \mathcal{O}'$, $\hat{f}(\xi) \in \mathcal{M}$, where $\hat{f}$ is the Fourier transform of f. Then $\hat{f}$ is a multiplier in $\mathcal{S}$ and in $\mathcal{M}$, while the PDO $\hat{f}(D)$ maps both $\mathcal{S}$ and $\mathcal{O}'$ into itself. Clearly, $\hat{f}(D)$ is a convolution operator. It can be directly shown that the convolutor that corresponds to it coincides with f, that is $f \in \mathfrak{L}(\mathcal{S})$, $f \in \mathfrak{L}(\mathcal{O}')$.

The idea of the proof of (12″) is the following: every operator $\mathrm{con}_f\colon \mathcal{S} \to \mathcal{S}$, $f \in \mathfrak{L}(\mathcal{S})$ can be extended by continuity to a continuous operator $\mathrm{con}_f\colon \mathcal{O}' \to \mathcal{O}'$. Then, using the fact that $\delta(x) \in \mathcal{O}'$, and that $f * \delta = f$, we obtain (12″).

Continuity of $\mathrm{con}_f\colon \mathcal{S} \to \mathcal{S}$ means (see Section 1.1.) that $\forall s, l$ we can find s', l' such that

$$\|f * \varphi\|_{(l)}^{(s)} \leqslant \mathrm{const}\|\varphi\|_{(l')}^{(s')}.$$

Using the description (7′) of the spaces $\mathcal{S}'$ and $\mathcal{O}'$, it can be checked that the operators con_f, $f \in \mathfrak{L}(\mathcal{S})$ commute with PDO $a(D)$, $a(\xi) \in \mathcal{M}$, and in particular with the PDO $(1 + |D|^2)^{s/2}$ which grade the scale $\{H_{(l)}^{(s)}\}$ according to smoothness. In view of this, we could choose $l' = \lambda(l)$, $s' = s + a(l)$. But then in the last inequality s and s' can be interchanged, that is, $\forall\, s, l\; \exists\, s', l'$ such that

$$\|f * \varphi\|_{(l)}^{(s')} \leqslant \mathrm{const}\|\varphi\|_{(l')}^{(s)}.$$

These estimates mean precisely that the operator $\mathrm{con}_f\colon \mathcal{O}' \to \mathcal{O}'$ is continuous.

Thus, the space $\mathcal{O}'$ of rapidly descreasing distributions is simultaneously the space of convolutors both for rapidly decreasing functions (distributions) and for temperately growing ones.

We note that the differential operator $P(D)$ with constant coefficients is the operator of convolution with the distribution $P(D)\delta(x) \in \mathcal{O}'$, and the shift operator T_h is the operator of convolution with $\delta(x + h)$.

Let us emphasize that the ability to provide an explicit description of convolutors is one of the most remarkable properties of Schwarz spaces.

1.5. Convolution Equations on $\mathbb{R}^n$

Theorem. *Let $\Phi = \mathcal{S}, \mathcal{O}, \mathcal{S}', \mathcal{O}'$, and $A \in \mathcal{O}'$ (that is, $A \in \mathfrak{L}(\Phi)$). Then the following conditions are equivalent*:

(I) *For every $\psi \in \Phi$, the convolution equation*

$$A * \varphi = \psi \tag{13}$$

has a unique solution $\varphi \in \Phi$.

(II) *Equation* (13) *has the fundamental solution* $G \in \mathcal{O}'$:

$$A * G = G * A = \delta(x). \tag{14}$$

For $\Phi = O' \ni \delta(x)$ condition (II) follows from (I) trivially.

From condition (I) and from the Banach bounded inverse theorem, we deduce continuity of the operator

$$(\mathrm{con}_A)^{-1}: \Phi \to \Phi, \quad \Phi = \mathcal{S}, \mathcal{O}, \mathcal{S}'. \tag{15}$$

It is clear that this operator commutes with shifts, that is, it is a convolution operator: $(\mathrm{con}_A)^{-1} = \mathrm{con}_G$ for some $G \in \mathcal{O}'$. Using Lemma 1.3 we arrive at (14). Using Lemma 1.3 and commutativity of convolutions in $\mathcal{O}'$, we deduce (I) from (II).

In the number of equivalent conditions of the theorem we can also include (II'), For some $c > 0$ and μ, the symbol $\hat{A}(\xi)$ is bounded from below:

$$|\hat{A}(\xi)| > c(1 + |\xi|)^{\mu} \quad \forall \xi \in \mathbb{R}^n. \tag{16}$$

In fact, condition (II) means that the distribution A is an invertible element of the ring (with respect to convolution) $\mathcal{O}'$. Isomorphism (9) is a ring isomorphism. Then (II) is equivalent to the invertibility of the symbol $\hat{A}$ in the ring (with respect to multiplication) $\mathcal{M}$, that is, each derivative $D^{\alpha}\hat{A}^{-1}$ must grow not faster than some power of ξ. An elementary computation shows that under the condition $\hat{A}(\xi) \in \mathcal{M}$, the latter statement is equivalent to the existence of the bound (16).

Remarks. 1) Deciphering the conditions for continuity of the operator (15) for $\Phi = \mathcal{S}, \mathcal{O}, \mathcal{S}', \mathcal{O}'$, we can replace (I) by statements about continuity of the operator

$$(\mathrm{con}_A)^{-1}: C^{(s)}_{(l)} \to C^{(s')}_{(l')}, \quad H^{(s)}_{(l)} \to H^{(s')}_{(l')}$$

for some set of s, s', l, l', that is, by statements about the solubility of (13) in spaces of functions (distributions) of finite smoothness and of fixed growth (decay) at infinity.

2) The operators con_A, $A \in \mathcal{O}'$ map the spaces $C^{(\infty)}_{(l)}$, $H^{(\pm\infty)}_{(l)}$ into themselves for all $l \in \mathbb{R}$ (this is connected to the fact that distributions in $\mathcal{O}'$ decay faster than any power). The operator con_A acts on the scales $C_{(l)} = \{C^{(s)}_{(l)}\}$, $H_{(l)} = \{H^{(s)}_{(l)}\}$ as an operator of finite order, that is, there exists a $d(l)$ such that for all s the operator

$$\mathrm{con}_A: C^{(s)}_{(l)} \to C^{(s-\varkappa)}_{(l)}, \quad H^{(s)}_{(l)} \to H^{(s-\varkappa)}_{(l)}, \quad \varkappa > d(l),$$

is continuous. Invertibility of A in the ring $\mathcal{O}'$ means that the inverse operator is also a finite order operator, that is, there exists a function $\delta(l)$ such that (13) has a solution $\varphi \in C^{(s+\varkappa)}_{(l)}$ or $\varphi \in H^{(s+\varkappa)}_{(l)}$, $\varkappa < \delta(l)$, for any right-hand side $\psi \in C^{(s)}_{(l)}$, $\psi \in H^{(s)}_{(l)}$.

1.6. Differential Equations on $\mathbb{R}^n$. As we have already said above, a differential operator $P(D)$ with constant coefficients is a convolution operator with distribution $P(D)\delta(x) \in \mathcal{O}'$, and Theorem 1.5 can be applied to it. Moreover, condition (16) is simplified. The reason for this is that, according to the Seidenberg-Tarski theorem (see for example the appendix in Hörmander [1963]) for the polynomial $\hat{A}(\xi) = P(\xi)$, the inequality (16) follows from the fact that $\hat{A}(\xi) \neq 0$. Thus, we have

Theorem. *The differential equation with constant coefficients*

$$P(D)\varphi(x) = \psi(x) \tag{17}$$

is uniquely soluble in any of the spaces $\mathscr{S}$, $\mathcal{O}$, $\mathscr{S}'$, $\mathcal{O}'$ if and only if the symbol $P(\xi)$ does not vanish anywhere on $\mathbb{R}^n$.

1.7. Some General Remarks. In the results we described above, we encounter a state of affairs that is going to recur frequently. It so happens that in some function spaces there is a natural (either constructive or axiomatic via invariance conditions) definition of convolutors. It turns out, furthermore, that the space of convolutors can be characterized more or less effectively. Moreover, solubility of equations in convolutions turns out to be equivalent to the existence of the fundamental solution, which is a convolutor. Let us note that many solubility proofs are in essence based on obtaining estimates of the fundamental solution, which ensure that it really is a convolutor in the corresponding space of right hand sides or of the initial data. Basically, that is exactly what Petrovskij [1938] does in his work on the correctness of the Cauchy problem.

Furthermore, in view of the available description of convolutors, the above mentioned solubility condition becomes an effective one. In this context, the Seidenberg-Tarski theorem allows us in the case of differential operators to simplify the condition due to the fact that the necessary power estimates are automatically satisfied.

Let us remark on another important point, which occurs in the theory we described above: solubility conditions in scales coincide with these in "limit spaces" (in our case, in $\mathscr{S}$, $\mathcal{O}$, $\mathscr{S}'$, $\mathcal{O}'$).

Our discussion in this section was considerably simplified by the fact that the functional space was invariant with respect to shifts. In the following sections dealing with the Cauchy problem, we shall have invariance with respect to an "incomplete" set of shifts, which, however, is sufficient to follow through with the procedure indicated above.

§2. The Homogeneous Cauchy Problem in Rapidly Decreasing (Slowly Increasing) Functions (Distributions)

Let $P(D)$ be a differential operator of mth order with respect to differentiation in x_1. As we already stated above, the Cauchy problem is the problem of finding a (smooth) function $u(x)$ satisfying the equation

$$P(D)u(x) = f(x), \quad x_1 \geqslant 0,$$

and the initial conditions

$$D_1^k u|_{x_1=0} = \varphi_k(x_2, \ldots, x_n), \quad k = 0, \ldots, m-1.$$

If $\varphi_0 = \cdots = \varphi_{m-1} \equiv 0$, then the resulting problem is called the *homogeneous Cauchy problem*. In this case $u(x)$ and $f(x)$ can be extended by zero for $x_1 < 0$; let us denote these extensions by u_+ and f_+, so that the equation can be replaced by the following equation on the whole space:

$$P(D)u_+(x) = f_+(x).$$

The resulting problem is amenable to treatment using the construction of the previous section.

We shall find necessary and sufficient conditions for solubility of the homogeneous Cauchy problem in the space of rapidly decreasing (slowly increasing) functions (distributions). It should not be thought that solubility in decreasing functions is a weaker property of the operator as compared to solubility in increasing functions: though data is taken from a smaller space, the solution is sought for also in a smaller function space. Similarly, solubility in the space of distributions is not a stronger property than solubility in the space of smooth functions. Thus, an exhaustive study of the problems under consideration would involve examining solubility in four different types of spaces. In the case of equations on $\mathbb{R}^n$ these were the spaces $\mathscr{S}$, $\mathscr{O}$, $\mathscr{O}'$, $\mathscr{S}'$; invertibility conditions for convolution operators coincide in all of them.

In this section we shall obtain analogous results for the subspaces $\mathscr{S}_+$, $\mathscr{O}_+$, $(\mathscr{O}')_+$, $(\mathscr{S}')_+$ consisting of elements with support in the half-space $x_1 \geqslant 0$. In this case solubility conditions in the cases of growing and decreasing functions (distributions) will be different. In all solubility conditions for convolution operators, the passage from limit spaces to scales with respect to smoothness and growth exponent can be effected automatically; we shall not do it due to lack of space.

2.1. Notation. Let us distinguish one coordinate in $\mathbb{R}^n$, t, so that $x = (t, y)$, $t = x_1, y = (x_2, \ldots, x_n)$, (τ, η) are the dual coordinates, $\tau = \sigma + i\rho$, $\xi = (\sigma, \eta)$.

If Φ is a space of functions or distributions, let us set

$$\Phi_\pm = \{\varphi \in \Phi, \operatorname{supp} \varphi \in \mathbb{R}_\pm{}^n\}, \quad \mathbb{R}_\pm{}^n = \{(t, y), \pm t \geqslant 0, y \in \mathbb{R}^{n-1}\}. \tag{18}$$

We denote by $\Phi_\oplus$, $\Phi_\ominus$ the quotient spaces Φ/Φ_- and Φ/Φ_+.

For $a < b$ let us set $\Phi[a, b) = T_{(-a,0)}\Phi_+/T_{(-b,0)}\Phi_+$ and $\Phi(a, b] = T_{(-b,0)}\Phi_-/T_{(-a,0)}\Phi_-$.

Spaces of the type Φ_+ and $\Phi[a, b)$ correspond, respectively, to the homogeneous Cauchy problem in the half-space $t \geqslant 0$ or in the strip $a \leqslant t < b$. The space $\Phi_\oplus$ is connected with the problem which is adjoint to the homogeneous Cauchy problem.

2.2. Spaces $\mathscr{S}_+$ and $(\mathscr{O}')_+$. As the notion of support of a function (distribution) is defined in the spaces $\Phi = \mathscr{S}, \mathscr{O}, \mathscr{O}', \mathscr{S}'$, the general definition (18) makes sense in the indicated spaces. These same spaces can otherwise be defined as projective or inductive limits of the subspaces $C_{(l)+}^{(s)}$ or $H_{(l)+}^{(s)}$.

Analogously, quotient spaces $\mathcal{S}_\oplus, \ldots, (\mathcal{S}')_\oplus$ can be understood either as quotient spaces of the corresponding linear topological spaces, or as projective and inductive limits of the quotient spaces $C^{(s)}_{(l)\oplus}$ or $H^{(s)}_{(l)\oplus}$. In other words, equalities (2), (3), (7), (7′) will stay valid if both right and left hand side spaces are equipped with the symbols $+, -, \oplus, \ominus$.

The relations holding between subspaces and quotient spaces of the Hilbert spaces $H^{(s)}_{(l)}$ are as follows:

$$(H^{(s)}_{(l)+})' = H^{(-s)}_{(-l)\oplus}, \quad (H^{(s)}_{(l)\oplus})' = H^{(-s)}_{(-l)+}, \tag{19}$$

which induce duality of the corresponding "limit spaces" (see the remarks following (7)).

$$(\Phi_+)' = (\Phi')_\oplus, \quad (\Phi_\oplus)' = (\Phi')_+, \quad \Phi = \mathcal{S}, \mathcal{O}. \tag{19′}$$

Let us indicate another way of characterizing the subspaces $C^{(s)}_{(l)+}$. It can be verified directly that if $\varphi \in C^{(s)}_{(l)+}$, then the following equality holds:

$$|\varphi|^{(s)}_{(l)} = \sup_{\rho<0} \sup_{x\in\mathbb{R}^n, |\alpha|\leqslant s} |e^{\rho t}(1+|x|^2)^{1/2} D^\alpha \varphi(x)| \overset{\text{def}}{=} \sup_{\rho<0} |\varphi|^{(s)}_{(l)[\rho]}. \tag{20}$$

Conversely, if $\exp(\rho t)\varphi \in C^{(s)}_{(l)+}$ $\forall \rho < 0$, and the right hand side of (20) is finite, then $\varphi \in C^{(s)}_{(l)}$.

In connection with the description of images of Φ_+ type spaces under the Fourier transform, we shall have recourse to the following construction.

Let E be a Banach space of functions of $\xi = (\sigma, \eta)$ with norm $|\ |_E$. Let us denote by E^+ the space of functions of the variables (τ, η), $\tau = \sigma + i\rho$ with the following properties.

1) For each fixed $\rho \leqslant 0$, the functions $\varphi_\rho(\xi) = \varphi(\sigma + i\rho, \eta)$ belong to E and are continuous functions of $\rho \in (-\infty, 0]$ with values in E.

2) The functions $\varphi(\tau, \eta)$ are holomorphic in τ in the half plane $\operatorname{Im} \tau < 0$.

3) The norm $\sup_{\rho<0} |\varphi_\rho|_E \overset{\text{def}}{=} |\varphi, E^+|$ is finite.

Using the projective and inductive limit operations, we can define the spaces

$$\mathcal{S}^+ = C^{(\infty)+}_{(\infty)} = \bigcap_{s,l} C^{(s)+}_{(l)}, \quad \mathcal{M}^+ = \bigcap_s C^{(s)+}_{(-\infty)} = \bigcap_s \bigcup_l C^{(s)+}_{(l)}.$$

If $\varphi \in C^{(s)}_{(l)+}$, $l > n$, then the classical Fourier transform $\hat{\varphi}(\tau, \eta)$ is defined for $\operatorname{Im} \tau \leqslant 0$, and is holomorphic in τ for $\operatorname{Im} \tau < 0$. Applying the Parseval inequality (4) to the functions $\exp(\rho t)\varphi$, $\varphi \in C^{(s)}_{(l)+}$, using (20) and the definition of norm in $C^{(s)+}_{(l)}$, we obtain

$$|\hat{\varphi}, C^{(l')+}_{(s)}| < \text{const} |\varphi|^{(s)}_{(l)}, \quad l' + n < 1, \quad \varphi \in C^{(s)}_{(l)+}.$$

Hence it follows that $\mathcal{F}\mathcal{S}_+ \subset \mathcal{S}^+$. The reverse inclusion is proved with the help of the Cauchy integral, so that

$$\mathcal{F}\mathcal{S}_+ = \mathcal{S}^+. \tag{21}$$

It is immediately seen that functions in $\mathcal{M}^+$ are multipliers in $\mathcal{S}^+$. In view of the isomorphism (21), to each symbol $a \in \mathcal{M}^+$ there corresponds a PDO $a(D)$ (see Section 1.1) mapping $\mathcal{S}_+$ into itself.

To obtain a Hilbert scale in $\mathscr{S}_+$, let us introduce symbols in $\mathscr{M}^+$ that are holomorphic in τ:

$$\delta_s^{\pm}(\tau, \eta) = (\pm i\tau + \sqrt{1 + |\eta|^2})^s, \quad \delta_s^+ \in \mathscr{M}^+. \tag{22}$$

The PDO's $\delta_s^+(D)$ can be taken to be the operators that grade the smoothness of the scale $H_{(l)+}^{(s)}$:

$$H_{(l)+}^{(s)} = \{\varphi \in H_{(l)+}^{(-\infty)}, \delta_s^+(D)\varphi \in H_{(l)+}\}.$$

If $\|\delta_s^+(D)_\varphi\|_{(l)}$ is taken to be the norm on $H_{(l)}^{(s)}$, we have the Hilbert space equivalent of equality (20),

$$\|\varphi\|_{(l)}^{(s)} = \sup_{\rho<0} \|e^{\rho t}\delta_s^+(D)\varphi\|_{(l)} \overset{\text{def}}{=} \sup_{\rho<0} \|\varphi\|_{(l)[\rho]}^{(s)}, \tag{20'}$$

that is, $\varphi \in H_{(l)+}^{(s)}$ if and only if $\exp(\rho t)\varphi \in H_{(l)}^{(s)}$ $\forall \rho < 0$, and the right hand side of (20′) is finite.

According to the classical *Paley-Wiener* [1934] *theorem*,

$$\mathscr{F} H_+ = H^+,$$

and moreover, the isomorphism is an isometry (Parseval's equality holds). This isomorphism can be extended to the spaces $H_{(l)+}^{(s)}$ for any $s \in \mathbb{R}$ and integers $l \geqslant 0$:

$$\mathscr{F} H_{(l)+}^{(s)} = H_{(s)}^{(l)+}. \tag{23}$$

Hence it follows that

$$\mathscr{F}(\mathscr{O}')_+ = \mathscr{M}^+. \tag{24}$$

2.3. Convolutors in $\mathscr{S}_+$ and $(\mathscr{O}')_+$. A convolution operator in Φ_+, $\Phi = \mathscr{S}, \mathscr{O}, \mathscr{S}', \mathscr{O}'$, is any continuous operator, which commutes with shifts, the range of which is in Φ_+, that is, with operators $T_{(t,y)}$, $t \leqslant 0$.

For $\Phi = \mathscr{S}, \mathscr{O}$, let us denote by Φ_a the subspace $IT_{(a,0)}\Phi_+ = T_{(-a,0)}\Phi_-$, which consists of elements of Φ that are zero for $t \geqslant a$. Clearly, $\Phi_b \supset \Phi_a$ for $b > a$, and we can consider the inductive limit $\Phi_\infty = \bigcup_{a=0}^{\infty} \Phi_a$.

Proposition. *Let $\Phi = \mathscr{S}, \mathscr{O}$. For each convolution operator on Φ_+ there exists a distribution $f \in (\Phi_\infty)'$ such that* (10) *holds.*

Conversely, if for $f \in (\Phi_\infty)'$ the right hand side of (10) *is in Φ_+ for each $\varphi \in \Phi_+$, then* (10) *is a convolution operator.*

To prove the proposition, we consider, as we did in the case of Proposition 1.3, the distributions $f_{(t,y)}$, $t \geqslant 0$, which belong to $(\Phi_t)'$. Note that $f_{(0,y)} = 0$. The condition of commutativity with shifts stated above, is equivalent to saying that the functional $f_{(t,y)}$ is independent of y, and that its restriction to Φ_h, $h < t$, coincides with $f_{(h,y)}$. Hence it follows that the functionals $f_{(t,y)}$ define a unique distribution f on the inductive limit Φ_∞, that is, (10) holds. The converse statement follows from the closed graph theorem.

As in § 1, we denote the operator (10) by con_f, and the set of all convolutors by $\mathfrak{L}(\Phi_+)$.

Theorem. (i) $\mathfrak{L}(\mathcal{S}_+) = (\mathcal{O}')_+$

(ii) *Every convolution operator on* $(\mathcal{O}')_+$ *can be obtained by extending by continuity some convolution operator on* $\mathcal{S}_+$.

Taking (ii) into consideration, we shall write

$$\mathfrak{L}((\mathcal{O}')_+) = (\mathcal{O}')_+.$$

The proof of the theorem follows the argument of Theorem 1.4. Let us remark that for $f \in (\mathcal{O}')_+$, the operator con_f coincides with the PDO $\hat{f}(D)$.

2.4. Convolutors in $\mathcal{O}_+$ and $(\mathcal{S}')_+$. In the course of describing convolutors on $\mathcal{O}_+$, there arise spaces of distributions that behave differently with respect to the variables t and y. Let us introduce the necessary concepts. For $l_1, l \in \mathbb{R}$ let us set

$$H^{(s)}_{(l_1, l)} = (1 + t^2)^{-l_1/2}(1 + |y|^2)^{-l/2} H^{(s)}.$$

We define $C^{(s)}_{(l_1, l)}$ in a similar way. Let us set

$$\mathcal{K} = \bigcap_l H^{(-\infty)}_{(-\infty, l)} = \bigcap_l \bigcup_{s, l_1} H^{(s)}_{(l_1, l)}. \tag{25}$$

Remark. The space $\mathcal{K}$ consists of distributions in $\mathcal{S}'$, which decay in y faster than any power, and grow temperately in t. In the case $n = 1$, $\mathcal{K} = \mathcal{S}'(\mathbb{R})$. For $n > 1$, the space $\mathcal{K}$ is in a certain natural sense the tensor product

$$\mathcal{K}(\mathbb{R}^n) \approx \mathcal{S}'(R) \otimes \mathcal{O}'(\mathbb{R}^{n-1}).$$

We shall need the characterization of the image under the Fourier transform of distributions in $\mathcal{K}_+$. Let us denote by $\mathcal{L}^+$ the space of functions $\psi(\tau, \eta)$ that are holomorphic in τ for $\operatorname{Im} \tau < 0$ together with all their derivatives $D^\alpha_\eta \psi$, and that moreover their derivatives admit the estimates

$$|D^\alpha_\eta \psi(\tau, \eta)| < c_\alpha (1 + |\tau| + |\eta|)^{p_\alpha} |\operatorname{Im} \tau|^{-q_\alpha}, \quad \operatorname{Im} \tau < 0, \quad \eta \in \mathbb{R}^{n-1}. \tag{26}$$

Then

$$\mathcal{F}\mathcal{K}_+ = \mathcal{L}^+. \tag{27}$$

Let us explain the idea of the proof of the isomorphism (27).

If $f \in \mathcal{K}_+$, then for any integer $l > 0$ we can find s and an integer $k > 0$ such that $f \in H^{(s)}_{(-2k, l)+}$, that is $f = (1 + t^2)^k g$, $g \in H^{(s)}_{(0, l)+}$. From the classical Paley-Wiener theorem combined with (23), we derive the isomorphisms

$$\mathcal{F} H^{(s)}_{(0, l)+} = H^{(0, l)+}_{(s)}, \tag{23'}$$

where $H^{(0,l)}$ is defined to be the space of functions $\psi(\xi)$ belonging in L_2 together with the derivatives $D^\alpha_\eta \psi$, $|\alpha| \leqslant l$. Since $\mathcal{F}((1 + t^2)^k g)) = (1 + i\partial/\partial\tau)^k g$, elements of $\mathcal{F}\mathcal{K}_+$ are obtained from elements of the space $\bigcap_l \bigcup_s H^{(0, l)+}_{(s)}$ by differentiation in τ. This last space can effectively be defined as $\bigcap_l \bigcup_s C^{(0, l)+}_{(s)}$. Hence it follows easily that $\mathcal{F}\mathcal{K}_+ \subset \mathcal{L}^+$. The opposite inclusion is proved using the Cauchy integral.

In Section 2.3 we already defined convolutors on $\mathcal{O}_+$.

Theorem. (i) $\mathfrak{L}(\mathcal{O}_+) = \mathcal{K}_+$.

(ii) *Every convolution operator on* $(\mathcal{S}')_+$ *can be obtained by extension by continuity of some convolution operator on* $\mathcal{O}_+$.

Taking (ii) into consideration, we shall write

$$\mathfrak{L}((\mathcal{S}')_+) = \mathcal{K}_+.$$

Let us comment on the differences between the proof of this theorem and the proof of Theorem 2.3. As the Fourier transform cannot be defined in $\mathcal{O}_+$, the inclusion $\mathcal{K}_+ \subset \mathfrak{L}(\mathcal{O}_+)$ is verified directly by reducing convolutions with distributions in $\mathcal{K}_+$ to compositions of differential operators with operators of convolution with usual functions.

As far as the converse inclusion is concerned, (ii) is proved using the procedure of Theorem 1.4, from which it follows that $\mathfrak{L}(\mathcal{O}_+) \subset (\mathcal{S}')_+$. Thus the theorem is proved for $n = 1$.

For $n > 1$, let us restrict the convolution operator A to functions of the form $\varphi(t)\psi(y)$, $\varphi \in \mathcal{O}(\mathbb{R}_t)_+$, $\psi \in \mathcal{O}(\mathbb{R}_y^{n-1})$. Let us associate with each $\psi \in \mathcal{O}(\mathbb{R}^{n-1})$, the operator

$$\mathcal{O}(\mathbb{R}_t)_+ \to \mathcal{O}(\mathbb{R}_t) + (\varphi \to (A(\varphi\psi))(t, 0)).$$

This is a convolution operator, and by the argument above, it is given by a convolution with some distribution $F_\psi \in (\mathcal{S}'(\mathbb{R}_t))_+$. Thus, there arises the operator

$$\mathcal{O}(\mathbb{R}_y^{n-1}) \to (\mathcal{S}'(\mathbb{R}_t))_+(\psi \to F_\psi).$$

From a certain version of the kernel theorem it follows that this operator is given by a generalized kernel in $\mathcal{K}_+$, that coincides with the original convolutor (see Proposition 2.3 for $\Phi = \mathcal{O}$).

2.5. Convolution Equations. On the basis of the results of Sections 2.3–2.5, and following the plan of Sections 1.5 and 1.6, we can prove the following statements.

Theorem 1. *Let* $\Phi = \mathcal{S}, \mathcal{O}', \mathcal{S}'$, *and* $A \in \mathfrak{L}(\Phi_+)$. *The following conditions are equivalent*:

(I) *For any* $\psi \in \Phi_+$, *the convolution equation* (13) *has a unique solution* $\varphi \in \Phi_+$.

(II) *Equation* (13) *has a fundamental solution* $G \subset \mathfrak{L}(\Phi_+)$.

(II′) *If* $\hat{A}$ *is the symbol of* A, *then* $\hat{A}^{-1} \in \mathcal{F}\mathfrak{L}(\Phi_+)$.

If $\Phi = \mathcal{O}$, (II) *and* (II′) *are equivalent, and* (I) *follows from* (II).

Remark. In the case $\Phi = \mathcal{O}$, we cannot include the implication (I) $\Rightarrow$ (II) in the theorem, as in this case the Banach bounded inverse theorem cannot be used.

Now we have to give substance to condition (II′). We shall use the explicit description of $\mathcal{F}\mathfrak{L}(\Phi_+)$.

If $\Phi = \mathcal{S}, \mathcal{O}'$, then $\mathcal{F}\mathfrak{L}(\Phi_+) = \mathcal{M}^+$. For a symbol $\hat{A} \in \mathcal{M}^+$, condition (II′) is equivalent to the existence of the following lower bound.

$$|\hat{A}(\tau, \eta)| > c(1 + |\tau| + |\eta|)^\mu, \quad \operatorname{Im} \tau \leqslant 0, \quad \eta \in \mathbb{R}^{n-1} \tag{28}$$

for some $c > 0$ and μ.

If $\Phi = \mathcal{O}, \mathcal{S}'$, then $\mathcal{F}\mathfrak{L}(\Phi_+) = \mathcal{L}^+$. For a symbol $\hat{A} \in \mathcal{L}^+$, condition (II′) is equivalent to the existence of the following lower bound:

$$|\hat{A}(\tau, \eta)| > c(1 + |\tau| + |\eta|)^\mu |\operatorname{Im} \tau|^\nu, \quad \operatorname{Im} \tau < 0, \quad \eta \in \mathbb{R}^{n-1} \tag{28'}$$

for some $c > 0$, μ, ν.

Under the conditions of Theorem 1, all the remarks of Section 1.5 can be repeated word for word once all the spaces considered there are equipped with the + symbol.

In the case of differential operators, taking the Seidenberg-Tarski theorem into account makes it possible to replace the estimates (28) and (28′) with conditions for the non-vanishing of the symbol in the corresponding domains. Thus, we have

Theorem 2. (i) *The differential equation* (17) *is uniquely soluble in* $\mathcal{S}_+$ *or* $(\mathcal{O}')_+$ *if and only if*

$$P(\tau, \eta) \neq 0, \quad \operatorname{Im} \tau \leqslant 0, \quad \eta \in \mathbb{R}^{n-1}. \tag{29}$$

(ii) *The differential equation* (17) *is uniquely soluble in* $(\mathcal{S}')_+$ *if and only if*

$$P(\tau, \eta) \neq 0, \quad \operatorname{Im} \tau < 0, \quad \eta \in \mathbb{R}^{n-1}. \tag{29'}$$

Let us sum up. As we showed in § 1, in the case of $\mathbb{R}^n$, the set of convolutors did not depend on whether we were considering functions (distributions) of temperate growth, or rapidly decreasing ones. In the case of the half-space $t \geqslant 0$, for functions (distributions) of temperate growth we obtain a larger set of convolutors than for rapidly decreasing functions: in the first case power growth law of convolutions in the variable t is allowed. This situation leads, in particular, to the fact that solubility conditions for differential operators in spaces of functions (distributions) of temperate growth, is weaker than in the case of rapidly decreasing ones. In the language of scales, it is connected to the fact that in one case we are dealing with the solubility of the Cauchy problem in spaces of functions where any asymptotic power law is possible, while in the other one, only some are.

2.6. Convolution Equations in a Finite Strip. In Petrovskij's theory, necessary and sufficient conditions for correctness of the Cauchy problem on a finite time interval were considered. That is, equation (17) was studied in the strip $\{a \leqslant t < b, y \in \mathbb{R}^{n-1}\}$. In this section we shall extend the theory developed above to convolution equations in the spaces $\Phi[a, b)$ (see Section 2.1) for $\Phi = \mathcal{S}, \mathcal{O}, \mathcal{O}', \mathcal{S}'$. First of all let us remark that $\Phi[a, b)$ can be defined in two ways: as quotient spaces of linear topological spaces $T_{(-c, 0)}\Phi_+$ or as (respectively) inductive and projective limits of the spaces $C^{(s)}_{(l)}[a, b)$ or $H^{(s)}_{(l)}[a, b)$. Both these methods lead to the same spaces.

The scales $\{H^{(s)}_{(l)}[a, b)\}$ and $\{H^{(s)}_{(l)}(a, b]\}$ are conjugate to each other. Conjugacy of these scales induces the following conjugacies:

$$(\Phi[a, b))' = \Phi'(a, b], \quad (\Phi(a, b])' = \Phi'[a, b), \quad \Phi = \mathcal{S}, \mathcal{O}.$$

As we already stated above, the PDO's $\delta_k^+(D)$ map the scale $\{H^{(s)}_{(l)+}\}$ into itself, and define on it a smoothness gradation. As PDO's and shift operators commute, the operators $\delta_k^+(D)$ map into themselves all shifts of the scale $\{H^{(s)}_{(l)+}\}$, and therefore map into itself the scale of quotient spaces $\{H^{(s)}_{(l)+}[a, b)\}$, defining on it a smoothness gradation. In a similar manner, the operators $\delta_k^-(D)$ realize a gradation on $\{H^{(s)}_{(l)+}(a, b]\}$.

Let us move on now to the definition of shift operators in a finite strip. Shift operator $T_{(-t, y)}$ for $t \geqslant 0$ are well defined in quotient spaces.

$$T_h\colon \Phi[a, b) \to \Phi[a, b), \quad h = (h_1, \dots, h_n), \quad h_1 \leqslant 0. \tag{30}$$

For $h_1 < a - b$ this is the null operator.

The concept of a continuous operator can be extended from the space Φ to the space $\Phi[a, b)$, $\Phi = \mathscr{S}, \mathcal{O}, \mathscr{S}', \mathcal{O}'$.

We call any continuous operator $A\colon \Phi[a, b) \to \Phi[a, b]$ a convolution operator on $\Phi[a, b)$, $\Phi = \mathscr{S}, \mathcal{O}, L', \mathcal{O}'$, if it commutes with the shifts (30). Let us go on to describe the corresponding convolutors. We shall start with the analogs of Propositions 1.3, 2.3 for convolution operators in a strip. For that, let us note as a function φ runs through the space $\Phi[a, b)$, the functions $IT_x\varphi$, $x = (t, y)$, $a \leqslant t \leqslant b$, run through $\Phi[0, b - a)$. Using the argument of Proposition 1.3, we can prove

Proposition. *Let $\Phi = \mathscr{S}, \mathcal{O}$. For every convolution operator on $\Phi[a, b)$ there exists a distribution $f \in \Phi'(0, b - a]$ such that* (10) *holds for $x = (t, y)$, $a \leqslant t \leqslant b$.*

Conversely, if for $f \in \Phi'(0, b - a]$ for each $\varphi \in \Phi[a, b)$ the right-hand side of (10) *is in $\Phi[a, b)$, then* (10) *defines a convolution operator.*

As above, we denote the operator (10) by con_f, and the set of all convolutors by $\mathfrak{L}(\Phi[a, b))$. It is clear that

$$\mathfrak{L}(\Phi[a + c, b + c)) = \mathfrak{L}(\Phi[a, b)) \quad \forall c \in \mathbb{R}, \quad \Phi = \mathscr{S}, \mathcal{O}. \tag{31}$$

Thus, in our description of convolutors we may assume without loss of generality that $a = 0$.

Theorem 1.

$$\mathfrak{L}(\mathscr{S}[0, b)) = \mathfrak{L}(\mathcal{O}[0, b)) = \mathcal{O}'[0, b). \tag{32}$$

According to Theorems 2.3 and 2.4, the operator con_f, $f \in (\mathcal{O}')_+$ maps the spaces $\mathscr{S}_+$, $\mathcal{O}_+$, and their shifts, into themselves. Therefore it can be extended to a continuous operator

$$\mathrm{con}_f\colon \Phi[0, b) \to \Phi[0, b), \quad f \in (\mathcal{O}')_+, \quad \Phi = \mathscr{S}, \mathcal{O}.$$

From (10) it can be seen that if the support of f belongs to the half-space $t \geqslant b$, con_f acts as the null operator. Thus, we have shown that $\mathcal{O}'[0, b) \subset \mathfrak{L}(\mathscr{S}[0, b))$, $\mathcal{O}'[0, b) \subset \mathfrak{L}(\mathcal{O}[0, b))$.

The inclusion $\mathcal{O}'[0, b) \supset \mathfrak{L}(\mathcal{O}[0, b))$ follows from the proposition, while the inclusion

$$\mathfrak{L}(\mathscr{S}[0, b)) \subset \mathcal{O}'[0, b)$$

is proved by using the argument of Theorem 1.4.

Using the relations

$$\Phi(a, b] = I\Phi[-b, -a), \quad \Phi'[a, b) = I(\Phi[-b, -a))',$$

it is possible to show that

$$\mathfrak{L}(\mathscr{S}(0, b]) = \mathfrak{L}(\mathscr{O}(0, b]) = \mathscr{O}'(-b, 0]$$

and

$$\mathfrak{L}(\mathscr{S}'[0, b)) = \mathfrak{L}(\mathscr{O}'[0, b)) = \mathscr{O}'[0, b).$$

Following the procedure of § 1 we can prove

Theorem 2. *Let $A \in \mathscr{O}'[0, b)$, and $\Phi = \mathscr{S}, \mathscr{O}, \mathscr{O}', \mathscr{S}$. The following conditions are equivalent:*

(I) *The convolution equation* (13) *is uniquely soluble in $\Phi[a, a + b]$ for all $a \in \mathbb{R}$.*

(I′) *The convolution equation* (13) *is uniquely soluble in $\Phi[a, a + b)$ for some $a \in \mathbb{R}$.*

(II) *There exists a $G \in \mathscr{O}[0, b)$ such that $A * G - \delta(x) = G * A - \delta = 0$ in the sense of $\mathscr{O}'[0, b)$.*

Let $A \in \mathscr{O}'[0, b)$. Let us denote by $\hat{A}(\tau, \eta)$ the Fourier transform of a representative of the residue class of A. Condition (II) of our theorem, unlike the analogous condition of Theorem 1.5, does not, in general, allow us to obtain effective necessary and sufficient conditions on $\hat{A}$ ensuring (I). However, we can extract separately a necessary and a sufficient condition from (II).

According to (II) there exists a symbol $\hat{G} \in \mathscr{M}^+$ such that

$$\hat{A}(\tau, \eta)\hat{G}(\tau, \eta) - 1 \in \mathscr{F}(T_{(-b,0)}(\mathscr{O}')_+) = e^{-i\tau b}\mathscr{M}^+.$$

From the definition of the space in the right hand side it follows that for some $c > 0$ and μ

$$|\hat{A}(\tau, \eta)\hat{G}(\tau, \eta) - 1| < c(1 + |\tau| + |\eta|)^\mu e^{b \operatorname{Im} \tau}.$$

Hence, trivially, we have

A necessary condition for (II) *to hold.* $\exists\, c_1, c_2$ such that

$$\{\hat{A}(\tau, \eta) = 0\} \Rightarrow \{\operatorname{Im} \tau > c_1 \ln(1 + |\tau| + |\eta|) + c_2\}. \tag{33}$$

A sufficient condition for (II) *to hold.* $\exists\, c > 0, \mu, \rho \leqslant 0$ such that

$$|\hat{A}(\tau, \eta)| > c(1 + |\tau| + |\eta|)^\mu, \quad \operatorname{Im} \tau \leqslant \rho. \tag{34}$$

If $\rho = 0$, inequality (34) is the invertibility condition of A in $(\mathscr{O}')_+$. For $\rho < 0$ this condition ensures invertibility in $(\mathscr{O}')_+$ of the distribution $A_\rho \overset{\text{def}}{=} e^{\rho t}A$, as the symbol $\hat{A}_\rho(\tau, \eta)$ equals $\hat{A}(\tau - i\rho, \eta)$. Taking into account the fact that spaces $\Phi[a, b)$ $b < \infty$ are invariant with respect to multiplication by $\exp(\rho t)$ for any ρ, we obtain the sufficiency of (34).

Let us discuss the connection between (33) and (34). Condition (34) means that the symbol $\hat{A}(\tau, \eta)$ has no zeroes in the half-plane $\operatorname{Im} \tau \leqslant \rho$ for some $\rho \leqslant 0$, and admits a power law lower bound in that half-plane. Condition (33) means that, in general, the symbol $\hat{A}(\tau, \eta)$ has zeroes in the lower half-plane, but the manifold of these zeroes separates (as η is increased) from any straight line $\operatorname{Im} \tau = \text{const.}$ very slowly.

In the case of differential operators, that is, $A = P(D)\delta(x)$ and $\hat{A} = P$, as we already stated above, condition (34) is equivalent to

$$P(\tau, \eta) \neq 0, \quad \operatorname{Im} \tau \leqslant \rho, \quad \eta \in \mathbb{R}^{n-1}. \tag{35}$$

On the other hand, the imaginary part of the algebraic function $\tau(\eta)$ which solves the equation $P(\tau, \eta) = 0$, cannot according to the Seidenberg-Tarski theorem, go to $-\infty$ slower than some power of $|\eta|$. Therefore (35) follows from condition (33). Thus, we have proved

Theorem 3. *Let $\Phi = \mathscr{S}, \mathscr{O}, \mathscr{S}', \mathscr{O}'$. The differential equation* (17) *is uniquely soluble in the space $\Phi[a, b)$ for any $a, b \in \mathbb{R}$, $a < b$ (or for some $a < b$) if and only if there exists a ρ for which* (35) *holds.*

Definition. We shall say that a polynomial $P(\tau, \eta)$ satisfies the *homogeneous correctness condition of Petrovskij* if there exists a ρ for which (35) holds.

This condition is different from the Petrovskij condition (see the Introduction) in that there is no assumption of solubility of the polynomial with respect to the highest power of τ.

Remark. As we already stated in the introduction, Petrovskij [1938] had a "scissors" cut between conditions (33) and (35); the Seidenberg-Tarski theorem shows that in the case of polynomials these conditions coincide.

2.7. A Problem Dual to the Homogeneous Cauchy Problem. In view of the conjugacy relations (19) we mentioned above, operators on quotient spaces

$$\operatorname{con}_B \colon \Phi_\oplus \to \Phi_\oplus, \quad \operatorname{con}_B \colon (\Phi')_\oplus \to (\Phi')_\oplus, \quad \Phi = \mathscr{S}, \mathscr{O}$$

can be interpreted as operators adjoint to

$$\operatorname{con}_A \colon (\Phi')_+ \to (\Phi')_+, \quad \operatorname{con}_A \colon \Phi_+ \to \Phi_+, \qquad A = IB.$$

Following this direction of thought, it is possible to obtain a description of convolutors:

$$\mathfrak{L}(\mathscr{S}_\oplus) = (\mathscr{K}')_-, \quad \mathfrak{L}(\mathscr{O}_\oplus) = (\mathscr{O}')_- \tag{36}$$

and to prove analogs of theorems of Section 2.5.

The results indicated above can also be obtained directly, by constructing a theory of convolutors in quotient spaces as of operators that commute with shifts; in fact, such a theory was constructed in the previous section, in the arguments of which we can set $a = -\infty$, $b = 0$. Then we have a theory of convolution operators in Φ_-; the theory in Φ_+ is obtained by using the reflection operator I.

Similarly, analogous theorems (of the type of Theorems 1, 2 of Section 2.5) for equations on quotient spaces can be obtained either following the scheme of Section 1.5, or from conjugacy considerations. We do not go into these details.

§ 3. The Non-homogeneous Cauchy Problem in Slowly Increasing (Rapidly Decreasing) Functions (Distributions)

In § 2, the Cauchy problem with zero initial data was reduced to a problem on the whole space in functions (distributions) supported in $t \geqslant 0$. This problem was investigated in the framework of convolution equations. We can treat the non-homogeneous problem in a similar manner.

Let $P(D) = \sum_{j=0}^{m} P_j(D_y)(D_t)^j$ be a differential operator with constant coefficients, and let $u(t, y)$ be a smooth solution of the Cauchy problem

$$P(D_t, D_y)u(t, y) = f(t, y), \quad t \geqslant 0, \tag{37}$$

$$D_t^k u(t, y)|_{t=0} = \varphi_k(y), \quad k = 0, \ldots, m-1. \tag{38}$$

Let us extend u, f smoothly for $t < 0$, using the same symbols u and f for these extensions. Let us set $u_+ = \theta_+ u$, $f_+ = \theta_+ f$, where $\theta_+(t)$ is the characteristic function of the semi-infinite line $t \geqslant 0$. Using the fact that $\frac{d}{dt}\theta_+(t) = \delta(t)$, we obtain for u_+ the equation

$$P(D_t, D_y)u_+ = F_+, \tag{39}$$

$$F_+ = f_+ + \frac{1}{i}\sum_{j=0}^{m-1}\sum_{l=0}^{m-1-j} P_{j+l+1}(D_y)\varphi_l(y)D_t^j\delta(t). \tag{40}$$

Let $\Phi = \mathscr{S}, \mathscr{O}$, and let p be a negative integer. Let us set

$$\Phi_{[+]} = \{\varphi_+ = \theta_+(t)\varphi, \varphi \in \Phi\}, \tag{41}$$

$$\Phi_{[+]}^{\{p\}} = \left\{\varphi = \varphi_+ + \sum_{j=0}^{|p|-1} \varphi_j(y)D_t^j\delta(t), \varphi_+ \in \Phi_{[+]}, \varphi_j \in \Phi(\mathbb{R}^{n-1}), j = 0, \ldots, |p| - 1\right\}. \tag{42}$$

Then the problem (37), (38) for $u, f \in \Phi(\mathbb{R}_+^n)$, and $\varphi_0, \ldots, \varphi_m \in \Phi(\mathbb{R}^{n-1})$ becomes the problem of invertibility of the operator

$$P(D)\colon \Phi_{[+]}^{\{p\}} \to \Phi_{[+]}^{\{p-m\}} \tag{43}$$

for $p = 0$. Setting

$$\Phi_{[+]}^{\{-\infty\}} = \bigcup_{p \leqslant 0} \Phi_{[+]}^{\{p\}}, \tag{42'}$$

we shall consider a more general problem of invertibility of convolution operators in the space (42′). This problem will be solved using the procedure of §§ 1, 2. First of all we shall describe the class $\mathfrak{L}(\Phi_{[+]}^{\{-\infty\}})$ of corresponding convolutors for $\Phi = \mathscr{S}, \mathscr{O}$. As we shall see below, it consists of the elements of $\mathfrak{L}(\Phi_+)$ that satisfy additional smoothness conditions in t for $t > 0$. These conditions are the analogs of the transmission (smoothness) conditions of Boutet de Monvel [1966], and of Vishik and Eskin [1964], in the classes $\mathscr{O}'$ and $\mathscr{K}$. We shall show that the invertibility of the operator (43) is equivalent to $P(D)$ having a fundamental solution in $\mathfrak{L}(\Phi_{[+]}^{\{-\infty\}})$, while this latter condition is equivalent to the invertibility of the symbol $P(\tau, \eta)$ in the space of the images of the Fourier transform, $\mathscr{F}\mathfrak{L}(\Phi_{[+]}^{\{-\infty\}})$. Using a characterization of this space, we shall obtain simple necessary and sufficient conditions on the symbol, under which (43) is an isomorphism.

At the end of the section, using the procedure in Section 2.7, we shall consider equations of the type (39), (40) in the strip $a \leqslant t < b$, obtaining thus, in the case of one equation of higher order, a modification of Petrovskij's [1938] theorem.

3.1. Convolutors on $\mathfrak{L}(\Phi_{[+]}^{\{-\infty\}})$. As we shall see below (see (47), (47')), the spaces $\Phi_{[+]}^{\{-\infty\}}$, $\Phi = \mathscr{S}, \mathscr{O}$, are constructed from Banach spaces with the help of inductive and projective limit operations. Starting with (47), (47') it is possible to introduce the concept of a continuous operator on $\mathscr{S}_{[+]}^{\{-\infty\}}$, $\mathscr{O}_{[+]}^{\{-\infty\}}$ in a way similar to the one used for $\mathscr{O}$, $\mathscr{S}'$, and so on in Section 1.1.

We call a continuous operator on $\Phi_{[+]}^{\{-\infty\}}$, $\Phi = \mathscr{S}, \mathscr{O}$, a convolution operator on that space, if it commutes with differential operators, and if its restriction to Φ_+ is a convolution operator on that space.

Let us emphasize that unlike the case of spaces Φ, Φ_+, in the definition of convolution operators on $\Phi_{[+]}^{\{-\infty\}}$, we replaced the condition of commuting with shifts, by a condition of commuting with differential operators; these conditions are similar, and in the case of spaces invariant relative to shifts, the second condition follows from the first one. Naturality of the condition of commuting with differential operators in $\Phi_{[+]}^{\{-\infty\}}$ is related to the fact, that differentiations (unlike shifts) are defined on all elements of this space and map it into itself.

We shall study convolutors on $\mathscr{S}_{[+]}^{\{-\infty\}}$ using the procedure of § 1.2. Beforehand, we must introduce a space of distributions $\Psi_+ \supset \mathscr{S}_{[+]}^{\{-\infty\}}$ that contains $\delta(x)$ and onto which convolution operators acting on $\mathscr{S}_{[+]}^{\{-\infty\}}$ must be extended. It is clear that distributions in the space Ψ_+ must satisfy two conditions: 1) same decay estimates must hold for them and for distributions in $\mathscr{S}_{[+]}$, 2) they have the same smoothness in t as elements of $\mathscr{S}_{[+]}^{\{-\infty\}}$.

In the case $n = 1$, the original space $\mathscr{S}_{[+]}^{\{-\infty\}}(\mathbb{R})$ has all these properties. Intuitively it is clear that in the case $n > 1$, we must take for Ψ_+ the tensor product.

$$\mathscr{S}_{[+]}^{\{-\infty\}}(\mathbb{R}) \otimes \mathfrak{L}(\mathscr{S}(\mathbb{R}^{n-1})), \quad \text{that is} \quad \mathscr{S}_{[+]}^{\{-\infty\}}(\mathbb{R}) \otimes \mathscr{O}'(\mathbb{R}^{n-1}).$$

To give precise meaning to this definition, we are obliged to introduce special bigraded scales.

Let us denote by $p_\oplus$ the canonical projection of Φ onto $\Phi_\oplus = \Phi/\Phi_-$. The homomorphism

$$p_\oplus \colon H_{(l)+}^{(p,s)} \to H_{(l)\oplus}^{(p,s)} \tag{44}$$

defines an additional filtration by smoothness, of the left hand side space. (We remind the reader that we denote by $H^{(p,s)}$ a space of functions of smoothness p in t and of smoothness s in y). In view of the above, for any integer $q \geqslant p$ let us set

$$H_{(l)+}^{(\{p,q\},s)} = \{\varphi \in H_{(l)+}^{(p,s)},\ p_\oplus \varphi \in H_{(l)\oplus}^{(q,s)}\}. \tag{45}$$

For $p = 0$ the mapping (44) is an isomorphism. For $p > 0$ this mapping is injective and has a cokernel, since functions in the left hand space together with their derivatives in t up to order $p - 1$ are zero at $t = 0$. If $p < 0$, functions from $H_{(l)+}^{(p,s)}$ can take any values for $t \to +0$, so that the mapping (44) is surjective. Its kernel consists of distributions supported in the plane $t = 0$. These are of the form

$$\sum_{j=0}^{r-1} g_j(y) D_t^j \delta(t), \quad g_j \in H_{(l)}^{(s)}(\mathbb{R}^{n-1}), \quad 0 \leqslant j \leqslant r - 1, \tag{46}$$

where $r = |p|$. For spaces (45) natural embeddings are defined so that scales of spaces (45) can be considered together with their projective and inductive limits with respect to various indices. We have:

$$\mathscr{S}_{[+]}^{\{-\infty\}} = \bigcup_p H_{(\infty)+}^{(\{p,\infty\},\infty)} = \bigcup_p \bigcap_{q,s,l} H_{(l)+}^{(\{p,q\},s)}, \tag{47}$$

$$\mathscr{O}_{[+]}^{\{-\infty\}} = \bigcup_{p,l} H_{(l)+}^{(\{p,\infty\},\infty)} = \bigcup_{p,l} \bigcap_{q,s} H_{(l)+}^{(\{p,q\},s)}. \tag{47'}$$

Let us return to the problem of convolutors in $\mathscr{S}_{[+]}^{\{-\infty\}}$. Let us set

$$U_{[+]} = \bigcup_p U_{[+]}^{\{p\}}, \quad U_{[+]}^{\{p\}} = \bigcap_{q,l} H_{(l)+}^{(\{p,q\},-\infty)}. \tag{48}$$

The space (48) satisfies the requirements for an admissible space of convolutors on (47); for $n = 1$, spaces (47) and (48) coincide.

Theorem 1. (i) *We have the equality*

$$\mathfrak{L}(\mathscr{S}_{[+]}^{\{-\infty\}}) = U_{[+]}. \tag{49}$$

(ii) *Each convolution operator in $\mathscr{S}_{[+]}^{\{-\infty\}}$ can be extended by continuity to a continuous operator in $U_{[+]}$.*

Note that in view of the statement of the theorem, $U_{[+]}$ is a ring with respect to convolution, so that it is natural to say that

$$\mathfrak{L}(U_{[+]}) = U_{[+]}. \tag{49'}$$

Proof of the theorem is by the following argument. If we take a convolution operator A on $\mathscr{S}_{[+]}^{\{-\infty\}}$, then its restriction to $\mathscr{S}_+$ is defined. This restriction (see Theorem 2.3) is an operator of convolution with some distribution $f \in (\mathscr{O}')_+$. The operator $\mathrm{con}_f\colon \mathscr{S}_+ \to \mathscr{S}_+$ can be extended by continuity (see Theorem 2.3) to a continuous operator $\tilde{A}$ on the space $(\mathscr{O}')_+$ which contains $\mathscr{S}_{[+]}^{\{-\infty\}}$. We have to show that $A = \tilde{A}$ on $\mathscr{S}_{[+]}^{\{-\infty\}}$. Operator $A - \tilde{A}$ equals 0 on $\mathscr{S}_+$ and commutes with differential operators; a direct examination of the generalized kernel of this operator shows that it equals 0 on $\mathscr{S}_{[+]}^{\{-\infty\}}$.

Remark. The space $U_{[+]}$ can be interpreted as the tensor product

$$U_{[+]} = \mathscr{S}_{[+]}^{\{-\infty\}}(\mathbb{R}) \oplus \mathscr{O}'(\mathbb{R}^{n-1}),$$

while equality (49) can be formally rewritten in the form

$$\mathfrak{L}(\mathscr{S}_{[+]}^{\{-\infty\}}(\mathbb{R}) \oplus \mathscr{S}(\mathbb{R}^{n-1})) = \mathfrak{L}(\mathscr{S}_{[+]}^{\{-\infty\}}(\mathbb{R})) \oplus \mathfrak{L}(\mathscr{S}(\mathbb{R}^{n-1})).$$

Theorem 2. *Let* $\Phi_{[+]} = \mathscr{S}^{\{-\infty\}}_{[+]}, U_{[+]}$. *The following conditions are equivalent for a distribution* $A \in U_{[+]}$:

(I) *Equation* (13) *is uniquely soluble in* $\Phi_{[+]}$.

(II) *Equation* (13) *has a fundamental solution in* $U_{[+]}$.

As $U_{[+]} \subset (\mathcal{O}')_+$, distributions $A \in U_{[+]}$ can be associated with convolution operators with symbols $\hat{A}(\tau, \eta) \in \mathscr{F} U_{[+]} \subset \mathscr{M}^+$. Because of that, we can include in the set of equivalent conditions of the theorem also

(II′) *The symbol* $\hat{A}(\tau, \eta)$ *is an invertible element of the ring* $\mathscr{F} U_{[+]}$.

In order to make Theorem 2 effective, we must provide an explicit description of the space $\hat{U}^+ \stackrel{\text{def}}{=} \mathscr{F} U_{[+]}$.

3.2. The Spaces $U_{[+]}$ and $\hat{U}^+$. According to definition (45), $H^{(\{p,q\},s)}_{(l)+} \supset H^{(q,s)}_{(l)+}$; let us fix a complement to the space on the right.

The Complement Lemma. Every element $\varphi \in H^{(\{p,q\},s)}_{(l)+}$ *can be represented in the form*

$$\varphi = \varphi' + \varphi'', \quad \varphi' \in H^{(q,s)}_{(l)+}, \quad \varphi'' \in (D_t - i)^{-q} g, \tag{50}$$

where g *has the form* (46) *with* $r = q - p$.

For $q = 0$ the decomposition (50) follows from the fact that (44) is a surjective mapping with kernel given by (46). For $q \neq 0$ we can write the corresponding decomposition for $(D_t - i)^q \varphi \in H^{(\{p-q,0\},s)}_{(l)+}$, and then to apply to this decomposition the operator $(D_t - i)^{-q}$.

The filtration of the space (45) induces the filtration

$$U^{\{p\}}_{[+]} \subset U^{\{p-1\}}_{[+]} \subset \cdots \subset U^{\{-\infty\}}_{[+]} \stackrel{\text{def}}{=} U_{[+]}$$

and we have isomorphisms

$$(D_t - i)^k U^{\{p\}}_{[+]} = U^{\{p-k\}}_{[+]}.$$

Similarly,

$$(D_t - i)^k \mathscr{S}^{\{p\}}_{[+]} = \mathscr{S}^{\{p-k\}}_{[+]}.$$

For $p < 0$, $k > 0$ these isomorphisms follow from our definitions. For $k < 0$ they may serve as definitions of spaces (42) for $p > 0$. Using the complement lemma, we can prove the

Gradation Lemma. *Let* $\Psi_{[+]} = \mathscr{S}^{\{-\infty\}}_{[+]}, U_{[+]}$. *Then* $\varphi \in \Psi_{[+]}$ *if and only if for every* q *we have the expansion*

$$\varphi = \sum_{j=0}^{q-p} \varphi_{p-j}(y)(D_t - i)^{p-j}\delta(t) + \varphi_{(q)}, \quad \varphi_{(q)} \in \Psi^{\{q\}}_{[+]}, \tag{51}$$

where $\varphi_{p-j}(y)$ *belongs, respectively, to* $\mathscr{S}(\mathbb{R}^{n-1})$ *and* $\mathcal{O}'(\mathbb{R}^{n-1})$.

Let us remark that the distributions $\varphi_{p-j}(y)(D_t - i)^{p-j}\delta(t)$ belong to $\Psi^{\{p-j-1\}}_{[+]}$ and do not belong to $\Psi^{\{p-j\}}_{[+]}$. In other words, the subspaces of such distributions grade our filtration. Comparison of expansions (51) for different q shows that coefficients of φ_{p-j} do not depend on q. The smallest p for which $\varphi_p(y) \neq 0$ is called the order of φ in t and is denoted by $\deg_t \varphi$.

Let us denote by $\hat{\mathscr{S}}^{\{-\infty\}+} \subset \hat{U}^+ \subset \mathscr{M}^+$ the spaces of images under the Fourier transform of distributions from $\mathscr{S}^{\{-\infty\}}_{[+]}$ and $U_{[+]}$. From the gradation lemma we obtain

Proposition 1. *Functions $\psi(\tau, \eta) \in M^+$ belong, respectively, to $\hat{\mathscr{S}}^{\{-\infty\}+}$, $\hat{U}^+$ if and only if for any sufficiently large q we have the expansion*

$$\psi(\tau, \eta) = \sum_{j=0}^{q-p+1} \psi_{p-j}(\eta)(\tau - i)^{p-j} + \psi_{(q)}(\tau, \eta), \tag{52}$$

moreover:

a) *if $\psi \in \hat{\mathscr{S}}^{\{-\infty\}+}$, then $\psi_{p-j} \in \mathscr{S}(\mathbb{R}^{n-1})$, $\psi_{(q)} \in M^+$ and $\forall t, s\ \exists\ c_{lsq}$ such that*

$$|\eta^\beta D_\eta^\alpha \psi_q(\tau, \eta)| < c_{lsq}(1 + |\tau| + |\eta|)^{-q}, \quad |\beta| \leqslant l, \quad |\alpha| \leqslant s, \quad \operatorname{Im} \tau \leqslant 0;$$

b) *If $\psi \in \hat{U}^+$, then $\psi_{p-j} \in \mathscr{M}(\mathbb{R}^{n-1})$ and $\forall s\ \exists\ \mu_{sq}, c_{sq}$ such that*

$$|D_\eta^\alpha \psi_q(\tau, \eta)| < c_{sq}(1 + |\eta|)^{\mu_{sq}}(1 + |\tau|)^{-q}, \quad |\alpha| \leqslant s, \quad \operatorname{Im} \tau \leqslant 0.$$

Thus, the symbols of distributions from $\mathscr{S}^{\{-\infty\}}_{[+]}$ and $U_{[+]}$ are asymptotic Laurent series in τ with coefficients in $\mathscr{S}$ or $\mathscr{M}$. As is easily seen, spaces from Proposition 1 are rings with respect to multiplication, and moreover the elements of $\hat{U}^+$ are multiplicators over $\hat{\mathscr{S}}^{\{-\infty\}+}$.

Proposition 2. *A function $\psi \in \hat{U}^+$ is an invertible element of this ring if and only if*

(a) *ψ is an invertible element of the larger ring $\mathscr{M}^+$;*

(b) *the highest order coefficient $\psi_p(\eta)$ of expansion* (52) *is an invertible element of the ring $\mathscr{M}(\mathbb{R}^{n-1})$.*

Let us remark on the proof of this proposition. Condition (b) guarantees invertibility of ψ in the ring of formal series $\sum \psi_{p-j}(\eta)(\tau - i)^{p-j}$, therefore a expansion of form (52) can be written down for ψ^{-1}. Condition (a) guarantees the required estimate for the remainder.

In view of § 2, (a) is equivalent to the estimate

$$|\psi(\tau, \eta)| > \mathrm{const}(1 + |\tau| + |\eta|)^{\mu}, \quad \operatorname{Im} \tau \leqslant 0, \tag{53}$$

while, in view of § 1, condition (b) is equivalent to the estimate

$$|\psi_p(\eta)| > \mathrm{const}(1 + |\eta|)^{\mu'}, \quad \eta \in \mathbb{R}^{n-1}. \tag{53'}$$

From (53), (53′) and the Seidenberg-Tarski theorem we obtain the following

Theorem. *The differential operator with constant coefficients $P(\tau, \eta) = \sum_{j\geqslant 0}^{m} P_j(\eta)\tau^j$ has a fundamental solution in $U_{[+]}$ if and only if*

$$P(\tau, \eta) \neq 0, \quad \operatorname{Im} \tau \leqslant 0, \quad \eta \in \mathbb{R}^{n-1}, \tag{54}$$

$$P_m(\eta) \neq 0, \quad \eta \in \mathbb{R}^{n-1}. \tag{54'}$$

3.3. Convolution Equations in $\mathcal{O}_{[+]}^{\{-\infty\}}$. Definition of convolution operators on $\mathcal{O}_{[+]}^{\{-\infty\}}$ was given in Section 3.1. To describe the convolutors we have, first of all, to single out a space $V_{[+]} \supset \mathcal{O}_{[+]}^{\{-\infty\}}$ that contains $\delta(x)$ and such that the operator (10) can be extended by continuity from $\mathcal{O}_{[+]}^{\{-\infty\}}$ to $V_{[+]}$.

On an intuitive level, $\mathcal{O}_{[+]}^{\{-\infty\}}$ is $\mathcal{O}_{[+]}^{\{-\infty\}}(\mathbb{R}) \oplus \mathcal{O}(\mathbb{R}^{n-1})$, therefore for $V_{[+]}$ we should take $\mathcal{O}_{[+]}^{\{-\infty\}}(\mathbb{R}) \oplus \mathcal{S}'(\mathbb{R}^{n-1})$. In a precise manner this space is defined in the following way:

$$V_{[+]} = \bigcup_p V_{[+]}^{\{p\}}, \quad V_{[+]}^{\{p\}} = \bigcup_l \bigcap_q H_{(l)+}^{(\{p,q\},-\infty)}. \tag{55}$$

Elements of the space (55) have the same smoothness in t as elements of $\mathcal{O}_{[+]}^{\{-\infty\}}$, however, unlike convolutors in $\mathcal{O}_+$, they have power law growth in y. Therefore prospective convolutors must be sought among elements of $V_{[+]} \cap \mathcal{K}_+$, where $\mathcal{K}_+$ is defined by (25). To describe these elements, we shall need the spaces $H_{(l,\lambda)}^{(p,s)}$, elements of which have different power asymptotics in the variables t and y. Let us set

$$H_{(l,\lambda)}^{(p,s)} = \{\varphi \in H_{(l,\lambda)}^{(-\infty)}, (1+|D_t|^2)^{p/2}(1+|D_y|^2)^{s/2}\varphi \in H_{(l,\lambda)}\}.$$

By analogy with (45) we can also introduce the spaces $H_{(l,\lambda)+}^{(\{p,q\},s)}$. Let us set

$$\mathfrak{N}_+ = \bigcup_p \mathfrak{N}_+^{\{p\}}, \quad \mathfrak{N}_+^{\{p\}} = \bigcap_{q,\lambda} H_{(-\infty,\lambda)+}^{(\{p,q\},-\infty)}. \tag{56}$$

Remark. For $n = 1$, the space (56) coincides with $\mathcal{O}_{[+]}^{\{-\infty\}}$.

Theorem 1. (i) *We have the equality*

$$\mathfrak{L}(\mathcal{O}_{[+]}^{\{-\infty\}}) = \mathfrak{N}_+.$$

(ii) *Each convolution operator on $\mathcal{O}_{[+]}^{\{-\infty\}}$ can be extended by continuity to a continuous operator on $V_{[+]}$.*

Let us note that in view of the statements of the theorem it is natural to consider that

$$\mathfrak{L}(V_{[+]}) = \mathfrak{N}_+.$$

The proof of the inclusion of $\mathfrak{N}_+$ in the space of convolutors on $\mathcal{O}_{[+]}^{\{-\infty\}}$ is done directly by using expansions of type (51) for $\mathcal{O}_{[+]}^{\{-\infty\}}$ and for $\mathfrak{N}_+$. As far as the reverse inclusion is concerned, we first prove (see the remark following Theorem 1 in Section 3.1) that a convolution operator A on $\mathcal{O}_{[+]}^{\{-\infty\}}$ is an operator con_f, where the distributions $f \in \mathcal{K}_+$ correspond to the restriction of A to $\mathcal{O}_+$. Verification of the inclusion $f \in \mathfrak{N}_+$ uses a version of the kernel theorem (compare with Theorem 2.4).

Theorem 2. *Let $A \in \mathfrak{N}_+$. Existence of a fundamental solution $G \in \mathfrak{N}_+$ is a sufficient condition for unique solubility of* (13) *on $\mathcal{O}_{[+]}^{\{-\infty\}}$, $V_{[+]}$, and a necessary and sufficient condition for solubility on $\mathfrak{N}_+$.*

Since $\mathfrak{N}_+ \subset \mathcal{K}_+$, the Fourier transform maps $\mathfrak{N}_+$ into some set $\mathfrak{L}^+$, while the existence of a fundamental solution is equivalent to the condition

(II′) The symbol $\hat{A}(\tau, \eta)$ is an invertible element of $\mathscr{F}\mathfrak{N}_+ \overset{\text{def}}{=} \mathfrak{L}^+$.

Let us describe the space $\mathfrak{L}^+$. First of all let us notice that for spaces $\Psi_{[+]} = \mathcal{O}_{[+]}^{\{-\infty\}}$, $V_{[+]}$, $\mathfrak{N}_+$, expansions of the form (51), with φ_{p-j} belonging, respectively, to $\mathcal{O}(\mathbb{R}^{n-1})$, $\mathscr{S}'(\mathbb{R}^{n-1})$, and $\mathcal{O}'(\mathbb{R}^{n-1})$, exist. Hence follows

Proposition 1. *A function $\psi(\tau, \eta) \in \mathscr{S}^+$ belongs to $\mathfrak{L}^+$ if and only if we have the expansion* (52), *moreover*: $\psi_{p-j} \in M(\mathbb{R}^{n-1})$, $\psi_{(q)} \in \mathfrak{L}^+$ *and* $\forall s\ \exists\ v_q, \mu_{sq}, c_{sq}$ *such that*

$$|D_\eta^\alpha \psi_{(q)}(\tau, \eta)| < c_{sq}(1 + |\eta|)^{\mu_{sq}} |\operatorname{Im} \tau|^{v_q}(1 + |\tau|)^{-q}, \quad |\alpha| \leqslant s.$$

Proposition 2. *A function $\psi \in \mathfrak{L}^+$ is an invertible element of this ring if and only if*

(a) *ψ is an invertible element of $\mathfrak{L}^+$*;

(b) *the highest order coefficient $\psi_p(\eta)$ is an invertible element of $\mathscr{M}$.*

Condition (a) is equivalent to the estimate

$$|\psi(\tau, \eta)| > \operatorname{const}(1 + |\tau| + |\eta|^\mu |\operatorname{Im} \tau|^r, \quad \operatorname{Im} \tau > 0. \tag{56′}$$

From (56), (54′) and the Seidenberg-Tarski theorem we have

Theorem 3. *A differential operator with constant coefficients $P(D_t, D_y) = \sum p_j(D_y) D_t^j$ has a fundamental solution in $\mathfrak{N}_+$ if and only if*

$$P(\tau, \eta) \neq 0, \quad \operatorname{Im} \tau < 0, \tag{57}$$

and condition (54′) *holds*,

Comparing Theorem (3) and Theorem 3.2, we see that the solubility condition for the inhomogeneous Cauchy problem includes the solubility condition for the homogeneous problem as well as the condition of non-vanishing of the coefficient of the highest power in τ of $P(\tau, \eta)$.

3.4. Relation to the Classical Formulation of the Cauchy Problem. Let

$$\Phi = \mathscr{S}, \quad \mathcal{O}, \quad A \in \mathfrak{L}(\Phi_{[+]}^{\{-\infty\}}) \quad \text{and} \quad m = \deg_t A.$$

Then the following operators are continuous:

$$\operatorname{con}_A\colon \Phi_{[+]}^{\{p\}} \to \Phi_{[+]}^{\{p-m\}} \quad \forall p \in Z. \tag{58}$$

If A satisfies the conditions of Theorem 2 of Section 3.1 and of Theorem 2 of Section 3.3, then operators (58) have a continuous inverse that is a convolution operator with $G \in \mathfrak{L}(\Phi_{[+]}^{\{-\infty\}})$, $\deg_t G = -m$,

$$(\operatorname{con}_A)^{-1} = \operatorname{con}_G\colon \Phi_{[+]}^{\{p-m\}} \to \Phi_{[+]}^{\{p\}}. \tag{59}$$

Let us consider particular cases of isomorphism (58), (59).

(a) $p = 0$, $\deg_t A = m > 0$. According to the definition of the space $\Phi_{[+]}^{\{-m\}}$, the isomorphism (58) means that there exists a unique solution to the problem

$$A * u_{[+]} = f_{[+]} + \sum_{j=0}^{m-1} f_j(y) D_t^j \delta(x), u_{[+]}, \quad f_{[+]} \in \Phi_{[+]}^{\{0\}}, \quad f_j \in \Phi(\mathbb{R}^{n-1}), \tag{60}$$

$$j = 0, 1, \ldots, m-1.$$

In view of the gradation lemmas of Sections 3.2, 3.3, the distribution A can be represented in the form $A = P + A_0$, where $A \in U_{[+]}^{\{0\}}$ (or $\mathfrak{N}_+^{\{0\}}$), while $P = \sum_{j=0}^{m} p_j(D_y)D_t^j\delta(t)$ is a differential operator. Then (60) can be rewritten in the form (see (39), (40))

$$\theta_+(Pu + A_0 * u_{[+]}) + \frac{1}{i}\sum_{j=0}^{m-1}\sum_{l=0}^{m-1-j} p_{j+l+1}(D_j)D_t^l u(0, y)D_t^j\delta(t)$$

$$= \theta_+ f + \sum_{j=0}^{m-1} f_j(y)D_t^j\delta(t), \tag{61}$$

where $u \in \Phi$ is an arbitrary smooth extension of $p_{\oplus}u_{[+]}$, and moreover the left hand side does not depend on the choice of the extension. As first terms in both the right hand and in the left hand side belong to $\Phi_{[+]}$, while the rest are distributions concentrated in the plane $t = 0$, we deduce from (61) the equalities

$$\theta_+(Pu + A_0 * u_{[+]}) = \theta_+ f, \tag{62}$$

$$p_m(D_y)D_t^{m-1-j}u(0, y) + \sum_{l<m-1-j} p_{j+l+1}(D_y)D_t^l u(0, y) = if_j(y),$$

$$j = 0, \ldots, m-1. \tag{63}$$

In the case of differential operators $A_0 \equiv 0$, that is, (62) becomes (37). Next we show that suitably choosing the functions $f_j(y) \in \Phi = \mathcal{S}, \mathcal{O}$ in (60), we obtain a solution u_+ that satisfies the Cauchy conditions (38) as $t \to +0$. For that we take as f_j functions obtained from (63) if we substitute $\varphi_k(y)$ for $D_t^k u(0, y)$. Clearly, if $\varphi_k(y) \in \Phi, f_j(y) \in \Phi$. Now let us show that with such a choice of f_j in (60) we obtain a solution of the problem (37), (38). In fact, from (63) it follows that

$$P_m(D_y)(D_t^{m-1-j}u(0, y) - \varphi_{m-1-j}(y)) = \sum_{t<m-1-j} P_{j+l-1}(D_y)(D_t^l u(0, y) - \varphi_l(y)).$$

For $j = m - 1$, we get

$$P_m(D_y)(u(0, y) - \varphi_0(y)) = 0.$$

If condition (54′) is satisfied, the operator $P_m(D_y)$ is invertible on $\Phi = \mathcal{S}, \mathcal{O}$, whence $u(0, y) = \varphi_0(y)$. Consecutively setting $j = m - 2$, etc., we arrive at the relations (37).

Thus, the general theorems of this section lead to a solubility theorem for the original problem (37), (38) with initial data in $\mathcal{S}, \mathcal{O}$. From that, using the scheme of § 1, we can derive Petrovskij [1938] type solubility theorems in spaces C_l^s.

(b) $\deg_t A = -m, m > 0, p = -m$. In this case the isomorphism (58) means that for any right hand side $f \in \Phi_{[+]}$, we can pick functions $u_+ \in \Phi_{[+]}^{\{m\}}, u_j(y) \in \Phi(\mathbb{R}^{n-1})$, $j = 0, \ldots, m-1$, such that

$$A * \left(u_+ + \sum_{j=0}^{m-1} u_j(y)D_t^j\delta(t)\right) = f_+, \tag{64}$$

that is, we have a problem in half-space with potentials on the boundary.

(c) $m = \deg_t A = p = 0$. In this case the equation $A * u_{[+]} = f_{[+]}$ is uniquely soluble in $\Phi_{[+]}$.

3.5. Convolution Equations in a Finite Strip. If in the construction of spaces in this section, we start in (44) not with $H_{(l)+}^{(p,s)}$, but with $H_{(l)}^{(p,s)}[a, b)$, we obtain spaces in a strip, which are naturally denoted by

$$\mathscr{S}^{\{-\infty\}}[a, b), \quad \mathscr{O}^{\{-\infty\}}[a, b), \quad U[a, b), \quad V[a, b).$$

A convolution operator on $\mathscr{S}^{\{-\infty\}}[a, b)$, $\mathscr{O}^{\{-\infty\}}[a, b)$ is a continuous operator, the restriction of which to $\mathscr{S}[a, b)$, $\mathscr{O}[a, b)$ is a convolution operator. Using the procedure of Section 2.6 we can show that

$$\mathfrak{L}(\mathscr{S}^{\{-\infty\}}[a, b)) = \mathfrak{L}(\mathscr{O}^{\{-\infty\}}[a, b)) = U[0, b - a). \tag{65}$$

Each convolution operator on $\mathscr{S}^{\{-\infty\}}[a, b)$, $\mathscr{O}^{\{-\infty\}}[a, b)$ can be extended by continuity to $U[a, b)$, $V[a, b)$, so that

$$\mathfrak{L}(U[a, b)) = \mathfrak{L}(V[a, b)) = U[0, b - a).$$

As in the case of the homogeneous Cauchy problem, it can be proved that solubility of the convolution equation in spaces named above is equivalent to the existence of a fundamental solution $G \in U[0, b - a)$.

The expansion

$$A = A_p(y)(D_t - i)^p\delta(t) + A_{p-1}(y)(D_t - i)^{p-1}\delta(t) + \cdots + A_{(q)},$$

is also valid for distributions $A \in U[0, c)$. Here $A_p, A_{p-1}, \ldots \in \mathscr{O}'(\mathbb{R}^{n-1})$, $A_{(q)} \in U^{(q)}[0, c)$, and the number p and the distributions $A_p, A_{p-1}, \ldots$ are independent of q.

Invertibility of $A \in U[0, c)$ (see Proposition 2 of Sections 3.2, 3.3) is equivalent to invertibility of A in the bigger ring $\mathscr{O}'[0, c)$ and the invertibility of the highest order term $A_p(y)$ in $\mathscr{O}'(\mathbb{R}^{n-1})$.

As we already mentioned in Section 2.6, in the case of differential operators, that is, for $A = P(D)\delta$, an ineffective invertibility condition in $\mathscr{O}'[0, c)$ becomes the effective Petrovskij homogeneous correctness condition. In the case of inveribility in $U[0, c)$ it is supplemented by condition (54′).

Theorem. *Let $\Phi[a, b) = \mathscr{S}^{\{-\infty\}}[a, b), \mathscr{O}^{\{-\infty\}}[a, b), U[a, b), V[a, b)$. The differential equation* (17) *is uniquely soluble in the space $\Phi[a, b)$ for any $a < b$ (or for some $a < b$) if and only if the polynomial P satisfies the Petrovskij homogeneous correctness condition and the additional condition* (54′).[1]

Let us make an interim summary. We embedded the inhomogeneous Cauchy problem into the problem of invertibility of a class of convolution operators. These operators form a subset of operators corresponding to the homogeneous Cauchy problem. Symbols of these operators can be expanded in an asymptotic

[1] In the case of operators P that are soluble with respect to the highest order derivative in time, the Petrovskij correctness condition holds (see the Introduction).

Laurent series in the variable dual to t. They constitute a generalization of symbols with a transmission (smoothness) condition that arise in the study of elliptic problems (see Vishik and Eskin [1964], Boutet de Monvel [1966]), to the inhomogeneous case. Invertibility conditions for convolution operators corresponding to the inhomogeneous Cauchy problem, consist of invertibility conditions for the homogeneous Cauchy problem and of conditions for the invertibility of the coefficient of the highest order time derivative in the ring of convolution operators with a smaller number of variables. In the case of differential operators, condition (54′) is added to the Petrovskij homogeneous correctness condition.

§4. Boundary Value Problems for Convolution Equations

As a prototype for the theory of §§ 1, 2, together with the theory of the Cauchy problem for partial differential equations with constant coefficients, we can take the theory of integral equations on the semi-infinite real line (in, say, L_p) with a difference kernel supported on the semi-axis $t \geqslant 0$:

$$u(t) + \int_0^\infty a(t-\theta)u(\theta)\,d\theta = f(t) \in L_p(0, \infty). \tag{66}$$

Extending u and f by zero for $t \leqslant 0$, we obtain an equation on the entire axis which can be solved with Fourier integrals. The theory of equations (66) with kernel $a(t)$ with support in the entire axis can serve as a model of theory of boundary value problems for convolution equations. In the early thirties, a method of solution of such equations was suggested by Wiener and Hopf. An exhaustive theory of equations (66) with kernel $a(t) \in L_1$ was developed by Krein [1958]. He showed that the left hand side defines (for example, in L_p) a Fredholm operator if and only if the kernel $A(t) = a(t) + \delta(t) \in W$ (W is the Wiener ring) admits the (canonical) factorization $A = A_+ * A_-$, where $A_\pm$ are invertible elements of the subrings $W_\pm \subset W$ consisting of elements of W concentrated in $\pm t \geqslant 0$. In the present section we construct the analog of this theory in spaces of smooth functions and of distributions. For simplicity, we restrict ourselves to the case of the semi-infinite real line, and not of a half-space.

Setting $A = \delta(t) + a(t)$, we rewrite (66) in the form

$$p_\oplus(A * u) = f, \tag{67}$$

where $p_\oplus$ is the canonical projection operator onto the quotient space.

We shall seek a solution $u \in \Phi_+$ of equation (67) for an arbitrary right hand side $f \in \Phi_\oplus$. As A we shall take distributions, for which the left hand side of (67) makes sense and defines a continuous operator from Φ_+ into $\Phi_\oplus$. We shall call these distributions *Wiener-Hopf convolutors*. Let us start by describing them in the case $\Phi = \mathcal{S}, \mathcal{O}$.

Remark. For the case of homogeneous distributions, the highest order symbol of which is different from zero outside of the origin (elliptic convolution equations), there exists a well developed theory (see Vishik and Eskin [1964]). We shall investigate equation (67) for more general classes of symbols, in which, in general, gradation with respect to homogeneity is impossible.

4.1. Wiener-Hopf Convolutors. Let Φ be one of the spaces of § 1 on the real line. The left shift operator T_h, $h \geqslant 0$, maps Φ_- into itself and induces a continuous operator $T_h: \Phi_\oplus \to \Phi_\oplus$.

We shall say that an operator $A: \Phi_+ \to \Phi_\oplus$ commutes with left shifts if

$$\{\varphi, T_h\varphi \in \Phi_+\} \Rightarrow \{AT_h\varphi = T_hA\varphi\}. \tag{68}$$

We call an continuous operator A from Φ_+ into $\Phi_\oplus$ a convolution operator if it commutes with left shifts in the sense of (68). An exact analog of Proposition 2.3 holds for such operators. Let $(\Phi_\infty)'$ be the space dual to the inductive limit $\bigcup_{h\geqslant 0} IT_h\Phi_+$. Then for any convolution operator $A: \Phi_+ \to \Phi_\oplus$ there exists a distribution $f \in (\Phi_\infty)'$ such that (10) holds for $t > 0$.

Conversely, if for $f \in (\Phi_\infty)'$ and for any $\varphi \in \Phi_\oplus$ the projection $p_\oplus$ maps the right hand side of (10) into $\Phi_\oplus$, then (10) defines a convolution operator from Φ_+ into $\Phi_\oplus$.

Let us denote the convolution operator corresponding to the right hand side of (10) by $p_\oplus \mathrm{con}_f$, and the family of $f \in (\Phi_\infty)'$ for which this operator is continuous, by $\mathfrak{L}(\Phi_+, \Phi_\oplus)$. As we already said above, we call the elements of $\mathfrak{L}(\Phi_+, \Phi_\oplus)$ Wiener-Hopf convolutors in Φ.

Let us remark that for $\Phi = \mathscr{S}, \mathscr{O}$, and $f \in \mathfrak{L}(\Phi_+, \Phi_\oplus)$ the operators

$$p_\oplus \mathrm{con}_f: (\Phi')_+ \to (\Phi')_\oplus, \quad p_\oplus \mathrm{con}_g: \Phi_+ \to \Phi_\oplus$$

are adjoint to each other if $g = If$. Therefore it is natural to take

$$\mathfrak{L}((\Phi')_+, (\Phi')_\oplus) = I\mathfrak{L}(\Phi_+, \Phi_\oplus), \quad \Phi = \mathscr{S}, \mathscr{O}. \tag{69}$$

Theorem.

$$\text{(i)}\ \ \mathfrak{L}(\mathscr{S}_+, \mathscr{S}_\oplus) = \{f \in \mathscr{S}', p_\oplus f \in (\mathscr{O}')_\oplus\},$$

$$\text{(ii)}\ \ \mathfrak{L}(\mathscr{O}_+, \mathscr{O}_\oplus) = \{f \in \mathscr{S}', p_\ominus f \in (\mathscr{O}')_\ominus\}.$$

Let us clarify the meaning of the equalities above. In case (i) the Wiener-Hopf convolutors must decrease rapidly from the right: this is necessary in order that the convolution product decay rapidly for $t > 0$. In case (ii) the Wiener-Hopf convolutors decay rapidly from the left: this is necessary in order that the convolution with functions temperately growing for $t > 0$, exist. From equalities (i), (ii) it follows that

$$\mathfrak{L}(\mathscr{S}_+, \mathscr{S}_\oplus) \cap \mathfrak{L}(\mathscr{O}_+, \mathscr{O}_\oplus) = \mathscr{O}', \tag{70}$$

that is, distributions in $\mathscr{O}'$ are Wiener-Hopf convolutors both on rapidly decreasing and on slowly growing functions. Moreover, these distributions preserve the fixed order of growth or decay of functions. It can be shown that

$$\bigcap_l \mathfrak{L}(C^{(\infty)}_{(l)+}, C^{(\infty)}_{(l)\oplus}) = \bigcap_l \mathfrak{L}(H^{(\infty)}_{(l)+}, H^{(\infty)}_{(l)\oplus}) = \mathscr{O}'. \tag{70'}$$

Remark. From the theroem and from definition (69) it follows that

$$\mathfrak{L}((\mathcal{O}')_+, (\mathcal{O}')_\oplus) = \mathfrak{L}(\mathcal{S}_+, \mathcal{S}_\oplus), \tag{71}$$

$$\mathfrak{L}((\mathcal{S}')_+, (\mathcal{S}')_\oplus) = \mathfrak{L}(\mathcal{O}_+, \mathcal{O}_\oplus). \tag{71'}$$

These equalities can be arrived at directly, having established the possibility to extend by continuity convolution operators $p_\oplus \mathrm{con}_f$: $\mathcal{S}_+, \mathcal{O}_+ \to \mathcal{S}_\oplus, \mathcal{O}_\oplus$ to continuous operators from $(\mathcal{O}')_+, (\mathcal{S}')_+$ to $(\mathcal{O}')_\oplus, (\mathcal{S}')_\oplus$.

4.2. Wiener-Hopf Convolutors with a Transmission[2] Property. As we saw already in § 3, in the study of the inhomogeneous Cauchy problem arise spaces of functions that are smooth in the interior of a half-space and have singularities on the boundary. In spaces $\mathfrak{L}(\Phi_+)$ we singled out the convolutors that can be extended to such spaces. In this section we shall consider the analogous problem for Wiener-Hopf convolutors. We shall single out *Wiener-Hopf convolutors with a transmission property*. We note that this kind of spaces of distributions first appeared in the work of Vishik and Eskin [1964] and of Boutet de Monvel [1966].

Let $\Phi = \mathcal{S}, \mathcal{O}$, and, as in § 3, let $\Phi_{[+]} = \theta_+(t)\Phi$, where $\theta_+(t)$ is the characteristic function of the semi-infinite real line $t > 0$. Differential operators act on the space $\Phi_{[+]}$. The action of these (unbounded) operators is induced by their action on $(\mathcal{S}')_+ \supset \Phi_{[+]}^{\{-\infty\}}$.

A convolution operator from $\Phi_{[+]}$ into $\Phi_\oplus$ is a continuous operator which

(1) induces a convolution operator from Φ_+ into $\Phi_\oplus$;

(2) commutes with differential operators.

Let us denote by $\mathfrak{L}(\Phi_{[+]}, \Phi_\oplus) \subset \mathfrak{L}(\Phi_+, \Phi_\oplus)$ the corresponding space of convolutors.

Theorem. *A distribution $f \in \mathfrak{L}(\Phi_+, \Phi_\oplus)$, $\Phi = \mathcal{S}, \mathcal{O}$, belongs to $\mathfrak{L}(\Phi_{[+]}, \Phi_\oplus)$ if and only if $p_\oplus f \in \Phi_\oplus$.*

Taking into account Theorem 4.1, the proposition above can be rewritten as the equalities

$$\mathfrak{L}(\mathcal{S}_{[+]}, \mathcal{S}_\oplus) = \{f \in \mathcal{S}', p_\oplus f \in \mathcal{S}_\oplus\}, \tag{72}$$

$$\mathfrak{L}(\mathcal{O}_{[+]}, \mathcal{O}_\oplus) = \{f \in \mathcal{S}', p_\ominus f \in (\mathcal{O}')_\ominus, p_\oplus f \in \mathcal{O}_\oplus\}. \tag{73}$$

Comparing (72), (73) with Theorem 4.1, we see that the transmission condition does not change the character of growth or decay of the convolutor; it only adds it a smoothness condition for $t > 0$.

From (72) and (73) it follows that

$$\mathfrak{L}(\mathcal{S}_{[+]}, \mathcal{S}_\oplus) \cap \mathfrak{L}(\mathcal{O}_{[+]}, \mathcal{O}_\oplus) = \{f \in \mathcal{O}', p_\oplus f \in \mathcal{S}_\oplus\} \stackrel{\text{def}}{=} U. \tag{74}$$

[2] The term "transmission property" is equivalent to the "smoothness condition" introduced by Vishik and Eskin.

It can be shown that distributions on U are convolutors on functions of fixed growth (decay), that is

$$\bigcap_l \mathfrak{L}(C^{(\infty)}_{(l)[+]}, C^{(\infty)}_{(l)\oplus}) = \bigcap_l \mathfrak{L}(H^{(\infty)}_{(l)[+]}, H^{(\infty)}_{(l)\oplus}) = U. \tag{74'}$$

4.3. Equations on the Semi-infinite Axis (67) Corresponding to Convolutors from U. Next we formulate the main result of this section, having to do with the solubility of equation (67) in the case of convolutors having the transmission property. An operator from Φ_+ into $\Phi_\oplus$ that corresponds to the left hand side of (67) can have a kernel or a cokernel. To account for these, we pass from the spaces Φ_+, $\Phi_\oplus$ to scales $\{\Phi^{(q)}_{[+]}\}$, $\{\Phi^{(r)}_{\oplus}\}$, which contain the original spaces. The first of these scales was defined in Section 3.1. Let us define the second one.

Let U be the space on the right hand side of (74). Then $U_- = (\mathcal{O}')_-$ and $U_+ = \mathcal{S}^{\{-\infty\}}_{[+]}$. For any integer r, let us set

$$\mathcal{S}^{\{r\}}_{\oplus} = U/\delta^+_{-r}(\mathcal{O}')_-, \tag{75}$$

where $\delta^+_{-r}(D)$ is the PDO corresponding to the symbol $(D_t - i)^r$. For $r = 0$, the space (75) is identified with $\mathcal{S}_\oplus$. There are surjectives maps of quotient spaces

$$\to \mathcal{S}^{\{r-1\}}_{\oplus} \to \mathcal{S}^{\{r\}}_{\oplus} \to \mathcal{S}^{\{r+1\}}_{\oplus} \to \cdots, \tag{76}$$

where kernels of all the maps are one-dimensional. Let us denote by $p^{\{r\}}_{\oplus}$ the canonical projection of U onto $\mathcal{S}^{\{r\}}_{\oplus}$.

Lemma. *For any integer r the mapping*

$$p^{\{r\}}_{\oplus}: \mathcal{S}^{\{r\}}_{[+]} \to \mathcal{S}^{\{r\}}_{\oplus}$$

is an isomorphism. The inverse operator has the form

$$\varphi \mapsto \delta^+_{-r}(D)\theta_+(t)\delta^+_r(D)\varphi.$$

Let us associate with $A \in U$ a continuous operator in the scales

$$p^{\{r\}}_{\oplus}\operatorname{con}_A: \mathcal{S}^{\{-\infty\}}_{[+]} \to \mathcal{S}^{\{r\}}_{\oplus}, \quad A \in U. \tag{77}$$

If in the construction above we substitute for $UV = \{f \in \mathcal{S}', p_\oplus f \in \mathcal{O}_\oplus\}$, then we obtain the spaces $\mathcal{O}^{\{r\}}_{\oplus}$. To each distribution $A \in U$ corresponds a continuous operator

$$p^{\{r\}}_{\oplus}\operatorname{con}_A: \mathcal{O}^{\{-\infty\}}_{[+]} \to \mathcal{O}^{\{r\}}_{\oplus}. \tag{77'}$$

In a similar way, replacing U by $U_{(l)} = \{f \in H^{(-\infty)}_{(l)}, p_\oplus f \in H^{(\infty)}_{(l)\oplus}\}$ we define spaces $U^{\{q\}}_{(l)[+]}$ and $U^{\{q\}}_{(l)[+]}$. For $A \in U$, the following mapping between scales is defined:

$$p^{\{-\infty\}}_{\oplus}\operatorname{con}_A: U^{\{-\infty\}}_{(l)[+]} \to U^{\{-\infty\}}_{(l)\oplus}. \tag{78}$$

Theorem. *For a distribution $A \in U$ the following conditions are equivalent:*
(I) *A is an invertible element of U, that is, $A * G = A * G = \delta(t)$ for some $G \in U$.*
(II) *A admits a canonical factorization in U, that is*

$$A = A_+ * A_-, \tag{79}$$

where $A_\pm$ are invertible elements of $U_\pm$, that is, $A_\pm * G_\pm = A_\pm * G_\pm = \delta(t)$ for some $G_\pm \in U_\pm$.

(III) *For any $l \in \mathbb{R}$ the mapping (78) is a finite order scale isomorphism, that is, there is an integer c, such that for any r the mappings*

$$p_{\oplus}^{\{r\}}\mathrm{con}_A\colon U_{(l)[+]}^{\{r+c\}} \to U_{(l)\oplus}^{\{r\}} \tag{80}$$

are isomorphisms between spaces.

The remaining part of this section is devoted to a discussion of the theorem stated above. In Section 4.4 we describe invertible elements of U and expalin the equivalence between (I) and (II). In Section 4.5 we present the method of Wiener-Hopf and prove equivalence of (II) and (III). In the end of the section we discuss various cases of the isomorphisms (78), (80).

4.4. Factorization of Distributions in U. We start by describing the invertible elements of the algebra U. According to definition (74),

$$A \in U \Leftrightarrow A = A_- + a, \quad A_- \in (\mathcal{O}')_-, \quad a \in \mathcal{S}. \tag{81}$$

if we denote by $\hat{U}$ the Fourier-image of U, then

$$\hat{A} \in \hat{U} \Leftrightarrow \hat{A} = \hat{A}_- + \hat{a}, \quad \hat{A}_- \in M^-, \hat{a} \in \mathcal{S}, \tag{81$'$}$$

that is, symbols of distributions from U, up to rapidly decaying functions, are functions with power law growth that are holomorphic in the upper half-space.

Let us clarify the invertibility conditions for A in U. As from invertibility in U follows invertibility in the larger ring $\mathcal{O}'$, we have the following lower bound for the symbol $\hat{A}$:

$$|\hat{A}(\sigma)| > c_1(1 + |\sigma|)^\mu, \quad \sigma \in \mathbb{R}. \tag{82}$$

If $G = G_- + g$, $G_- \in (\mathcal{O}')_-$, $g \in \mathcal{S}$ is the inverse of the distribution A written in the form (81), then

$$A_- * G_- - \delta(t) = -(A_- * g + G_- * a + a * g).$$

The left hand side of that equality belongs to $(\mathcal{O}')_-$, while the right hand side is in $\mathcal{S}_-$. Passing to Fourier-images, we obtain

$$\hat{A}_-(\tau)\hat{G}_-(\tau) - 1 = 0(|\tau|^{-\infty}), \quad |\tau| \to \infty, \quad \operatorname{Im}\tau \geqslant 0.$$

Hence we deduce a lower bound for large $|\tau|$:

$$|\hat{A}_-(\tau)| > c_R(1 + |\tau|)^\nu, \quad \operatorname{Im}\tau \geqslant 0, \quad |\tau| > R. \tag{83}$$

As in the representation (81) the distribution A_- is defined mod $\mathcal{S}_-$, the symbol $\hat{A}_-(\tau)$ is defined up to a function that is holomorphic in the upper half-plane and decays faster than any power. Therefore the estimate (83) is independent of the choice of A_-.

Conditions (82), (83) are in fact also sufficient invertibility conditions for A in U. Let (83) be satisfied. As the function $\hat{A}_-(\tau)$ can have only a finite number of zeroes inside a half-disc of radius R, $\hat{A}_-(\tau) = P(\tau)\hat{B}_-(\tau)$, where $P(\tau)$ is a polynomial, while $\hat{B}_-(\tau)$ is an invertible element of $\mathcal{M}^-$. Thus

$$\hat{A}(\sigma) = \hat{B}_-(\sigma)\hat{C}(\sigma), \quad \hat{C}(\sigma) = P(\sigma) + \hat{a}(\sigma)/B_-(\sigma), \tag{84}$$

and, in view of (82),

$$|\hat{C}(\sigma)| > c_2(1 + |\sigma|)^{\mu'}, \tag{82'}$$

Next we describe a subclass $\hat{\mathcal{K}} \subset \hat{U}$, to which the symbol $\hat{C}$ belongs; moreover, condition (82) will turn out to be the invertibility condition in that class.

Let us denote by $\mathcal{K}$ a subspace of U:

$$\mathcal{K} = \{A \in \mathcal{O}', p_\oplus A \in \mathcal{S}_\oplus, p_\ominus A \in \mathcal{S}_\ominus\}.$$

Hence it follows that

$$C \in \mathcal{K} \Leftrightarrow C = C_+ + a = C_- + a', \quad C_\pm \in \mathcal{S}^{\{+\infty\}}_{[\pm]}, a, a' \in \mathcal{S}.$$

From the description of the Fourier-image of the space $\mathcal{S}^{\{-\infty\}}_{[+]}$ presented in § 3, it follows that $\hat{C} \in \hat{\mathcal{K}}$ if and only if $\hat{C} \in \mathcal{M}$, and for every $z \in \mathbf{C}$, $\operatorname{Im} z \neq 0$, $\hat{C}$ can be expanded in the asymptotic series

$$\hat{C}(\sigma) = \sum_{j=0}^{\infty} c_{p-j}(\sigma - z)^{p-j},$$

and series for the derivatives are obtained by differentiating this series. The estimate (82′) ensures invertibility in $\hat{\mathcal{K}}$. Thus we have established

Proposition. *For $A \in U$, the following conditions are equivalent:*
(I) *A is an invertible element of U;*
(II) *estimates* (82), (83) *are valid for the symbol $\hat{A}$;*
(III) *A admits the representation*

$$A = B_- * C, \; B_-, C \text{ are invertible elements of } (\mathcal{O}')_-, \mathcal{K}. \tag{85}$$

Definition of the canonical factorization has already been given in condition (II) of Theorem 4.3. Distribution $A_\pm$ of (79) is defined up to a constant. In fact, let $A = B_+ * B_-$, where $B_\pm$ is an invertible element of $U_\pm$. Then

$$\hat{C}(\tau) \overset{\text{def}}{=} \frac{\hat{A}_+(\tau)}{\hat{B}_+(\tau)} = \frac{\hat{B}_-(\tau)}{\hat{A}_-(\tau)} \in \mathcal{M}^+ \cap \mathcal{M}^-,$$

that is, $\hat{C}$ is an entire function that grows not faster than a polynomial. By Liouville's theorem $\hat{C}$ is a polynomial. In a similar way it is proved that $1/\hat{C}$ is a polynomial, whence $\hat{C} \equiv \text{const}$.

Let us set $d_+ = \deg A_+ (A_+ \in U_+)$. As A_+ is defined up to a constant, d_+ is uniquely determined by the canonical factorization. This integer is called the index of the factorization.

Let us sketch the proof of equivalence of (I) and (II) in Theorem 4.3. In view of the proposition, it suffices to establish the existence of canonical factorization for invertible elements $A \in \hat{\mathcal{K}}$ (that is, of ones that satisfy (82)):

$$A = A_+ * A_-, \quad A_\pm \in \mathcal{K}_\pm, \quad (A_\pm * G_\pm = G_\pm * A_\pm = \delta(t)). \tag{86}$$

Without loss of generality, we shall assume that $\deg A = 0$; otherwise we can represent A in the form $((D_t - i)^p\delta(t)) * B$, where $\deg B = 0$. If $\deg A = 0$, then $\hat{A}(-\infty) = \hat{A}(+\infty) \neq 0$ (due to (82)). In this case the increment of $\arg \hat{A}(\sigma)$ as $-\infty < \sigma < \infty$ is an integer multiple of 2π, say, $2m\pi$. This also is the increment of the symbol $(\sigma + i)^m(\sigma - i)^{-m}$, which already is factorized.

Thus let us consider the factorization (86) of A such that $\deg A = 0$ and the increment of $\arg \hat{A}(\sigma)$ is zero; we shall take $\hat{A}(\pm\infty) = 1$. Then $\ln \hat{A}(\sigma)$ is a single-valued function on the compactified real line. This function can be expanded in power series of $1/\sigma$, that is, $\ln \hat{A}(\sigma) \in \hat{\mathscr{K}}$, $\deg \ln \hat{A}(\sigma) \leqslant -1$. Such a function can be represented as

$$\ln \hat{A}(\sigma) = B_+(\sigma) + B_-(\sigma), \quad B_\pm \in \hat{\mathscr{K}}_\pm, \quad \deg B_\pm \leqslant -1,$$

hence

$$\hat{A}(\sigma) = \hat{A}_+(\sigma)\hat{A}_-(\sigma), \quad \hat{A}_\pm(\sigma) = \exp(B_\pm(\sigma)) \in \hat{\mathscr{K}}_\pm.$$

Passing to the inverse Fourier transform, we obtain the desired factorization (86).

4.5. The Wiener-Hopf Method. Let A be an invertible element of U. Then A admits the canonical factorization (79), and let d_+ be the factorization index. As operator con_{A_-}, $A_- \in (\mathcal{O}')_-$ commutes with projection operators $p^{\{r\}}_\oplus$, we have the commutative diagram

$$\begin{array}{ccc}
U^{\{r+d_+\}}_{(l)[+]} & \xrightarrow{p^{\{r\}}_\oplus \mathrm{con}_A} & U^{\{r\}}_{(l)\oplus} \\
\Big\downarrow \mathrm{con}_{A_+} & & \Big\uparrow \mathrm{con}_{A_-} \\
U^{\{r\}}_{(l)[+]} & \xrightarrow{p^{\{r\}}_\oplus} & U^{\{r\}}_{(l)\oplus}
\end{array}$$

If $A_\pm$ is an invertible element of $U_\pm$, the vertical mappings are space isomorphisms. According to Lemma 4.3 the lower mapping is an isomorphism, and therefore the upper one is also an isomorphism. Thus, we have shown the implication (II) $\Rightarrow$ (III) of Theorem 4.3, and moreover the number c in (80) equals the factorization index.

Remark. From what we said above it follows that for $f \in U_{(l)}$ the equation

$$A * u - f \in \delta^+_{-r}(D)(\mathcal{O}')_- \tag{87}$$

has a unique solution $u \in U^{\{r+d_+\}}_{(l)[+]}$, which is given by the explicit formula

$$u = G_+ * \delta^+_{-r}\theta_+\delta^+_r(D)G_- * f,$$

where $G_\pm$ is the distribution from (79).

Let us now sketch the proof of necessity of the invertibility condition for A for solubility of equation (87). Without loss of generality we may assume that $r = 0$, $d_+ = 0$ (otherwise we can replace u by $\delta^+_{d_+}(D)u$). Thus, for all l, and in particular for $l = 0$, the operator

$$p_\oplus \,\mathrm{con}_A \colon H^{(\infty)}_{[+]} \to H^{(\infty)}_\oplus$$

has a bounded inverse, and therefore there exists an s such that we have the estimate

$$\int_0^\infty |u(t)|^2\,dt \leqslant c\int_0^\infty |\delta_s^-(D)(A*u)(t)|^2\,dt\ \forall u \in H_{[+]}^{(\infty)}. \tag{88}$$

Without loss of generality we may assume that $s = 0$ (otherwise we can replace A by $\delta_s^-(D)A$). We have to show that conditions (82), (83) on the symbol $\hat{A}$ follow from estimate (88).

Let us consider in detail the derivation of (83). For that let us substitute into (88) the auxiliary function

$$v_z(t) = \sqrt{2\,\mathrm{Im}\,z}\,\theta_+(t)e^{izt} \in \mathscr{S}_{[+]},$$

corresponding to an arbitrary point $z \in \mathbf{C}$, $\mathrm{Im}\,z > 0$; the factor $\sqrt{2\,\mathrm{Im}\,z}$ is included for normalization: $\|v_z\| = 1$. A direct calculation shows that

$$\hat{v}_z(\sigma) = \frac{i\sqrt{\mathrm{Im}\,z}}{\sqrt{\pi}(\sigma - z)}, \quad (A_- * v_z)(t) = -\hat{A}_-(z)v_z(t).$$

Substituting it (with $s = 0$) into (88) we obtain

$$c^{-1} \leqslant |\hat{A}_-(z)| + \|a * v_z\|, \tag{89}$$

where

$$\|a * v_z\| = \|\hat{a}(\sigma)\hat{v}_z(\sigma)\| \leqslant (\pi\,\mathrm{Im}\,z)^{-1/2}\|a\|.$$

Substituting this estimate into (89), we get that

$$|\hat{A}_-(z)| > \frac{1}{2c} \quad \text{for} \quad \mathrm{Im}\,z \geqslant 4c^2\|a\|^2/\pi. \tag{90}$$

On the other hand, as $|\hat{v}_z(\sigma)| < \mathrm{const}|\sigma - z|^{-1}$,

$$\|a * v_z\| \leqslant c(h)/|\mathrm{Re}\,z|, \quad a \in \mathscr{S}, \quad 0 \leqslant \mathrm{Im}\,z \leqslant h;$$

comparing this inequality with (90), we arrive at (83).

In conclusion, let us go back to those corollaries from Theorem 4.3, that are relevant to the question of solubility of the original equation (67) in the space $\Phi_{[+]}^{\{q\}}$, $\Phi = \mathscr{S}, O, H_{(l)}^{(\infty)}$:

a) Let the factorization index $d_+ > 0$. For $r = -d_+$, theorem 4.3 states that for each $f \in \Phi = U, V, U_{(l)}$ there exists unique solution $u \in \Phi_{[+]}$ of the equation

$$A * u - f \in \delta_{d_+}^+(D)(\mathscr{O}')_-.$$

Let $z_1(t), \ldots, z_{d_+}(t)$ be solutions corresponding to $f = \delta_{j-1}^+(D)\delta(t)$, $j = 1, \ldots, d_+$. The functions $z_j(t)$ are linearly independent and belong to the kernel of the operator $p_\oplus\,\mathrm{con}_A$. Conversely, it can be shown that every element in the kernel of $p_\oplus\,\mathrm{con}_A$ is a linear combination of the z_j. Thus, for $d_+ > 0$ equation (67) has a d_+-dimensional kernel. In order to make this problem well defined, we must impose additional (boundary) conditions on u:

$$B_j(u) = c_j, \quad j = 1, \ldots, d_+,$$

where B_j are linear functionals, and c_j are arbitrary constants. The solubility condition for the resulting problem (*Lopatinskij condition*) consists of the following:

$$\det \|B_j(z_k)\| \neq 0. \tag{91}$$

If

$$B_j(u) = (B_j(D)u)(0),$$

where $B_j(D)$ are pseudo-differential operators, then the Lopatinskij condition (91) takes the form

$$\det \left\| \int_{-\infty}^{\infty} B_j(\sigma)\sigma^{k-1}A_+^{-1}(\sigma)\,d\sigma \right\| \neq 0.$$

b) Let $d_+ < 0$. For $r = 0$, Theorem 4.3 states that for any right hand side $f \in \Phi_{[+]}$ we can find $u \in \Phi_{[+]}^{\{d_+\}}$, such that $A * u - f \in (\mathcal{O}')_-$. Every distribution $u \in \Phi_{[+]}^{\{d_+\}}$ has the form $u = u_{[+]} + \sum c_j D_t^j \delta(t)$, $u_{[+]} \in \Phi_{[+]}$. Thus we obtain solubility of the problem with potentials

$$A * (u_{[+]} + \textstyle\sum c_j D_t^j \delta(t)) - f \in (\mathcal{O}')_-.$$

Hence it follows that

$$u_{[+]} + \textstyle\sum c_j D_t^j \delta(t) = G_+ * \theta_+(G_- * f).$$

The coefficients c_j are equal to zero if and only if

$$(D_t^j(G_- * f))(+0) = 0, \quad j = 0, \ldots, |d_+| - 1.$$

Thus, for $d_+ < 0$, equation (67) has a solution $u_{[+]} \in \Phi_{[+]}$ if and only if the right hand side f satisfies $|d_+|$ additional conditions, that is, the problem has a $|d_+|$-dimensional cokernel.

Chapter 2
Exponential Correctness Classes for the Cauchy Problem

A classical result states that that Cauchy problem for the heat equation can be solved for initial data growing as $\exp(a|x|^2)$. For the wave equation (because of finite domain of dependence) there are in fact no restrictions on the growth rate of the initial data. On the other hand, when solving the Cauchy problem for the Schrödinger equation, it is impossible to depart from initial data with power law growth. How is this distinction expressed in terms of symbols of differential operators?

Initially the problem of determination of maximal exponential growth admissible for the Cauchy data was investigated for parabolic equations and systems thereof (Ladyzhenskaya [1950], Ejdel'man [1964]). Shilov (see Gel'fand and Shilov [1958]) studied the general problem of correctness classes for the Cauchy problem of the form $\exp(a|x|^p)$ for arbitrary systems of differential equations with constant coefficients. He introduced a characteristic of the symbol (the type), that determines the character of the correctness class. Analysis shows that this characteristic is unstable, and Shilov's classes are intrinsically connected with the constant coefficient case.

Below, a different procedure for the determination of correctness classes is adopted. At the level of constant coefficients, the difference lies in the fact that together with the Cauchy problem we consider a problem with exponentially growing right hand side, controlling the character of the change of exponential asymptotics with time. As the result we obtain more stable classes, and the passage to the variable coefficients case becomes possible, so that as a particular case we recover the parabolic theory. Another distinguishing feature lies in that we study anisotropic classes, that contain principally different estimates in different directions (usually the analysis is restricted to isotropic classes of the form $\exp(h(|x))$, where h is a convex function of one variable; more frequently just $\exp(a|x|^p)$ is considered).

The correctness class for the Cauchy problem is determined by the behavior of the symbol of the differential operator in a complex domain (in all the variables, and not only in a variable that is dual to the time, as in Petrovskij's theory). In the version we have adopted everything is determined by the maximal tubular domain in which the symbol is different from zero. The scheme for studying exponential correctness classes is the same as the one used in the case of power law classes (Chapter 1): we study convolution equations in spaces of test functions and of distributions with exponential asymptotic behavior. In particular, we investigate subspaces of $\mathscr{S}'(\mathbb{R}^n)$, elements of which are distributions with support in some fixed convex set, or the dual question about Hardy type spaces of functions holomorphic in tubular domains that take generalized boundary values in $\mathscr{S}'$. These equations have been thoroughly studied in cases of the light cone and of future-past tubes (in fact for arbitrary cones and radial tubular domains) connected with problems of mathematical physics (Vladimirov [1978], Schwarz [1952]).

§ 1. Convolution Equations in Spaces of Exponentially Decaying Functions and Distributions

As our guide, we take the Cauchy problem in exponentially decaying (growing) functions (distributions). All four of the possible combinations can be fruitfully studied. Conjugacy considerations prompt us to group them in twos. Let us start with exponentially decaying functions. We preface this study by reminding the reader of some notions from the theory of convex functions.

1.1. Convex Functions. Let $\mu(x)$ be a function on $\mathbb{R}^n_x$ that takes either finite values or the value $+\infty$. Let dom μ be the closure of the set of x on which $\mu(x)$ is finite (the support of μ). Let us denote by E_μ its epigraph in $\mathbb{R}^{n+1}_{x_0,x}$: the set $\{(x_0, x)\colon x_0 \geqslant \mu(x)\}$. In what follows we shall assume that μ is a convex function, that is, E_μ is a convex set, and that E_μ is a closed set, that is, that μ is lower semicontinuous.

An *asymptotic cone* V_μ for E_μ, or simply for μ, is the maximal convex cone a translation of which is contained in E_μ. If U is a closed set in $\mathbb{R}^n$, then we say that its *indicator function* μ_U is a function that is zero at points $x \in U$ and $+\infty$ otherwise. The function μ_U is convex; dom $\mu_U = U$.

Let $\mathbb{R}^n_\xi$ be the space dual to $\mathbb{R}^n_x$, and let $\langle \xi, x\rangle$ be the duality pairing. Let us denote by $\hat{\mu}(\xi)$ the *conjugate in the sense of Young function* to $\mu(x)$:

$$\hat{\mu}(\xi) = \sup_{x \in \operatorname{dom} \mu} (-\mu(x) + \langle \xi, x\rangle). \tag{1}$$

This function will be convex. We have that $\hat{\hat{\mu}} = \mu$. Let us note that if μ_U is the indicator function, then $\hat{\mu}_U(\xi) = \sup_{x \in U} \langle \xi, x\rangle$ is a degree 1 homogeneous function, and therefore $E_{\hat{\mu}_U}$ is a cone.

With respect to conjugacy in the sense of Young, the concepts of asymptotic cone V_μ and of the support dom μ, are conjugate in the sense that if $E_\nu = V_\mu$, then $\hat{\nu}$ is the indicator function of dom $\hat{\mu}$.

We shall also make use of two other cones connected with convex functions: dom V_μ, the projection of V_μ on $\mathbb{R}^n_x$, and $V(\operatorname{dom} \mu)$, the asymptotic cone of dom μ. It is clear that $V(\operatorname{dom} \mu) \supset \operatorname{dom} V_\mu$. If ν is the indicator function of $V(\operatorname{dom} \mu)$, then $\hat{\nu}$ is the indicator function of $\operatorname{dom}(V_{\hat{\mu}})$. Let us note that if V is a cone, $\hat{\mu}_V$ is the indicator function of the cone $\hat{V}$, where the dual cone $\hat{V}$ is the set of all ξ such that $\langle \xi, x\rangle \geqslant 0$ for all $x \in V$.

Let us denote by L_μ the maximal subspace, a translation of which is contained in E_μ (or, which amounts to the same, in V_μ). Then L_μ is necessarily contained in $\mathbb{R}^n_x(x_0 = 0)$. Let L'_μ be the subspace orthogonal to L_μ in the dual space $\mathbb{R}^n_\xi$. Then dom $\hat{\mu} \subset L'_\mu$, and L'_μ is generated by dom $\hat{\mu}$. Thus dom $\hat{\mu}$ is of full dimension if and only if E_μ contains no straight lines.

Examples. 1) Let $\mu(x_1, \dots, x_n) = (x_2^2 + \dots + x_n^2)/4x_1$ for $x_1 > 0$ and $\mu = +\infty$ for $x_1 \leqslant 0$. Then E_μ is a cone and the function conjugate in the sense of Young, $\hat{\mu}$, is the indicator function of the paraboloid $\{\xi_1 \leqslant -(\xi_2^2 + \dots + \xi_n^2)\}$.

2) Let $\mu = \mu_U$, where U is defined by the conditions $x_1 > 0$, $x_1^2 - x_2^2 - \dots - x_n^2 > 0$, that is, U is one half of the spherical cone. Then $\hat{\mu}_U$ is also the indicator function of one half of the spherical cone $\xi_1 < 0$, $\xi_1^2 - \xi_2^2 - \dots - \xi_n^2 > 0$.

Let us describe the set of the functions we can base ourselves on in applications to the Cauchy problem.

3) Let $\chi(\eta)$ be a convex function, and let $\hat{\nu}[\chi](\rho, \eta)$ be the indicator function of the set $\{\rho \leqslant -\chi(\eta)\}$, (t, y) are the dual variables. Then the conjugate function in the sense of Young $\nu[\chi] = \infty$ for $t \leqslant 0$, and $\nu[\chi] = t\hat{\chi}(y/t)$ for $t > 0$, where $\hat{\chi}$ is the conjugate function of χ.

4) Let us truncate the function $v[\chi]$ for $t < a$, $a > 0$, that is, let us consider the function $v[\chi, a]$ that coincides with $v[\chi]$ for $t \geqslant a$ and equals $+\infty$ for $t < a$. Then $\operatorname{dom} \hat{v}[\chi, a] = \operatorname{dom} \hat{v}[\chi]$ and $\hat{v}[\chi, a](\rho, \eta) = a(\rho + \chi(\eta))$ for $\rho \leqslant -\chi(\eta)$.

1.2. Hölder Scales Corresponding to Exponentially Growing Weights. Let $\mu(x)$ be a convex function satisfying conditions of Section 1.1, and such that $\operatorname{dom} \mu$ is of full dimension. Let us consider the spaces $C^{(s)}_{(l),\mu}$ with the norm

$$|\varphi|^{(s)}_{(l),\mu} = \sup_{x \in \mathbb{R}^n, |\alpha| \leqslant s} |\exp(\mu(x))(1 + |x|)^l D^\alpha \varphi(x)|, \tag{2}$$

Functions in this space are equal to zero for $x \notin \operatorname{dom} \mu$. Let us set

$$\mathcal{S}_\mu = C^{(\infty)}_{(\infty),\mu} = \bigcap_{s,l} C^{(s)}_{(l),\mu}. \tag{3}$$

If $\mu \equiv 0$, we obtain $\mathcal{S}(\mathbb{R}^n)$; if $\mu = 0$ for $x_1 > 0$ and $\mu = +\infty$ for $x_1 \leqslant 0$, we obtain $\mathcal{S}_+$. Let U be a convex set, and let μ_U be its indicator function. Then $\mathcal{S}_{\mu_U} \overset{\text{def}}{=} \mathcal{S}_U$ consists of elements of $\mathcal{S}(\mathbb{R}^n)$ with support in U. For $\mu(x)$ of Example 1 ($\mu(x) = (x_2^2 + \cdots + x_n^2)/4x_1$ for $x_1 > 0$), functions in $\mathcal{S}_\mu$ for a fixed $x_1 > 0$ decay as $\exp(-a(x_2^2 + \cdots + x_n^2))$, $a > 0$.

The norm (2) is equivalent (compare with (20) of Chapter 1) to the norm

$$\sup_{x \in \mathbb{R}^n, |\alpha| < s, \xi \in \operatorname{dom} \hat{\mu}} |\exp(-\hat{\mu}(\xi)) \exp(\langle \xi, x \rangle)(1 + |x|)^l D^\alpha \varphi(x)|$$
$$\overset{\text{def}}{=} \sup_{\xi \in \operatorname{dom} \hat{\mu}} \exp(-\hat{\mu}(\xi)) |\varphi|^{(s)}_{(l)[\xi]}. \tag{2'}$$

The same argument as in Section 2.2, Chapter 1, leads to the following proposition.

Proposition. *The space $\mathcal{S}_\mu$ is dual relative to the Fourier transform to the space $\mathcal{S}^\mu$ of functions $\psi(\xi)$ that are infinitely differentiable on the tubular set*

$$D(\hat{\mu}) = \{\xi \in \mathbf{C}^n, \operatorname{Im} \xi \in \operatorname{dom} \hat{\mu}, \hat{\mu}(\operatorname{Im} \xi) < \infty\},$$

holomorphic in complex directions in $D(\hat{\mu})$ (these correspond to L'_μ), and, furthermore, for all α, N and for some c,

$$|D^\alpha \psi(\xi)| < c \exp(\hat{\mu}(\operatorname{Im} \xi))(1 + |\xi|)^N, \quad \xi \in D(\hat{\mu}). \tag{4}$$

If E_μ contains no straight lines, and only in this case, the space $\mathcal{S}^\mu$ consists of functions holomorphic in all the variables. This will happen, in particular, for $\mathcal{S}^U \overset{\text{def}}{=} \mathcal{S}^{\mu_U}$ if U contains no straight lines. If then V is the asymptotic cone of U (it contains no straight lines) then $D(\hat{\mu})$ is the closure of the radial tubular domain $\mathbb{R}^n - i\hat{V}$, where $\hat{V}$ is the cone dual to V. If (and only if) the cone V_μ coincides with the ray $\{x_0 > 0, x = 0\}$, the space $\mathcal{S}^\mu$ consists of entire functions.

1.3. Hilbert Scales Corresponding to Exponentially Growing Weights. Let us impose an additional restriction upon the functions μ. Let us demand that the cone $\operatorname{dom} V_\mu$ is of full dimension. This condition is stronger than the one imposed before, that is, that $\dim(\operatorname{dom} \mu) = n$. The present condition is equivalent to requiring that $\operatorname{dom} \hat{\mu}$ (or $V(\operatorname{dom} \hat{\mu})$) contain no straight lines. Hence it follows that by an affine transformation dom $\hat{\mu}$ can be mapped into a direct product of some number of semi-infinite straight lines.

Lemma. *Under the conditions imposed above on μ, there exists a polynomial $\delta(\xi)$ such that $\delta(\xi) \neq 0$ for $\xi \in D(\hat{\mu})$, and $|\delta(\xi)| > c(1 + |\xi|)$ for $\xi \in D(\hat{\mu})$.*

In view of the remark above the statement of the lemma, its proof is reduced (by an affine transformation of $D(\hat{\mu})$) to the case of direct product of $\mathbb{R}^k$ with $(n - k)$ copies of half-planes, and in that case the polynomial can be constructed directly.

Obviously, for all s $\delta^s(\xi)$ is a multiplicator on $\mathscr{S}^\mu$, and therefore the pseudo-differential operator $\delta^s(D)$ leaves $\mathscr{S}_\mu$ invariant. Let us consider the spaces $H^{(s)}_{(l),\mu}$ with the norm

$$\|\varphi\|^{(s)}_{(l),\mu} = \left(\int |\delta^s(D)\varphi|^2(1 + |x|^2)^l \exp(2\mu(x))\, dx\right)^{1/2}. \tag{5}$$

Proposition. *The scales $C^{(s)}_{(l),\mu}$ and $H^{(s)}_{(l),\mu}$ are equivalent. In particular, $\mathscr{S}_\mu = H^{(\infty)}_{(\infty),\mu} = \bigcap_{s,l} H^{(s)}_{(l),\mu}$.*

Let us set

$$(\mathscr{S}')_\mu = \bigcup_{s,l} H^{(s)}_{(l),\mu} = H^{(-\infty)}_{(-\infty),\mu}; \quad (\mathscr{O}')_\mu = \bigcap_{l} \bigcup_{s} H^{(s)}_{(l),\mu}.$$

We use the notation "with primes" in order to establish the correspondence with the analogous notation from Chapter 1, and in the meantime we do not discuss the question of realization of these spaces as paired by duality (we do not introduce $\mathscr{O}_\mu$ either). The space $\mathscr{M}^\mu$ of partially holomorphic functions admitting the estimate (4) with N depending on α, is defined in a natural way. With respect to the Fourier transform, we have the duality

$$\mathscr{F}(\mathscr{O}')_\mu = \mathscr{M}^\mu.$$

A number of words about the proofs are in order. The estimate of Hilbert norms in terms of the Hölder ones is obtained in an elementary fashion (and so that no specific properties of the weights $\exp\mu$ are used). The possibility of estimating the Hölder norms in terms of the Hilbert norms is, on the other hand, non-trivial. Let us introduce an equivalent Hölder norm[1]

$${}'|\varphi|^{(s)}_{(l),\mu} = \sup_{x \in \operatorname{dom}\mu} |(\exp\mu(x))(1 + |x|)^l \delta^s(D)\varphi(x)|.$$

This can be done, since $C^{(0)}_{(0),\mu} \subset (\mathscr{S}')_\mu$ (by the part of the equivalence we have already proved). For $\xi \in \operatorname{dom}\hat{\mu}$ let us set

$${}'|\varphi|^{(s)}_{(l),[\xi]} = {}'|\varphi|^{(s)}_{(l),\langle\xi,x\rangle}, \quad \|\varphi\|^{(s)}_{(l)[\xi]} = \|\varphi\|^{(s)}_{(l),\langle\xi,x\rangle}.$$

---

[1] For these norms monotonicity in s, l is hard to prove. However, from the said above it follows that there exists ε such that $'|\ |$ can be estimated in terms of the analogous norm with indices $s + \varepsilon$, $l + \varepsilon$. All our arguments can be automatically extended to Banach scales with this property, so that, in particular, limit spaces are well defined.

For some k we have the estimate

$$'|\varphi|^{(s)}_{(l),[\xi]} \leqslant c\|\varphi\|^{(s+k)}_{(l),[\xi]} \tag{6}$$

with constant c independent of ξ. Let us set

$$|[\varphi]|^{(s)}_{(l),\mu} = \sup_{\xi \in \operatorname{dom} \hat{\mu}} \exp(-\hat{\mu}(\xi))\|\varphi\|^{(s)}_{(l),[\xi]}. \tag{7}$$

It is clear that

$$|[\varphi]|^{(s)}_{(l),\mu} \leqslant \|\varphi\|^{(s)}_{(l),\mu}. \tag{8}$$

From (8) and (6) it follows directly that

$$'|\varphi|^{(s)}_{(l),\mu} \leqslant c|[\varphi]|^{(s+k)}_{(l),\mu} \leqslant c\|\varphi\|^{(s+k)}_{(l),\mu}. \tag{9}$$

1.4. Convolutors on the Spaces $\mathscr{S}_\mu$ and $(\mathcal{O}')_\mu$. Let us consider shift operators $T_{-h}\varphi(x) = \varphi(x-h)$, where h are internal points of the cone dom V_μ. Recall that in view of our assumptions dom V_μ is of full dimension. These operators leave the spaces $\mathscr{S}_\mu$, $(\mathcal{O}')_\mu$ invariant. Note that they are dual to operators of multiplication by $\exp(-i\langle\xi, h\rangle)$. Continuous operators on $\mathscr{S}_\mu$, $(\mathcal{O}')_\mu$ that commute with the indicated shifts, are called convolution operators on the corresponding spaces. As in Chapter 1, it is proved that every convolution operator on $\mathscr{S}_\mu$ can be represented in the form (1.10), and that every operator of the form (1.10) is a convolution operator. In this manner we introduce the space of convolutors $\mathfrak{L}(\mathscr{S}_\mu)$.

We shall obtain a description of the space of convolutors if yet another restriction is imposed on the function $\mu(x)$. We shall call a function $\mu(x)$ *almost conic* if for some c the supremum in (1) is attained for $|x| \leqslant c$. Let us denote the set of all such μ by $\mathscr{L}$.

An example of a function $\mu \in \mathscr{L}$ is provided by a function $\mu = \mu[V, a]$, such that E_μ coincides with a translation of a cone V to the point a. Here the supremum is attained for $x = a'$, $a = (a_0, a')$. A function $\mu \in \mathscr{L}$ can be represented in the form $\inf_{a \in K} \mu[V, a]$, where K is a compact set, $V = V_\mu$. In terms of $\hat{\mu}$ it means that the restriction of $\hat{\mu}$ to dom $\hat{\mu}$ can be extended to a convex function on all of $\mathbb{R}^n$, and, moreover, it must be possible to project the asymptotic cone of the extended function on all of $\mathbb{R}^n$ (that is, it must have no vertical generating lines).

Let $\hat{\nu}$ be the indicator function of dom $\hat{\mu}$, and let $\nu(x)$ be the function conjugate to it in the sense of Young. The epigraph E_ν is the asymptotic cone of the epigraph E_μ.

Theorem. *Let $\mu \in \mathscr{L}$. Then convolution operators on $\mathscr{S}_\mu$ can be extended by continuity to convolution operators on $(\mathcal{O}')_\mu$. We have that*

$$\mathfrak{L}(\mathscr{S}_\mu) = \mathfrak{L}((\mathcal{O}')_\mu) = (\mathcal{O}')_\nu. \tag{10}$$

Furthermore $(\mathcal{O}')_\nu$ is a ring with respect to convolution.[2]

---

[2] The restriction $\mu \in \mathscr{L}$ is not needed in the proof of the inclusion $(\mathcal{O}')_\nu \subset \mathfrak{L}(\mathscr{S}_\mu)$.

The argument of the proof is as before. Inclusion in one way follows from the fact that elements of the ring with respect to multiplication $\mathscr{M}^\mu$ are multiplicators on $\mathscr{S}_\mu$. The proof of the reverse inclusion uses the extension of the convolution operator to $(\mathscr{O}')_\mu$ and a uniform bound on the family $\delta(x - y)\mu(y)$, where the parameter y runs through dom μ.

1.5. Convolution Equations on $\mathscr{S}_\mu$ and $(\mathscr{O}')_\mu$. Using the procedure of Chapter 1, we prove

Theorem. *The equation $A * \varphi = \psi$ is uniquely soluble on $\mathscr{S}_\mu$, $(\mathscr{O}')_\mu$ if and only if it has the fundamental solution $G \in (\mathscr{O}')_\nu$: $A * G = \delta(x)$. This is equivalent to having for some $c > 0$, N, the following estimate for the (symbol) Fourier transform of the convolutor, $\hat{A}(\xi)$,*

$$|\hat{A}(\xi)| > c(1 + |\xi|)^N, \quad \xi \in D(\hat{\mu}). \tag{11}$$

In particular, we must have $\hat{A}(\xi) \neq 0$ on $D(\hat{\mu})$.

Corollary. *The differential equation $P(D)\varphi = \psi$ on $\mathscr{S}_\mu$, $(\mathscr{O}')_\mu$ is uniquely soluble if and only if the distance $d(\xi)$ from $\xi \in D(\hat{\mu})$ to the surface of zeroes $\{P(\xi) = 0\}$ admits a power law lower bound (in terms of $c(1 + |\xi|^N)$).*

The reduction is achieved by estimating the modulus of the polynomial in terms of the distance to the surface of zeroes (see Lemma 4.1.1 in Hörmander [1963]).

If $D(\hat{\mu})$ is a semi-algebraic set (that is, one defined by a system of algebraic equations and inequalities) then the desired estimate follows from the condition $P(\xi) \neq 0$ according to the Seidenberg-Tarski theorem. That is what happens in the case $D(\hat{\mu}) = \mathbb{R}^n$ (that is, for the space $\mathscr{S}(\mathbb{R}^n)$) and in the case $D(\hat{\mu}) = \mathbb{R}^{n-1} \times \mathbb{C}_-$, where $\mathbb{C}_-$ is the lower half-plane (for $\mathscr{S}(\mathbb{R}^n)_+$). An important example of a semi-algebraic set is provided by the maximal tubular set on which a given polynomial is different from zero.

1.6. Connection with the Cauchy Problem. Let dom $\hat{\mu}$ be an unbounded set. Then it contains a ray. We can choose variables $\xi = (\rho, \omega)$ such that the ray is given by the condition $\omega = 0$, $\rho > 0$. In the conjugate picture, the asymptotic cone V_μ will be contained in the half-space $\{t \geqslant a\}$, where (t, y) are the variables dual to (ρ, ω). If here dom μ is contained in a half-space $t \geqslant a$, then a convolution equation in $\mathscr{S}_\mu$, $(\mathscr{O}')_\mu$ can be interpreted as a homogeneous Cauchy problem in spaces of functions with respective exponential asymptotics.

If the ray $\{\omega = 0, \rho > 0\}$ coincides with the asymptotic cone of the set dom $\hat{\mu}$, then V_μ is the half-space $\{t \geqslant 0\}$ and generically dom μ is a half-space. The case of dom $\mu = \mathbb{R}^n$ is interpreted in a natural way as the homogeneous Cauchy problem with zero data at $t = -\infty$.

A meaningful set of examples will result if we consider convolution equations $A * \varphi = \psi$ on spaces $\mathscr{S}_\mu$ for $\mu = \nu[\chi, a]$ (see Example 4). If here A is an invertible element of $(\mathscr{O}')_{\nu[\chi]}$, then all such equations are soluble. We remind the reader that we are dealing here with functions, which decrease for a fixed $t > a$ as $\exp(-c\hat{\chi}(y/t))$.

If the asymptotic cone $V(\operatorname{dom}\hat{\mu})$ is not a ray, then any of its directions can be taken as the direction dual to time. Consequently, dom V_μ will lie in the intersection of dual half-spaces. In this case, when $\dim V(\operatorname{dom}\hat{\mu}) = n$, convolutors in $\mathfrak{L}(\mathcal{S}_\mu)$ are naturally called hyperbolic. Here dom μ is generically a space-like domain with respect to dom V_μ.

Let the symbol of a differential operator $P(D_t, D_y)$ satisfy the homogeneous correctness condition in the sense of Petrovskij: $P(\tau, \eta) \neq 0$ for $\rho = \operatorname{Im}\tau < c$, $\eta \in \mathbb{R}^{n-1}$. Let us denote by $\Omega_P \subset \mathbb{R}^n_{(\rho,\omega)}$ the maximal set for which (i) $P(\tau, \xi) \neq 0$ for $(\operatorname{Im}\tau, \operatorname{Im}\xi) \in \Omega_P$, (ii) $\{(\rho, \omega): \rho < c, \omega = 0\} \subset \Omega_P$. Let $D_P = \mathbb{R}^n + i\Omega_P$ be the corresponding tubular set in $\mathbb{C}^n$ (in which the symbol is different from zero). The set Ω_P is convex, and has the form $\{\rho < \chi_P(\omega)\}$, where χ_P is a convex function.

Theorem. *If the support* dom $\hat{\mu} \subset \Omega_\rho$ *is at a distance admitting a power lower bound from the boundary* Ω_ρ, *then the equation* $P(D_t, D_y)u = f$ *is uniquely soluble in* $\mathcal{S}_\mu$. *If* $\mu \in \mathcal{L}$, *then the indicated condition on* μ *is necessary for solubility of the differential equation in* $\mathcal{S}_\mu$. *In particular, the equation is soluble in* $\mathcal{S}_\mu$ *for* dom $\hat{\mu} = \{\rho \leqslant -\chi_\rho(\omega) - \varepsilon\}$, $\varepsilon > 0$.

The last claim used the semi-algebraicity of Ω_P. The most important class of μ for which we have solubility in $\mathcal{S}_\mu$ is

$$\mu(t, y) = v[\chi_P, a] - \varepsilon t = t\hat{\chi}_P(y/t) - \varepsilon t, \quad t > a, \quad \varepsilon > 0.$$

Examples. 1) Let us consider the heat equation, that is, an operator with symbol $P(\tau, \eta) = \tau - i(\eta_1)^2 - \cdots - i(\eta_{n-1})^2$. For this symbol the domain Ω_P is the paraboloid $\rho < -|\omega|^2$. That means that the heat equation is soluble, in particular, in spaces $\mathcal{S}_\mu$ for $\mu(t, y) = -\varepsilon t + |y|^2/4t$ for $t > 0$; $\mu = \infty$ for $t \leqslant 0$. Solubility will persist if we make a translation in t and consider, for example, $\mu(t, y) = \varepsilon t + |y|^2/4(t + a)$ for $t > 0$; $\mu = \infty$ for $t \leqslant 0$.

2) Let us consider the wave equation $P(\tau, \eta) = \tau^2 - \Sigma\eta_j^2$. Then Ω_P is one half of the spherical cone $\rho^2 - |\omega|^2 > 0$, $\rho < 0$. Let us use the notation $V = \{t^2 - |y|^2 > 0, t > 0\}$. We denote by $\mathcal{S}_{V,\varepsilon}$ the space of functions obtained from $\mathcal{S}_V$ (the functions in $\mathcal{S}$ with support in V) by multiplication by $\exp \varepsilon t$. Let $\mathcal{S}_{V,\varepsilon}[a, \infty)$ be the subspace of functions that vanish for $t \leqslant a$. The results we formulated above, as applied to the wave equation, establish solubility in the spaces $\mathcal{S}_{V,\varepsilon}[a, \infty)$, $a > 0$, $\varepsilon > 0$.

3) In the case $P(\tau, \eta) = \tau + \Sigma\eta_j^2$ (symbol of the Schrödinger operator) Ω_P coincides with a ray, and therefore we cannot transcend the bounds of power weights in space variables.

1.7. Remarks About the Problem in a Strip and the Inhomogeneous Cauchy Problem. The convexity property will be preserved if outside a convex set we take the value of μ to be $+\infty$ ("truncate"). Let $\xi = (\rho, \omega)$, (t, y) are the dual variables and $V(\operatorname{dom}\hat{\mu})$ contains the ray $\{\omega = 0, \rho = 0\}$. The function $\mu[a, \infty)$ is obtained by truncating μ *for* $t \leqslant a$. For example, $v[\chi, a] = v[\chi][a, \infty)$. We set $\mathscr{S}_\mu[a, b) = \mathscr{S}_\mu[a, \infty)/\mathscr{S}_\mu[b, \infty)$. In particular, $\mathscr{S}_{v[\chi]}[a, \infty) = \mathscr{S}_{v[\chi, a]}/\mathscr{S}_{v[\chi, b]}$. The spaces $(\mathcal{O}')_\mu[a, b)$ are defined similarly. In these quotient spaces, a theory of convolutors can be developed following the line of Section 2.6, Chapter 1.

In particular, in the notation of Theorem 1.4, $(\mathcal{O}')_v[0, b-a) \subset \mathfrak{L}(\mathscr{S}_\mu[a, b)) \subset (\mathscr{S}')_v[0, b-a)$. In particular, this is the case for $\mu = v[\chi]$. If dom $\hat{\mu}$ has the form $\rho < -\chi(\omega)$, and the polynomial $P(\tau, \xi) \neq 0$ for $\operatorname{Im} \tau < -c - \chi(\operatorname{Im} \xi)$ for some c, then the equation $P(D_t, D_y)u = f$ is uniquely soluble in $\mathscr{S}_\mu[a, b)$. If the set dom $\hat{\mu}$ is semi-algebraic, then this condition on μ, P is necessary for solubility. In particular, we have solubility for $\chi = \chi_P$.

By analogy with § 3 of Chapter 1, we introduce spaces connected with the inhomogeneous Cauchy problem (for exponentially decreasing data). Let $\mu(a) = \mu|_{t=a}$. Let us denote by $\mathscr{S}_\mu^{\{-\infty\}}[a, b)$ the space of distributions of the form $\psi_+ + \sum_{k=0}^{l} \psi_k D_t^k \delta(t)$, where $\psi_+ = \theta_+(t-a)\psi$, $\psi \in \mathscr{S}_\mu/\mathscr{S}_\mu[b, \infty)$, $\psi_k \in \mathscr{S}_{\mu(a)}$.

Assume that $P(\tau, \eta)$ is a polynomial soluble with respect to the highest power of τ, and that the equation $P(D_t, D_y)u = f$ is soluble in $\mathscr{S}_\mu[a, b)$. Then it is soluble in $\mathscr{S}_\mu^{\{-\infty\}}[a, b)$. For $\mu \in \mathscr{L}$ solubility conditions in $\mathscr{S}_\mu[a, b)$, $\mathscr{S}_\mu^{\{-\infty\}}[a, b)$ coincide. This statement can be interpreted as the solubility of the Cauchy problem for data decaying as $\exp(-\mu(a, y))$. Let us state these results for $\mu = v[\chi]$.

Theorem. *Let $P(\tau, \eta)$ be a polynomial correct in the sense of Petrovskij, and let $\chi(\eta)$ be a convex function such that $\chi(\eta) \geqslant \chi_P(\eta) - c$, $0 \leqslant a < b < \infty$. Then the equation $P(D_t, D_y)u = f$ is soluble in the spaces $\mathscr{S}_{v[\chi]}[a, b)$, $(\mathscr{S}')_{v[\chi]}[a, b)$, and $\mathscr{S}_{v[\chi]}^{\{-\infty\}}[a, b)$. In particular, we can take $\chi = \chi_P$. Conversely, if we have solubility in any of these spaces, and if χ is a semi-algebraic function, then it satisfies the indicated condition.*

§ 2. Convolution Equations in Exponentially Growing Functions (Distributions)

2.1. Spaces of Exponentially Growing Functions (Distributions). While dealing with exponential decay, we introduced norms containing weights of the form $\exp \mu$, where μ is a convex function. With exponentially growing functions we connect weights, logarithms of which are concave. We keep the standard notation μ for (downward) convex functions. Then the functions $-\mu$ will be concave; they can take the value $-\infty$ as well as finite values. We also retain the restrictions on μ from the previous section: the asymptotic cone V_μ is of full dimension (dom $\hat{\mu}$ contains no straight lines), and μ is almost conic ($\mu \in \mathscr{L}$).

Let us introduce spaces $C_{(l), -\mu}^{(s)}$ with norm $|\ |_{(l), -\mu}^{(s)}$, defined by (1.2). Let us set $\mathscr{S}_{-\mu} = \bigcap_{s, l} C_{(l), -\mu}^{(s)}$. In these spaces, there is factorization with respect to values of a function outside of dom μ. We define the Hilbert norms $\|\ \|_{(l), -\mu}^{(s)}$ (of the space $H_{(l), -\mu}^{(s)}$) as dual to the norms $\|\ \|_{(-l), \mu}^{(-s)}$. We set $(\mathscr{S}')_{-\mu} = \bigcup_{s, l} H_{(l), -\mu}^{(s)}$. By definition, $(\mathscr{S}_\mu)' = (\mathscr{S}')_{-\mu}$.

Proposition. *For $\mu \in \mathscr{L}$, the scales $|\ |_{(l), -\mu}^{(s)}$ and $\|\ \|_{(l), -\mu}^{(s)}$ are equivalent.*

As we already mentioned, the non-trivial part here is the estimation from above of the Hölder norms in terms of the Hilbert ones. The proof proceeds by iterative widening of the set of μ for which the embedding holds. As the first step we consider μ of the form $\mu_V + \langle \xi, x\rangle$, where μ_V is the indicator function of some n-hedral angle. This case is easily established by the classical embedding theorems. Next we consider the case when E_μ is an arbitrary polyhedral angle: by a finite triangulation this case is reduced to the previous one. After that we can consider the case when E_μ is an arbitrary cone, and finally, when $\mu \in \mathscr{L}$. During these two final steps we use the fact that if embedding has already been proved for μ_α (uniformly in α) and $\mu = \inf_\alpha \mu_\alpha$, then embedding holds for μ as well.

Corollary.

$$\mathscr{S}_{-\mu} = H^{(\infty)}_{(\infty), -\mu}, \quad (\mathscr{S}_{-\mu})' = (\mathscr{S}')_\mu.$$

The theory of convolution operators and of convolution equations on $(\mathscr{S}')_{-\mu}$ is constructed by conjugacy:

$$\mathfrak{L}((\mathscr{S}')_{-\mu}) = (\mathscr{O}')_{I\nu},$$

where $E_\nu = V_\mu$ is the asymptotic cone of E_μ (see theorem 1.4). We do not go into details of this case.

2.2. Convolutors on $\mathscr{S}_{-\mu}$ and $(\mathscr{S}')_\mu$. The shift operators, T_h, where h is an interior point of V_μ act in the spaces $\mathscr{S}_{-\mu}$. Convolution operators are continuous operators that commute with these shifts. On $\mathscr{S}_{-\mu}$ they can be given the representation (1.10). We have that $\mathfrak{L}((\mathscr{S}')_\mu) = \mathfrak{L}(\mathscr{S}_{-I\mu})$. The description of the convolution operators depends essentially on whether the epigraph E_μ contains straight lines or not (that is, whether dom $\hat{\mu}$ is of full dimension or not). Let us denote by $\mathscr{L}^0$ the set of all weights $\mu \in \mathscr{L}$ for which E_μ contains no straight lines.

Theorem. *For $\mu \in \mathscr{L}^0$*

$$\mathfrak{L}(\mathscr{S}_{-\mu}) = \mathfrak{L}((\mathscr{S}')_{I\mu}) = (\mathscr{S}')_{I\nu}, \tag{12}$$

where $E_\nu = V_\mu$. In particular, if the cone $E_\nu = V_\nu$ contains no straight lines ($\hat{\nu}$ is the indicator function of a convex domain that contains no straight lines), then $(\mathscr{S}')_\nu$ is a ring with respect to convolution.

Let us discuss the connection with multipliers and with the Fourier transform. The functions in $\mathscr{S}_{-\mu}$ grow rapidly and can have no regular Fourier transform. On the other hand, distributions in $(\mathscr{S}')_\mu$, and, in particular, convolutors in $(\mathscr{S}')_\nu$ have as Fourier transforms holomorphic functions in tubular domains $D(\hat{\mu}) = \mathbb{R}^n + i(\text{dom } \hat{\mu})$.

Proposition. *For $\mu \in \mathscr{L}^0$ the space $(\mathscr{S}')_\mu$ is dual with respect to the Fourier transform to the space of functions $\psi(\xi)$ that are holomorphic in interior points of $D(\hat{\mu})$ and for which for some $c > 0$, k, N, we have*

$$|\psi(\xi)| < c \exp(\hat{\mu}(\operatorname{Im} \xi))\, d(\xi)^k (1 + |\xi|)^N,$$

where $d(\xi)$ is the distance from ξ to the boundary of $D(\hat{\mu})$.

2.3. Convolution Equations

Theorem. *The equation $A * \varphi = \psi$, $A \in \mathfrak{L}(\mathcal{S}_{-\mu}) = (\mathcal{S}')_{I\nu}$ on $\mathcal{S}_{-\mu}$ (or the adjoint equation $IA * \varphi = \psi$ on $(\mathcal{S}')_{I\mu}$) is uniquely soluble if and only if there exists a fundamental solution $G \in (\mathcal{S}')_{I\nu}$. This is equivalent to having for some $c_1, c_2 > 0$, k_1, k_2, N_1, N_2*

$$c_1\, d(\xi)^{k_1}(1 + |\xi|)^{N_1} < |\hat{A}(\xi)| < c_2\, d(\xi)^{k_2}(1 + |\xi|)^{N_2}$$

at interior points of $D(I\hat{\mu}) = D(I\hat{\nu})$, $\hat{A}$ being the Fourier transform of A.

Here the right inequality means that $\hat{A}(\xi)$ is a multiplier, while the left one ensures invertibility of $\hat{A}$ in the ring of multipliers.

Corollary. *The differential equation $P(D)u = f$ on $\mathcal{S}_{-\mu}$, $(\mathcal{S}')_{I\mu}$ is uniquely soluble if and only if $P(-\xi) \neq 0$, if ξ is an interior point of $D(\hat{\mu})$.*

As we see, in the case of exponentially growing weights, the standard simplification for differential equations of the solubility condition for convolution equations can be effected without any additional constraints on the domain $D(\hat{\mu})$.

Let us describe briefly how these results change for weights that do not satisfy condition $\mathcal{L}^0$. Let $\mathbb{R}^{n_1}$ be the maximal subspace, a translation of which is contained in E_μ, and let $\mathbb{R}^{n_2}$ be some complementing subspace; $x = (x_1, x_2)$ is the corresponding subdivision of the variables: $x_1 \in \mathbb{R}^{n_1}$, $x_2 \in \mathbb{R}^{n_2}$. In (2), let us replace the weight $(1 + |x|)^l$ by $(1 + |x_1|)^{l_1}(1 + |x_1|)^{l_2}$, and, correspondingly, let us introduce the spaces $C^{(s)}_{(l_1,l_2),\mu}$, as well as $H^{(s)}_{(l_1,l_2),\mu}$. Let us set $\mathcal{K}_\mu = \bigcap_{l_1} \bigcup_{l_2,s} H^{(s)}_{(l_1,l_2),\mu}$.

Proposition. $\mathfrak{L}(\mathcal{S}_{-\mu}) = \mathfrak{L}((\mathcal{S}')_{I\mu}) = \mathcal{K}_{I\nu}$.

A particular case of this proposition is the description of $\mathfrak{L}((\mathcal{S}')_+)$ (see Section 2.4 of Chapter 1). For $\mathcal{L}(\mathcal{S}_{-\mu})$ all the scheme of convolution equations construction goes through, but we shall not state the corresponding results.

2.4. Remarks Concerning the Problem in the Strip and Concerning the Inhomogeneous Cauchy Problem. The constructions of Section 1.7 can be automatically applied to the case of exponentially growing weights. With the same notation, and under the same conditions on μ (dom $\hat{\mu}$ contains the ray $\{\rho < 0, \omega = 0\}$) let us define

$$\mathcal{S}_{-\mu}(a, b] = \mathcal{S}_{-\mu(-\infty,b]}/\mathcal{S}_{-\mu(-a,a]}, \quad a < b.$$

We have that

$$(\mathcal{S}_{-\mu}(a, b])' = (\mathcal{S}')_\mu[a, b).$$

Let $\mu \in \mathcal{L}$. Then $\mathfrak{L}(\mathcal{S}_{-\mu}(a, b]) = \mathfrak{L}(\mathcal{S}_{-\mu}(a - b, 0])$. In particular, for $\mu \in \mathcal{L}^0$, $\mathfrak{L}(\mathcal{S}_{-\mu}(a, b]) = (\mathcal{S}')_{I\nu}(a - b, 0]$ in the notation of Theorem 2.2. We shall not discuss general convolution equations, restricting ourselves to differential equations. Let dom $\hat{\mu}$ have the form $\{\rho < -\chi(\omega)\}$ and let the symbol $P(\tau, \eta)$ never vanish for $\operatorname{Im} \tau > c + \chi(\operatorname{Im} \zeta)$ for some c. Then we have solubility in $\mathcal{S}_\mu(a, b]$.

In the semi-algebraic case this condition on χ is also necessary.

As in Section 1.7, we can define the space $\mathcal{S}^{\{-\infty\}}_{-\mu}(a, b]$, consisting of distributions of the form $\psi_- + \sum_{k=1}^{l} \psi_k D_t^k \delta(t)$, where $\psi_- = \theta_-(t - b)\psi$, $\psi \in \mathcal{S}_{-\mu}/\mathcal{S}_{-\mu(\infty,a]}$,

$\psi_k \in \mathscr{S}_{-\mu(b)}$. If the symbol of the differential operator is soluble with respect to the highest power of τ, and if the equation $P(D_t, D_y)u = f$ is soluble in $\mathscr{S}_{-\mu}(a, b]$, then it is soluble in $\mathscr{S}_{-\mu}^{\{-\infty\}}[a, b)$. In the case of $\mu \in \mathscr{L}$ the solubility conditions coincide. In the case of μ of the form $\nu[\chi]$ we obtain the

Theorem. *Let $P(\tau, \eta)$ be a polynomial that is correct in the sense of Petrovskij; let $\chi(\omega)$ be a convex function such that $\chi(\omega) > \chi_p(\omega) - c$ for some c, $-\infty < a < b \leqslant 0$. Then the equation $P(D_t, D_y)u = f$ is soluble in the spaces $\mathscr{S}_{-I\nu[\chi]}[a, b)$, $(\mathscr{S}')_{-I\nu[\chi]}[a, b)$, $\mathscr{S}_{-I\nu[\chi]}^{\{-\infty\}}[a, b)$. In particular, we can take $\chi = \chi_p$. Conversely, if χ is semi-algebraic, then the indicated condition on χ follows from solubility in one of these spaces.*

This statement can be given an interpretation in the language of homogeneous and inhomogeneous Cauchy problems with exponentially growing data. In order to solve the Cauchy problem in the positive direction, we changed the direction of time in the statement of the theorem (passing to $I\nu[\chi]$). In the case of the inhomogeneous problem, we prescribe data for $t = a$, $a < 0$, that grow as $\exp(|a|\hat{\chi}(y/a))$. Making a translation in time, we can prescribe such data for $t = 0$. Let us remark that the solution is considered in a time interval of length not exceeding $|a|$ (in $\mathscr{S}_{I\nu[\chi]}$ there is factorization with respect to values for $t \geqslant 0$). This corresponds to the well-known fact that the Cauchy problem with exponentially growing initial data has a solution with controllable growth only on a finite time interval.

We shall obtain the maximal admissible growth rate for $\chi = \chi_p$. In particular, for the heat equation (for $t = 0$) the growth rate $\exp(|y|^2/c)$, $c > 0$, is admissible, while for $t < c$ the solution grows as $\exp(|y|^2/(c - t))$.

Let us indicate how the classical results concerning isotropic correctness classes are obtained in this scheme. Let $\operatorname{dom} \chi_p = \mathbb{R}^{n-1}$, and let p be the smallest integer, for which the majorant $c_1|\eta|^p + c_2 \geqslant \chi_p(\eta)$ exists. We shall call this p the genus of the symbol P (this concept is different from the genus of Shilov). From homogeneity considerations for the function $\hat{\chi}(y) = c|y|^{p'}$, $p' = p/(p-1)$, we have for some c that $\chi(\eta) \geqslant \chi_p(\eta) - c'$. Let $\mu(t, y) = c|a - t|^{-1/(p-1)}|y|^{p/(p-1)}$, $\mu = +\infty$ for $t \leqslant 0$. Then in the spaces $\mathscr{S}_{-\mu}[0, a - \varepsilon)$, $\mathscr{S}_{-\mu}^{\{-\infty\}}[0, a - \varepsilon)$ we shall have solubility for all a. (The constant c is determined by the symbol P and does not depend on a). We emphasize again that solubility in the spaces $\mathscr{S}_{-I\nu[\chi_p]}^{\{-\infty\}}[a, -\varepsilon)$ can be interpreted as a sharp description of anisotropic correctness classes, that take into account various possibilities for growth rates of the initial data (and of the right hand sides) in different directions.

§ 3. Special Classes of Differential Operators and their Correctness Classes

3.1. Exponentially Correct Operators. Let $P(\tau, \eta)$, $\tau \in C^1$, $\eta \in \mathbb{R}^{n-1}$ be the symbol of a differential operator, that is correct in the sense of Petrovskij. Let us remind the reader that this means tht P can be solved with respect to the highest power of τ (say, τ^k) and $P(\tau, \eta) \neq 0$ for $\operatorname{Im} \tau < c$, $\eta \in \mathbb{R}^{n-1}$, for some c. In § 1 we associated with P the set $\Omega_p = \{\rho < -\chi_p(\omega)\} \subset \mathbb{R}^n_{\rho,\omega}$, for which $P(\tau, \zeta) \neq 0$ for $\operatorname{Im}(\tau, \zeta) \in \Omega_p$.

We shall call a polynomial P *semi-exponentially correct*, if the projection $\operatorname{dom} \chi_p$ of the set Ω_p on $\mathbb{R}^{n-1}_\omega$ is an unbounded set of full dimension. In this case the asymptotic cone $V(\operatorname{dom} \chi_p)$ does not reduce to a point.

A polynomial $P(\tau, \eta)$ is called *exponentially correct* (or, in different terminology, strongly correct) if the projection of Ω_p (that is, $\operatorname{dom} \chi_p$) coincides with $\mathbb{R}^{n-1}$. In other words, exponential correctness means that all the translates of P into the complex domain, $P_\omega(\tau, \eta) = P(\tau, \eta + i\omega)$ are correct in the sense of Petrovskij. In the case of semi-exponential correct polynomials this is true for some unbounded open set of $\omega \in \mathbb{R}^{n-1}$.

Two other versions of the definition of exponential correctness are:

(i) $d(\tau, \eta)$, the distance to the manifold of zeroes of P goes to infinity as $\operatorname{Im} \tau \to -\infty$, $\eta \in \mathbb{R}^{n-1}$.

(ii) $P^{(\alpha)}(\tau, \eta)P^{-1}(\tau, \eta) \to 0$ uniformly in $(\operatorname{Re} \tau, \eta)$ as $\operatorname{Im} \tau \to -\infty$, $\eta \in \mathbb{R}^{n-1}$, for any non-zero multi-index α.

Semi-exponential correctness is equivalent to the existence of $a, N \in \mathbb{R}^{n-1}$, for which $d(\tau, \eta + ia + i\sigma N) \to \infty$ for $\sigma \to \infty$, $\operatorname{Im} \tau \to -\infty$. The vectors a, N are characterized by $N \in V(\operatorname{dom} \chi_p)$; $a + i\sigma N \in \operatorname{dom} \chi_p$ for sufficiently large c.

The alternate formulations (i), (ii) of exponential correctness invite one to relate it to hypoellipticity.

Proposition. *Hypoelliptic symbols that are correct in the sense of Petrovskij, are exponentially correct.*

The converse is, naturally, untrue, as the example of hyperbolic polynomials shows (see below).

An important characteristic of $P(\tau, \eta)$ is V_p, the asymptotic cone of Ω_p. Every direction in V_p can be chosen to be the direction dual to time, so that the polynomial will preserve its correctness in the sense of Petrovskij.

In terms of solubility of the Cauchy problem, the characteristic feature of exponentially correct polynomials consists of the following. Let $\mathscr{S}_c(\mathbb{R}^n)$ be the space $\mathscr{S}_\mu$ corresponding to the weight $\mu(t, y) = \exp c|y|$; $\mathscr{S}_c[a, \infty)$ is the subspace of functions with support in $t \geqslant a$, $\mathscr{S}_c[a, b) = \mathscr{S}_c[a, \infty)/\mathscr{S}_c[b, \infty)$, $a < b$. Analogous definitions are given for $(\mathscr{S}')_c$. Below we assume that $b < \infty$.

Theorem. *The equation $P(D)u = f$ is soluble in $\mathscr{S}_c[a, b)$ for all c if and only if $P(\tau, \zeta)$ is exponentially correct.*

Analogous claims can be established for $c < 0$, and also for solubility in $(\mathscr{S}')_c[a, b)$ and in $\mathscr{S}_c^{(-\infty)}[a, \infty)$.

These statements follow directly from §§ 1, 2, but it is easier to prove them than the case of general weights. In analogous claims for semi-exponentially correct operators, the exponential weight can be prescribed only in certain directions in

$\mathbb{R}_y^{n-1}$. We do not quote the corresponding statements. Results of §§ 1, 2 allow us to formulate solubility theorems in $\mathscr{S}_\mu$, that take into account the possible character of exponential decay (growth) in a much more precise manner. We turn next to the consideration of special types of operators.

3.2. Hyperbolic Polynomials. Let us start with operators with a homogeneous symbol. Recall that a homogeneous symbol $P_0(\tau, \eta)$ is correct in the sense of Petrovskij if and only if it is hyperbolic, that is, when $P_0(\tau, \eta) = \prod_{j=1}^{k} (\tau - \lambda_j(\eta))$, where the roots $\lambda_j(\eta)$ are real. Homogeneous hyperbolic polynomials are always exponentially correct: for them $\Omega_{p_0} = V_{p_0}$ is the connected component of the set $\{P(\sigma, \eta) \neq 0, (\sigma, \eta) \in \mathbb{R}^n\}$.

A homogeneous symbol $P_0(\tau, \eta)$ is called *strictly hyperbolic* if it remains correct in the sense of Petrovskij after the addition of lower order terms. The condition of strict hyperbolicity is $\lambda_i(\eta) \neq \lambda_j(\eta)$, $i \neq j$ (the roots are simple). If P_0 is a homogeneous strictly hyperbolic symbol, then the inhomogeneous symbol $P = P_0 + Q$, $\deg Q < k = \deg P_0$ is also called strictly hyperbolic. This is equivalent to the following statement: for every c_1 there exists c_2, for which $|P(\tau, \eta)| > c_1(|\tau| + |\eta|)^{k-1}$ for $\operatorname{Im} \tau < c_2$, $\eta \in \mathbb{R}^{n-1}$. An inhomogeneous strictly hyperbolic symbol is exponentially correct. Here $V_p = \Omega_p = V_{p_0}$.

An exponentially correct polynomial is called hyperbolic if $\dim V_p = n$. Then its principal part is necessarily hyperbolic (but not necessarily strictly hyperbolic).

3.3. 2*b*-Parabolic Polynomials (in the Sense of Petrovskij). We turn now to the consideration of quasi-homogeneous exponentially correct polynomials. Let us prescribe the weight $2b$ to τ, and let us denote by $\deg_{2b}$ the weighted degree with respect to weights $(2b, 1, \ldots, 1)$. Let $P_0(\tau, \eta)$ be a quasi-homogeneous polynomial with respect to this system of weights, and $\deg_{2b} P_0 = m$. It will be called $2b$-*parabolic* if $P_0(\tau, \eta) \neq 0$ for $\operatorname{Im} \tau \leqslant 0, \eta \neq 0$. This, in terms of the roots $\lambda_j(\eta)$ (these are homogeneous algebraic functions), is equivalent to $\operatorname{Im} \lambda_j(\eta) > 0$ for $\eta \neq 0$. A $2b$-parabolic polynomial is exponentially correct. Here the cone V_{p_0} is reduced to a ray, while $\rho = \chi_p(\omega)$ is a homogeneous function of degree $2b$.

Adding any lower order term with respect to the weight $(2b, 1, \ldots, 1)$, that is, in passing to $P = P_0 + Q$, $\deg_{2b} Q < m = \deg_{2b} P_0$, we again obtain an exponentially correct polynomial. Such a polynomial is called (inhomogeneous) $2b$-parabolic. Equivalent conditions of $2b$-parabolicity are: for some c_1, c_2, we have that $|P(\tau, \eta)| \geqslant c_1(|\tau| + |\eta|^{2b})^k$ for $\operatorname{Im} \tau < c_2$, or for every c_1 there exists c_2, for which $|P(\tau, \eta)| \geqslant c_1(|\tau| + |\eta|^{2b})^{k-1}$ for $\operatorname{Im} \tau < c_2$.

The domain Ω_p has the same asymptotics as Ω_{p_0}; in particular, Ω_p contains a translate of Ω_{p_0} in the direction ρ. Hence it follows that if χ satisfies the conditions of Theorems 1.7, 2.4 for the symbol P_0, then for all $P = P_0 + Q$, $\deg_{2b} Q < m$, we have solubility in $\mathscr{S}_{\nu[\chi]}[a, b)$, $\mathscr{S}_{\nu[\chi]}^{\{-\infty\}}[a, b)$, $\mathscr{S}_{-I\nu[\chi]}[a, b)$, $\mathscr{S}_{-I\nu[\chi]}^{\{-\infty\}}[a, b)$. In the semi-algebraic case, there is no solubility for other χ. In particular, we can take $\nu[\chi_p] = t^{-1/(2b-1)}\hat{\chi}_p(y)$. Recall that $\hat{\chi}_p$ is a homogeneous function of degree $2b/(2b-1)$. We obtain thus a description of anisotropic correctness classes of the Cauchy problem. Isotropic classes correspond to weights of the form $\exp(ct^{-1/(2b-1)}|y|^{2b/(2b-1)})$. The use of tubular domains allows us to obtain sharp estimates of the Green's function for $2b$-parabolic equations, and frequently also to compute the asymptotic behavior by using the method of sharpest descent.

3.4. $(2b + 1)$-Hyperbolic Polynomials. As we pass from the even weight $2b$ to the odd one $2b + 1$, it turns out that for $b \neq 0$ there are no exponentially correct polynomials in the class of $(2b + 1, 1, \ldots, 1)$-homogeneous ones. However, there may be semi-exponentially correct ones. We shall call such a semi-exponentially correct polynomial $(2b + 1)$-hyperbolic. A polynomial is called *strictly* $(2b + 1)$-*hyperbolic* if it remains semi-exponentially correct under any change of its $(2b + 1, 1, \ldots, 1)$-lower order terms. It is clear, that the property of polynomial being $(2b + 1)$-hyperbolic, is a property of its $(2b + 1, 1, \ldots, 1)$-principal part. Hyperbolic polynomials are included in this scheme for $b = 0$.

Thus, let $P_0(\tau, \eta)$ be a quasi-homogeneous polynomial with respect to the weights $(2b + 1, 1, \ldots, 1)$, $\deg_{2b+1} P_0 = m$; $P_0 = \prod_{j=1}^{k} (\tau - \lambda_j(\eta))$ is called *strictly* $(2b + 1)$-*hyperbolic* if and only if (i) all the roots $\lambda_j(\eta)$ are real and simple: $\lambda_i(\eta) \neq \lambda_j(\eta)$, $i \neq j$, $\eta \neq 0$; (ii) for some $N \in \mathbb{R}^{n-1}$, the directional derivative $\langle \operatorname{grad} \lambda_j(\eta), N\rangle > 0$ for all j for $\eta \neq 0$. Then N is called a $(2b + 1)$-hyperbolicity direction of $P_0(\tau, \eta)$. These directions, moreover, form a cone $V = V(\operatorname{dom} \Omega_p)$. Necessarily, $\dim V = n$. The condition of strict $(2b + 1)$-hyperbolicity for $P = P_0 + Q$ is equivalent to the existence for every c, of c_1, c_2, such that $|P(\tau, \eta + ivN)| \geqslant c(|\tau| + |\eta|^{2b+1})^{k-1}$ for $v > c_1$, $\operatorname{Im} \tau < c_2$.

As far as the question of solubility of differential equations in the spaces $\mathscr{S}_\mu$, etc., is concerned, everything we said about $2b$-parabolic operators carries over to $(2b + 1)$-hyperbolic ones. However, the homogeneous functions χ_p of degree $2b + 1$ have a substantially different structure.

Let P_0 be a quasi-homogeneous strictly $(2b + 1)$-hyperbolic polynomial; let V be the cone of $(2b + 1)$-hyperbolicity directions. This is a solid cone that contains no straight lines for $b > 0$. Then $\operatorname{dom} \chi_{P_0} = V$, and χ_{P_0} is a homogeneous function of degree of homogeneity $2b + 1$. Then everywhere in $\mathbb{R}^{n-1}$ $\hat{\chi}_{P_0}$ is a finite homogeneous function of degree $(2b + 1)/2b$. For $(-y) \in V'$, where V' is the cone dual to V, the function $\hat{\chi}_{P_0}$ vanishes. On the other hand, if $\tilde{V} \supset\supset (-V')$ is a cone, then outside of $\tilde{V}$ the function $\hat{\chi}_{P_0}$ can be estimated from below by $c|y|^{(2b+1)/2b}$, $c > 0$. This means that for the solution of the Cauchy problem, only power law growth in y is allowed on $(-V')$, while outside of $(-V')$, exponential growth (as $\exp(a|y|^{(2b+1)/2b})$) is allowed as well. A simpler fact is: if $\mu(t, y) = c|y|$ outside of $\tilde{V}$ and $\mu = 0$ on $\tilde{V}$, then we have solubility in the spaces $\mathscr{S}_\mu[a, b)$.

As an example, let us consider $P(\tau, \eta) = \tau - (\eta_1^3 + \cdots + \eta_{n-1}^3)$. This is a 3-hyperbolic polynomial, and its 3-hyperbolicity cone V_+ is the positive n-hedron $\{\eta_j > 0, 1 \leqslant j \leqslant n - 1\}$. For $\eta \in V_+$, we have that $\chi_P(\eta) = (\eta_1^3 + \cdots + \eta_{n-1}^3)$. Correspondingly, $\chi_P(y) = c[(y_1)_+^{3/2} + \cdots + (y_{n-1})_+^{3/2}]$, where $c = 2\sqrt{3}/9$, $\lambda_+^{3/2} = \lambda^{3/2}$ for $\lambda > 0$, $\lambda_+^{3/2} = 0$ for $\lambda < 0$. In particular, $\hat{\chi}_P = 0$ in the purely negative n-hedron. An interesting situation arises already for $n = 2$, when the Cauchy problem can be solved with data growing as $\exp(a|y|^{3/2})$ for $y \to \infty$, but only power law growth is allowed for $y \to -\infty$.

3.5. Polynomials Parabolic with Respect to the Newton Polygon. We shall extend the concept of $2b$-parabolicity so as to include in one class $2b$-parabolic polynomials for different b, as well as their products. Let us fix a polynomial $P(\tau, \eta)$, and in the positive quadrant $\mathbb{R}^2_+$ let us consider the set of all integer pairs (α, γ) such that for some monomial $\tau^\alpha \eta^\beta$ in P with non-zero coefficient, $|\beta| = \gamma$. Let us supplement this set by projections on the axes (α), (γ), and the point $\{0\}$, and let us take the convex hull of this set. The resulting polygon N_P is called the *Newton polygon* of the polynomial P. The solubility condition in τ^k entails that there is no non-coordinate horizontal side. We also assume that there is no non-coordinate vertical side.

Under these conditions, we call a polynomial *N-parabolic*, if there exist constants $c_1 > 0, c_2$, such that for every $(\alpha, \gamma) \in N_P$, we have that $|P(\tau, \eta)| > c_1|\tau|^\alpha|\eta|^\gamma$ for $\operatorname{Im} \tau < c_2$. If N_p is the triangle $\{2b\alpha + \gamma \leqslant m, \alpha \geqslant 0, \gamma \geqslant 0\}$, we obtain the $2b$-parabolic polynomials. The set of N-parabolic polynomials is closed (i) relative to small perturbations by monomials $\tau^\alpha\eta^\beta$ $(\alpha, |\beta|) \in N_P$, (ii) relative to arbitrary perturbations by monomials $\tau^\alpha\eta^\beta$ $(\alpha, |\beta|)$ is an interior point of N_P, (iii) relative to multiplication. In particular, any products of $2b$-parabolic polynomials (with different b) are N-parabolic.

In the definition of a quasi-homogeneous $2b$-parabolic polynomial in terms of the roots $\lambda_j(\eta)$, we can discard the polynomial dependence in η (retaining homogeneity). We shall call the corresponding, polynomial in τ, symbols, $2b$-parabolic quasi-polynomials. In a similar way we can define hyperbolic quasi-polynomials. It turns out, that every $2b$-parabolic polynomial can be represented, modulo lower order terms, as a product of $2b_j$-parabolic quasi-polynomials. The b_j's encountered in the product have a simple geometric interpretation: $(2b_j, 1)$ are the direction vectors of non-coordinate sides of N_P (for more details, see Volevich and Gindikin [1968]).

Every N-parabolic polynomial is exponentially correct: here V_P coincides with a ray. Let b be the largest of b_j entering the decomposition above. Then the $2b$-parabolic quasi-polynomial in the decomposition of P is a polynomial; let us denote it by $P_{2b}(\tau, \eta)$. Let χ'_P be the function $\chi_{P_{2b}}$. It is homogeneous of degree $2b$. We have solubility in $\mathscr{S}_{v[\chi'_P]}[a, b)$, $\mathscr{S}_{-Iv[\chi'_P]}[a, b)$, etc.

3.6. Dominantly Correct Polynomials. We do not demand any longer that N_P have no vertical non-coordinate sides. Let $U_p \subset N_p$ be the set of all integer lattice points in N_p, such that $(\alpha, \gamma) \in U_p \Rightarrow (\alpha + \delta, \gamma) \in N_p$ for some $\delta > 0$. This corresponds to lower order monomials. Let us call a polynomial $P(\tau, \eta)$ *dominantly correct*, if for arbitrary $c > 0$, there exists c_1, such that for every $(\alpha, \gamma) \in U_P$, we have that $|P(\tau, \eta)| > c|\tau|^\alpha|\eta|^\gamma$ for $\operatorname{Im} \tau < c_1$. In this class belong all the $2b$-parabolic polynomials, and all the hyperbolic ones. It is closed relative to any variation in monomials $\tau^\alpha\eta^\beta$ with $(\alpha, |\beta|) \in U_P$. Every dominantly correct polynomial admits a decomposition (modulo lower order terms that correspond to U_P) $\tau^b H(\tau, \eta) R(\tau, \eta)$, where H is a strictly hyperbolic, and R is an N-parabolic (a product of $2b$-parabolic) quasi-polynomials (see Volevich and Gindikin [1968]).

By direct verification it is seen that dominantly correct polynomials are exponentially correct. Estimates of the correctness classes are obtained as in Section 3.5.

The concept of dominant correctness can be generalized (to semi-dominant correctness), so that from it follows semi-exponential correctness, and so that $(2b + 1)$-hyperbolic polynomials belong in the resulting class. We do not go into these details.

3.7. Pluriparabolic Polynomials. We introduce another class of exponentially correct polynomials. They will be, in some natural sense, hyperbolic in some variables, and parabolic in others. In this context, the coexistence of two types of variables in the same symbol occurs in a non-trivial way. Thus, let there be two types of variables $\tau = (\tau_1, \ldots, \tau_l)$, and $\eta = (\eta_1, \ldots, \eta_m)$. To the first ones we give weight $2b$, and to the second set, weight 1. Let the polynomial $P_0(\tau_1, \ldots, \tau_l; \eta_1, \ldots, \eta_m)$ be quasi-homogeneous with respect to this system of weights of degree k in τ (of quasi-homogeneous degree $2bk$). Assume that it possible to choose one of the τ-variables, say, τ_1, to be the variable dual to time, in such a way, that we have a polynomial that is correct in the sense of Petrovskij. We shall call P_0 *pluriparabolic* if for some $c > 0$ we have that $|P_0(\tau, \eta)| > c(|\mathrm{Im}\, \tau_1| + |\eta|^{2b})^k(|\tau| + |\eta|^{2b})^{k-1}$ for $\mathrm{Im}\, \tau_1 \leqslant 0$. If $\lambda_j(\tau', \eta)$ are the roots of P_0 in τ_1, $\tau' = (\tau_2, \ldots, \tau_l) \in \mathbb{R}^{l-1}$, then this condition is equivalent to the following: (i) $\lambda_j(\tau', 0)$ are real and simple ($P_0(\tau, 0)$ is a strictly hyperbolic polynomial), and (ii) for some $c > 0$ we have that $\mathrm{Im}\, \lambda_j(\tau', \eta) \geqslant c|\eta|^{2b}$, $\tau' \in \mathbb{R}^{l-1}$, $\eta \in \mathbb{R}^m$.

For a fixed η the symbol P_0 will be strictly hyperbolic, while for a fixed τ' it will be $2b$-parabolic. For $k = 1$, pluriparabolic polynomials have a simple structure: they can be represented as a sum of a hyperbolic polynomial of degree 1 and of a degree $2b$ homogeneous elliptic polynomial with positive definite imaginary part. For $k > 1$, the "intertwining" of the variables has a more complicated character. Let us demonstrate it by an example.

Let $Q(\tau)$ be a strictly hyperbolic polynomial, and let $V = V_Q$ be the corresponding cone. Let us consider a polynomial mapping F of degree $2b$ of $\mathbb{R}^m$ into $\mathbb{C}^l$, such that $\mathrm{Im}\, F(\eta)$ is an interior point of V for every $\eta \in \mathbb{R}^m$, $\eta \neq 0$ (F is V-elliptic). Then the polynomial $P(\tau, \eta) = Q(\tau - F(\eta))$ is pluriparabolic.

Pluriparabolic polynomials are exponentially correct, as are the polynomials obtained by adding any lower order terms with respect to the weights $(2b, \ldots, 2b, 1, \ldots, 1)$. The asymptotic cone of P coincides with the cone of its hyperbolic part, and each of its directions can be taken as the variable dual to time, retaining the pluriparabolicity. The spaces $\mathscr{S}_\mu$, in which pluriparabolic differential equations are soluble, are easily described. We do not present this description, noting that we have coexistence of a finite domain of dependence in hyperbolic variables, with parabolic behavior in the rest of the variables.

Chapter 3
The Cauchy Problem for Linear Equations with Variable Coefficients

In previous chapters, we treated the Cauchy problem for differential equations with constant coefficients in classes of functions (distributions) with power law or exponential decay (growth) as a particular case of the more general problem of solubility of convolution equations in certain function spaces, and, in that form, it was completely solved.

Formulation of the Cauchy problem for equations with variable coefficients in such generality cannot be justified, and in this case discussion from the very beginning is restricted to a specific class of operators.

Up till now two main approaches to the study of the Cauchy problem for equations with variable coefficients have been developed: the parametrix method originating with Hilbert and E. Levi, and the method of energy estimates due to Courant, Friedrichs, and H. Lewy [1928]. The natural target for the parametrix method is presented by parabolic equations, while the method of energy estimates is best applied to hyperbolic equations. The corresponding classes of higher order equations (systems) were singled out by Petrovskij in mid-thirties (see Petrovskij [1938], [1937]). These, respectively, are the $2b$-parabolic in the sense of Petrovskij, and strictly hyperbolic (hyperbolic in the sense of Petrovskij) equations (systems). For $2b$-parabolic systems Petrovskij constructed Green's matrices and obtained the solution of the Cauchy problem in the form of their convolution with the initial data. Petrovskij restricted himself to the case of coefficients depending on time only, remarking at the same time that this restriction is related to the use of Fourier's method, and does not follow from the nature of the problem. Later on, using the parametrix method, Ladyzhenskaya [1950] (in the case of a single high order equation) and Ejdel'man [1964] constructed Green's matrices for general parabolic systems with variable coefficients (see also the monograph of Solonnikov [1965]). As elements of these matrices decay exponentially in the space variables, their use made possible solving the Cauchy problem in classes of exponentially growing functions.

In the case of hyperbolic systems with variable coefficients, Petrovskij [1937] obtained estimates of the L_2 norms of solutions and of their derivatives up to some finite order in terms of L_2 norms of the right hand side and of the Cauchy data. Estimates of this nature ensure uniqueness of solutions, and, if combined with direct methods of construction of solutions, lead to existence theorems. Petrovskij [1937] approximated coefficients of the hyperbolic system by entire functions, and constructed the approximate solution by the Cauchy-Kovalevskaya method. Technically, Petrovskij's construction of energy integrals is very cumbersome as it uses a very rudimentary form of the theory of singular integral operators (which was to be developed twenty years later).

For general hyperbolic systems with variable coefficients (both linear and quasilinear) Ladyzhenskaya [1952] constructed convergent finite difference schemes, extended Petrovskij's estimates to these schemes, and obtained an existence theorem for the Cauchy problem without using the Cauchy-Kovalevskaya theorem.

In the case of a single high order hyperbolic equation, a simpler way of obtaining energy estimates was suggested by Leray [1953]; this is the so-called "separating" operator method.

In this chapter we present both these methods of solution of the Cauchy problem for equations with variable coefficients. We try to single out the natural classes of operators these methods can be applied to. It turns out that in the case of the parametrix method it is convenient to extend the treatment beyond the confines of differential operators, and to conduct the discussion in classes of pseudo-differential operators (PDO).

§ 1. The Homogeneous Cauchy Problem for Pseudodifferential Equations

1.1. The General Idea of the Method consists basically of solving an equation with variable coefficients by the method of successive approximations, where as the first approximation we take the solution of the equation with constant (frozen) coefficients. This approach succeeds in the cases when locally, in appropriate function spaces, the operator is a small perturbation of a similar operator with coefficients frozen at some point.

To make the above precise, let us consider the homogeneous Cauchy problem for a differential equation with variable coefficients

$$\begin{gathered} P(x; D_t, D_y)u(t, y) = \sum p_{j\alpha}(x) D_t^j D_x^\alpha u = f(t, y) \\ u(x, t) = f(x, t) = 0 \quad \text{for } t < 0, \quad (t, y) \in \mathbb{R}^n, \end{gathered} \tag{1}$$

where the symbol $P(x; \tau, \eta)$ satisfies the Petrovskij homogeneous correctness condition uniformly in x: $P(x; \tau, \eta) \neq 0$ for $\operatorname{Im} \tau \leqslant \gamma_0$. We seek a solution of (1) in the form

$$u(t, y) = (2\pi)^{-n/2} \int_{\operatorname{Im} \tau = \gamma} \exp(it\tau + iy \cdot \eta) P^{-1}(x; \tau, \eta) \hat{g}(\tau, \eta)\, d\xi, \tag{2}$$

where $\xi = (\operatorname{Re} \tau, \eta)$, while $\hat{g}$ is the Fourier-Laplace transform of some unknown density g, $g(t, y) = 0$ for $t < 0$. If the coefficients of P do not depend on x, then (2) gives the solution of the problem (1) for $g = f$. In the case of variable coefficients, let us substitute (2) into (1), more the differential operator inside the integral and differentiate using the Leibnitz-Hörmander formula. Then we obtain for g the following pseudo-differential equation (a precise definition of a PDO is given below):

$$g + r(x; D_t, D_y)g = f \tag{3}$$

with the symbol

$$r(x;\tau,\eta)=\sum_{|\alpha|=1}^{\operatorname{ord}P}\frac{1}{\alpha!}P^{(\alpha)}(x;\tau,\eta)D_x^\alpha P^{-1}(x;\tau,\eta). \tag{4}$$

We shall use the standard notation

$$a_{(\alpha)}^{(\beta)}(x;\xi)=\left(\frac{\partial}{\partial\xi}\right)^\beta\left(\frac{1}{i}\frac{\partial}{\partial x}\right)^\alpha a(x;\xi).$$

For $\alpha=0$ ($\beta=0$) we shall write $\alpha^{(\beta)}$ (respectively, $a_{(\alpha)}$). We want the symbol (4) to be bounded by a constant $c(\rho)$, such that the $c(\rho)\to 0$, $\rho=\operatorname{Im}\tau\to-\infty$. To attain this, it is sufficient to demand that the two following conditions hold:

$$P^{(\alpha)}(x;\tau,\eta)/P(x;\tau,\eta)\to 0,\quad \operatorname{Im}\tau\to-\infty \qquad \text{(uniformly in } x, \operatorname{Re}\tau,\eta)$$

$$|P_{(\alpha)}(x;\tau,\eta)P^{-1}(x;\tau,\eta)|<\text{const}.$$

The first condition, in view of Section 3.1, is the condition

(A) For any fixed $x\in\mathbb{R}^n$, the polynomial $P(x;\xi)$ is exponentially correct.

The second estimate follows from the condition of constancy of the strength:

(B) There exist constants c and γ_0 such that $\forall x', x''\in\mathbb{R}^n$

$$|P(x';\tau,\eta)P^{-1}(x'',\tau,\eta)|<c,\quad \xi\in\mathbb{R}^n,\quad \operatorname{Im}\tau\leqslant\gamma_0.$$

Under conditions (A), (B) not only the symbol (4), but also all its x derivatives tend to zero as $\operatorname{Im}\tau\to-\infty$. Hence we deduce that the PDO $r(x;D_t,D_x)$ maps $(H_{[\rho]})_+$, $\rho\leqslant\gamma_0$, into itself, and its norm $\to 0$ as $\rho\to-\infty$. But then for $\rho<\gamma_1<\gamma_0$ equation (3) can be solved in $(H_{[\gamma]})_+$ by a Neuman series. Having found g, we can use (2) to obtain the solution $u(x)$ of the original problem. PDO (2) is called the parametrix (more precisely, the right parametrix) of equation (1). Making operator (2) act on equation (1) on the left, we can again obtain an equation of the same type as (3); from uniqueness for this equation, uniqueness for problem (1) follows.

Thus, in the case of exponentially correct differential operators of constant strength the problem (1) has a unique solution in $(H_{[\rho]})_+$ (for $|\rho|$ sufficiently large). We want to study in detail the question of the dependence of the smoothness of the solution of (1) on the smoothness of the right hand side f. For that, it will be convenient for us to include operators satisfying conditions (A) and (B) into the algebra of PDO which contains both these operators and their parametrices.

1.2. Calculus of Pseudo-differential Operators Associated with the Homogeneous Cauchy Problem. The traditional calculus of PDO, which is oriented towards the study of local properties of differential operators, is developed in the scale of spaces $\{H^{(s)}\}$, and the operators themselves are considered up to lower order (smoothing) operators. For solutions of the homogeneous Cauchy problem we shall develop a calculus of PDO acting simultaneously in all the scales $\{H_{[\rho]}^{(s)}\}$ for all $\rho\leqslant\gamma_0$; moreover, these operators are considered up to (modulo) operators the norm of which in this scale goes to zero as $\rho\to-\infty$.

The Class of Symbols. A function $a(x; \tau, \eta)$ is called a symbol if it is defined and infinitely differentiable for $x \in \mathbb{R}^n$, $\eta \in \mathbb{R}^{n-1}$, $\tau \in \mathbb{C}$, $\operatorname{Im} \tau \leqslant \gamma_0(a)$, holomorphic in τ, and grows together with its derivatives not faster than a power of $|\tau| + |\eta|$, that is, for some $m = m(a)$

$$|a_{(\alpha)}(x; \tau, \eta)| < c_\alpha(1 + |\tau| + |\eta|)^m. \tag{5}$$

Furthermore, we shall assume that the symbol stabilizes for sufficiently large x, that is

$$a(x; \tau, \eta) = a(\tau, \eta) + a'(x; \tau, \eta), \tag{6}$$

$$a'(x; \tau, \eta) \equiv 0 \text{ for } |x| > X(a). \tag{6'}$$

Remark. All results of this section remain valid if (6′) is replaced by a weaker condition

$$\int |a'_\alpha(x; \tau, \eta)|\, dx < c_\alpha(1 + |\tau| - |\eta|)^m. \tag{6''}$$

Pseudo-differential Operators. Every symbol considered above can be made to correspond to a PDO:

$$a(x; D_t, D_y)u = (2\pi)^{-n/2} \int_{\operatorname{Im} \tau = \rho} a(x; \tau, \eta)\hat{u}(\tau, \eta) \exp(it\tau + iy\cdot\eta)\, d\xi. \tag{7}$$

The right hand side of (7) is defined for all $u \in H^{(\infty)}_{[\rho]}$ and $\rho \leqslant \gamma_0(a)$. It can be checked that this operator maps $H^{(\infty)}_{[\rho]}$ into itself. If $u \in (H^{(\infty)}_{[\gamma]})_+$, then for $\rho < \gamma$ the right hand side of (7) is independent of ρ and defines an operator that maps $(H^{(\infty)}_{[\gamma]})_+$, $\gamma \leqslant \gamma_0(a)$, into itself. By continuity this operator can be extended to a continuous operator mapping $(H^{(-\infty)}_{[\gamma]})$, $(H^{(-\infty)}_{[\gamma]})_+$ into themselves.

The Spaces $(H^\lambda_{[\gamma]})_+$. We related to each symbol a PDO acting in the limit spaces $H^{(\pm\infty)}_{[\gamma]}$, $(H^{(\pm\infty)}_{[\gamma]})_+$. Now we want to connect these spaces by a scale of spaces of functions of finite smoothness (distributions of finite order). In Chapter 1 we considered the space $(H^{(s)}_{[\gamma]})_+$ defined as the image of the PDO $\delta^+_{-s}(D)$. Were we to restrict ourselves to PDO's with homogeneous symbols, these spaces would have sufficed. However, for a precise description of operators with inhomogeneous symbols we require a wider and more flexible scale. This is the scale of spaces $(H^\delta_{[\gamma]})_+$, defined as images in $(H^{(-\infty)}_{[\gamma]})_+$ of PDO with constant coefficients:

$$(H^\delta_{[\gamma]})_+ = \{\varphi \in (H^{(-\infty)}_{[\gamma]})_+, \delta(D)\varphi \in (H_{[\gamma]})_+\}, \tag{8}$$

here we introduce the norm $\|u\|^\delta_{[\gamma]} = \|\delta(D)u\|_{[\gamma]}$.

In order that the spaces (8) are modules over $H^{(\infty)}$, we impose an additional restriction on the symbol δ: $\exists N = N(\delta)$ such that

$$\begin{gathered} |\delta(\tau', \eta')\delta^{-1}(\tau'', \eta'')| < c(1 + |\xi' - \xi''|)^N, \quad \forall \xi', \quad \xi'' \in \mathbb{R}^N, \\ \operatorname{Im} \tau' = \operatorname{Im} \tau'' \leqslant \gamma_0(\delta), \quad \xi' = (\operatorname{Re} \tau', \eta'), \quad \xi'' = (\operatorname{Re} \tau'', \eta''). \end{gathered} \tag{9}$$

An example of symbols for which (9) is satisfied is provided by exponentially correct polynomials and their real powers.

The class S^δ of symbols is defined by the conditions:

$$|a_{(\alpha)}(x;\tau,\eta)| < c_\alpha|\delta(\tau,\eta)|, \quad \forall\xi\in\mathbb{R}^n, \quad \operatorname{Im}\tau \leqslant \gamma_1; \tag{10}$$

$\exists N$ such that $\forall\xi' = (\operatorname{Re}\tau',\eta')$ and $\forall\xi'' = (\operatorname{Re}\tau'',\eta'')$, $\operatorname{Im}\tau' = \operatorname{Im}\tau'' = \rho$

$$|a_{(\alpha)}(x;\tau',\eta') - a_{(\alpha)}(x;\tau'',\eta'') < \varepsilon_\alpha(\rho)(1+|\xi'-\xi''|)^N|\delta(\tau',\eta')|,$$

$$\varepsilon_\alpha(\rho)\to 0, \quad \rho\to-\infty. \tag{11}$$

Remark. If condition (9) holds for δ (and we shall deal only with such δ), then it follows from (11) that an analogous estimate with ξ' and ξ'' exchanging places, is also valid.

Proposition 1. *If $a(x;\tau,\eta)\in S^\delta$, then the operator*

$$a(x;D_t,D_y)\colon (H^{\delta\lambda}_{[\gamma]})_+ \to (H^\lambda_{[\gamma]})_+, \quad \gamma\leqslant\gamma_0(\delta,\lambda),$$

is bounded, and its norm does not exceed a constant independent of γ.

Proposition 2 (Commutation formula). *Let $a_j(x;\tau,\eta)\in S^{\delta_j}$, $j=1,2$, and $a(x;\tau,\eta) = (a_1a_2)(x;\tau,\eta)$. Then the operator*

$$a_1(x,D)\cdot a_2(x,D) - a(x,D)\colon (H^{\delta_1\delta_2\lambda}_{[\gamma]})_+ \to (H^\lambda_{[\gamma]})_+,$$

is bounded, and its norm goes to zero as $\gamma\to-\infty$.

These statements are proved by techniques that are in principle the same as the classical PDO techniques used to obtain estimates in the scales $H^{(s)}$ (see for example Kohn and Nirenberg [1965]).

1.3. The Homogeneous Cauchy Problem for PDO and for Exponentially Correct Differential Operators of Constant Strength. Next we shall study the question of solubility of the equation

$$a(x;D_t,D_y)u = f\in(H^\lambda_{[\gamma]})_+, \tag{12}$$

where in the left hand side we have a PDO with a symbol satisfying a certain analog of conditions (A), (B) of section 1.1.

Condition (C). *$a(x;\tau,\eta)\in S^{a(x^0)}$ for any $x^0\in\mathbb{R}^n$, where $a(x^0) = a(x^0;\tau,\eta)$ and there exists $A>0$ such that*

$$|a(x;\tau,\eta)| > A|a(x^0;\tau,\eta)|, \quad \operatorname{Im}\tau\leqslant\gamma(a).$$

We immediately note that from (C) follows the constant strength condition

$$|a(x';\tau,\eta)| < c|a(x'';\tau,\eta)|, \quad \forall x', \quad x''\in\mathbb{R}^n, \quad \operatorname{Im}\tau\leqslant\gamma_0. \tag{13}$$

Remark. For $\alpha=\beta=0$ and $x=x^0$, it follows from (11) that the symbol $\delta(\tau,\eta) = a(x^0;\tau,\eta)$ satisfies (9), so that $a\in S^\delta$.

Let us quote two statements dealing with symbols that satisfy condition (C); these can serve to motivate the introduction of this class.

Proposition 1. *The symbols $a(x;\tau,\eta)$ and $a^{-1}(x;\tau,\eta)$ satisfy condition* (C) *simultaneously.*

Proposition 2. *The polynomial symbol $P(x; \tau, \eta)$ satisfies condition* (C) *if and only if it satisfies conditions* (A), (B) *of section* 1.1.

The first statement is obvious; let us sketch the proof of the second one. If conditions (A), (B) are satisfied for a polynomial symbol P, then, expanding it in the Taylor series in the variable ξ at a point $\xi'' = (\mathrm{Re}\,\tau'', \eta'')$ we find that ($\mathrm{Im}\,\tau' = \mathrm{Im}\,\tau''$)

$$P(x; \tau', \eta') - P(x; \tau'', \eta'') = \left[\sum_{|\alpha|>0} \frac{(\xi' - \xi'')^\alpha}{\alpha!} \frac{P^{(\alpha)}(x; \tau'', \eta'')}{P(x; \tau'', \eta'')}\right] P(x; \tau'', \eta''), \qquad (14)$$

whence (11) follows trivially for $\alpha = \beta = 0$. In view of the finite dimensionality of the space of polynomials, it is possible to choose a number of points $x^1, \ldots, x^J$ in such a way that

$$P(x; \tau, \eta) = \sum_{j=1}^{J} c_j(x) P(x^j; \tau, \eta).$$

With the help of this representation, the right hand side of (14) can be epxressed as a linear combination of similar expressions with constant coefficients. Hence (10), (11) follow for $\alpha > 0$, $\beta = 0$.

Using the Lagrange interpolation formula, we can express the symbols $P^{(\beta)}_{(\alpha)}$, $\beta > 0$, as linear combinations of the symbols $P_{(\alpha)}$, thus proving (11) for $\beta > 0$.

On the other hand, if (11) is satisfied, the square bracket in the right hand side of (14) does not exceed $c\varepsilon(\rho)$ for $|\xi' - \xi''| \leqslant 1$. In view of the Lagrange interpolation formula, the coefficients of the polynomial (in $(\xi' - \xi'')$ do not exceed $c'\varepsilon(\rho)$, that is, (A) is satisfied, while (B) follows from (13).

Theorem. *Let condition* (C) *be satisfied, and let $\delta(\tau, \eta) = a(x^0; \tau, \eta)$. Then for each $\lambda(\tau, \eta)$ that satisfies* (9) *there exists $\gamma_1 = \gamma_1(\lambda)$, such that for $\gamma < \gamma_1$ the mapping*

$$a(x, D)\colon (H^{\delta\lambda}_{[\gamma]})_+ \to (H^{\lambda}_{[\gamma]})_+. \qquad (15)$$

is an isomorphism of spaces.

The proof follows the construction of Section 1.1 and is based on Propositions 1, 2 of Section 1.2. We seek a solution of (12) in the form $u = a^{-1}(x, D)g$. $g \in (H^{\lambda}_{[\gamma]})_+$. According to Proposition 1, $u \in (H^{\lambda\delta}_{[\gamma]})$, and by Proposition 2, g satisfies the equation

$$g + [a(x; D)\cdot a^{-1}(x; D) - 1]g = g + Rg = f, \qquad (16)$$

where the norm of the operator $R\colon (H^{\lambda}_{[\gamma]})_+ \to (H^{\lambda}_{[\gamma]})_+$ goes to zero as $\gamma \to -\infty$. From that we deduce surjectivity of operator (15). Acting on (15) on the left by the operator $a^{-1}(x, D)$, we shall show injectivity of this operator.

Assume that $a(x; \tau, \eta)$ satisfies condition (C) and that in equation (12) the right hand side f belongs to $(H^{\infty}_{[\gamma_0]})_+$. Then with every symbol λ that satisfies (9) we can associate $\gamma(\lambda)$ such that the solution $u(x)$ belongs to $(H^{\lambda}_{[\gamma]})_+$ for $\gamma < \gamma(\lambda)$. In order to establish solubility of (12) in the space of C^∞ functions, we must pass to a finite strip.

For $a < b$ the quotient spaces

$$H^\mu[a, b) = T_{(-a,0)}(H^\mu_{[\gamma]})_+ / T_{(-b,0)}(H^\mu_{[\gamma]})_+$$

are defined.

The right hand side does not depend on γ either in the "stock" of its elements or in topology, so that the spaces are well defined. The PDO we have introduced map the scale of spaces $H^{(-\infty)}[a, b)$ into itself. From the theorem above we derive

Corollary 1. *Let condition* (C) *be satisfied for* $a(x; \tau, \eta)$, $\delta = a(x^0, \cdot)$, $-\infty < a < b < \infty$. *Then the mapping*

$$H^{\lambda\delta}[a, b) \to H^\lambda[a, b)(u \mapsto a(x; D)u)$$

is an isomorphism.

Corollary 2. *Under the conditions of the theorem, for any* $-\infty < a < b < \infty$, *the mapping*

$$H^{(\pm\infty)}[a, b) \to H^{(\pm\infty)}[a, b)(u \mapsto a(x; D)u)$$

is an isomorphism.

1.4. The Homogeneous Cauchy Problem in Slowly Growing and Slowly Decreasing Functions. In preceeding sections we considered PDO's in smoothness scales related to $H^{(-\infty)}$. In the remaining part of the section we shall deal with scales with weights. Let us start with the simpler case of power weights; moreover, we restrict ourselves to weights $l(x)$ that satisfy conditions of the type of (9),

$$|l(x')l^{-1}(x'')| \leqslant c(1 + |x' - x''|)^N (c = c(l), N = N(l))$$

and estimates for derivatives $l^{(\alpha)}(x) = D^{(\alpha)}l(x)$:

$$|l^{(\alpha)}(x)l^{-1}(x)| < c_\alpha; \quad x \in \mathbb{R}^n. \tag{17}$$

Let a symbol $\delta(\tau, \eta)$ satisfy condition (9) and let estimates of the type of (17) be satisfied for its derivatives,

$$|\delta^{(\alpha)}(\tau, \eta)\delta^{-1}(\tau, \eta)| < D_\alpha, \quad \eta \in \mathbb{R}^{n-1}, \quad \operatorname{Im} \tau \leqslant \gamma(\delta).$$

We remark that exponentially correct polynomials and their real powers satisfy the conditions indicated above. By analogy with (8) let us define

$$H^\delta_{l[\gamma]+} = \{\varphi \in (\mathscr{S}')_{[\gamma]+}, \delta(D)l(x)\varphi \in H_{[\gamma]+}\}, \tag{18}$$

and let us introduce in this space the norm $\|u\|^\delta_{l[\gamma]} = \|\delta(D)l(x)u\|_{[\gamma]}$. In definition (18) $\delta(D)$ and $l(x)$ can be interchanged. The norm thus arising will be eqivalent to the norm $\| \quad \|^\delta_{l[\gamma]}$.

With each pair of weights $\delta(\tau, \eta)$, $l(x)$ that satisfy the conditions indicated above, we can associate a class S^δ_l of symbols $a(x; \tau, \eta)$ that are infinitely differentiable in all variables, holomorphic in τ for $\operatorname{Im} \tau \leqslant \gamma(a)$ and satisfy the condition (compare with (10), (11))

$$|a^{(\beta)}_{(\alpha)}(x; \tau, \eta)| < \varepsilon_{\alpha\beta}(\operatorname{Im} \tau)|\delta(\tau, \eta)l(x)|, \quad \operatorname{Im} \tau \leqslant \gamma(a, \delta),$$

where

$$\varepsilon_{\alpha\beta}(\operatorname{Im}\tau)\to 0,\quad \operatorname{Im}\tau\to-\infty,\quad \alpha\geqslant 0,\quad \beta>0.$$

Proposition 1. *If $a(x;\tau,\eta)\in S_l^\delta$ then for any pair of weights $m(x)$, $\lambda(\tau,\eta)$ satisfying the condition above, the operator*

$$a(x;D_t,D_y)\colon (H_{lm}^{\delta\lambda})_{[\gamma]+}\to (H_m^\lambda)_{[\gamma]+},\quad \gamma\leqslant\gamma(\delta,\lambda,l,m). \tag{19}$$

is bounded.

Proposition 2 (commutation formula). *Let $a_j(x;\tau,\eta)\in S_{l_j}^{\delta_j}$, $j=1,2$, then there exists a symbol $a(x;\tau,\eta)\in S_{l_1l_2}^{\delta_1^j\delta_2}$ such that $a_1(x;D_t,D_y)\cdot a_2(x;D_t,D_y)=a(x;D_t,D_y)$. Also, the norm of the operator*

$$a(x,D_t,D_y)-(a_1a_2)(x;D_t,D_y)\colon (H_{l_1l_2m}^{\delta_1\delta_2\lambda})_{[\gamma]+}\to (H_m^\lambda)_{[\gamma]+}$$

goes to zero as $\gamma\to-\infty$.

Proofs of these statements make essential use of the Calderon-Vaillancourt theorem of boundedness of PDO in L_2 (see for example Taylor [1981]). It is exactly this theorem that allows us to renounce the symbol stabilization condition we used in Section 1.2.

Propositions 1 and 2 allow us to prove an analog of Theorem 1.3 for the spaces (18).

Theorem. *Let $a(x;\tau,\eta)\in S_l^\delta$, and assume that there is a constant A such that*

$$|a(x;\tau,\eta)|>A|\delta(\tau,\eta)l(x)|,\quad \operatorname{Im}\tau\leqslant\gamma(a).$$

then for any symbols λ, m satisfying the conditions indicated above, it is possible to find $\gamma(\lambda,m)$ such that for $\gamma<\gamma(\lambda,m)$ the mapping (19) *is a space isomorphism.*

Passing to a finite strip, $0\leqslant t\leqslant c$ for definiteness, we shall obtain isomorphisms for various "limit" spaces of the scale (18).

Corollary. *Let the conditions of the theorem be satisfied, and $\Phi=\mathscr{S},\mathcal{O},\mathcal{O}',\mathscr{S}'$. Then $\forall c>0$*

$$a(x;D)\colon \Phi[0,c)\to\Phi[0,c). \tag{20}$$

is an isomorphism.

Remark. As Φ in (20) we can take various spaces obtained from (18) by operations of taking projective and inductive limits. In particular, we can take

$$\bigcap_{l,s'}\bigcup_{s}H_{(l)}^{(s,s')};\quad \bigcup_{s,l}\bigcap_{s'}H_{(l)}^{(s,s')}. \tag{21}$$

The first of these can be interpreted as $\mathcal{O}'(\mathbb{R})\otimes\mathscr{S}(\mathbb{R}^{n-1})$, and the second one as $\mathscr{S}'(R)\otimes\mathcal{O}(\mathbb{R}^{n-1})$.

1.5. The Inhomogeneous Cauchy Problem in Slowly Increasing and Slowly Decreasing Functions. We shall say that a symbol $a(x;\tau,\eta)$ of Section 1.2 satisfies the transmission condition, if it belongs, in the variables τ, η, to a translation of the space $\hat{U}^+$ of § 3 of Chapter 1, that is, in some half-plane $\operatorname{Im} \tau \leqslant \gamma_0$ it can be expanded in an asymptotic Laurent series in τ:

$$a(x; \tau, \eta) = \sum_{j=0}^{\infty} a_{m-j}(x; \eta)(\tau - i\gamma_0)^{m-j}. \tag{22}$$

If $a_m(x; \eta) \not\equiv 0$, then m is called the order of a in t; $m = \operatorname{ord}_t a$. With each symbol we can associate a PDO. It can be checked that this operator maps the spaces (21) into themselves, as it does the analogous spaces in a strip,

$$a(x, D)\colon \Phi^{\{-\infty\}}[0, c) \to \Phi^{\{-\infty\}}[0, c), \quad \Phi = \mathscr{S}, \mathscr{O}. \tag{23}$$

Theorem. *Let the symbol $a(x; \tau, \eta)$ satisfy conditions of Theorem 1.4 and the transmission condition. Let us assume, in addition, that*

$$a_m(x; \eta) \equiv \text{const}, \quad m = \operatorname{ord}_t a. \tag{24}$$

Then the mappings (23) *are space isomorphisms.*

Let us explain the proof of this theorem in the case $\Phi = \mathscr{S}$. Let us denote by Ψ the first of the spaces in (21). Then, according to Section 1.4, the mapping

$$a(x, D)\colon \Psi[0, c) \to \Psi[0, c) \tag{23'}$$

is an isomorphism. As $\mathscr{S}^{\{-\infty\}}[0, c) \subset \Psi[0, c)$, and as symbols with the transmission condition leave $\mathscr{S}^{\{-\infty\}}[0, c)$ invariant, (23) can be considered as the restriction of the mapping (23′). Hence follows the injectivity of (23). To prove surjectivity, let us take an arbitrary element $f \in \mathscr{S}^{\{-\infty\}}[0, c)$. Then there exists an element $u \in \Psi[0, c)$, such that $a(x; D)u = f$. It remains only to prove partial hypoellipticity of the PDO a in the variable t:

$$\{u \in \Psi[0, c), a(x, D)u \in \mathscr{S}^{\{-\infty\}}[0, c)\} \Rightarrow \{u \in \mathscr{S}^{\{-\infty\}}[0, c)\}. \tag{25}$$

The idea of the proof of (25) is very simple. In view of (24), we can assume that $a(x; D) = (D_t - i\gamma_0)^m + (D_t - i\gamma_0)^{m-1}B$, where the operator B does not lower smoothness in t. Thus,

$$u = -(D_t - i\gamma_0)^{-1}Bu + (D_t - i\gamma_0)^{-m}f. \tag{26}$$

If the distribution u has smoothness s_0 in t for $t > 0$, then, by (26) it has smoothness $s_0 + 1$. Iterating on this argument, we obtain (25).

In fact, the isomorphism (25) means that all the mappings

$$a(x, D)\colon \Phi^{\{p\}}[0, c) \to \Phi^{\{p-m\}}[0, c), \quad m = \operatorname{ord}_t a. \tag{27}$$

are isomorphisms.

From definition (A) it follows that an exponentially correct symbol is soluble with respect to the highest power of τ, so that exponentially correct polynomials of constant strength satisfy the conditions of the theorem, while the mapping (27) is an isomorphism for $\Phi = \mathscr{S}, \mathscr{O}$. For $p = 0$, we deduce from the isomorphism (27) the correctness of the Cauchy problem in a strip with initial data in $\mathscr{S}$ or $\mathscr{O}$. The Cauchy problem with initial data in $C_{(l)}^{(\infty)}$ for any (even) l can be treated similarly.

1.6. PDO's with Holomorphic Symbols and Correctness Classes for the Cauchy Problem with Variable Coefficients. We saw in Chapter 2, that the study of operators in spaces with exponential asymptotics $\exp \mu$, where μ is a convex function, is reduced to the study of their simultaneous action in spaces with norms containing weight functions $\exp(\langle \xi, x \rangle)$, where ξ runs through the convex set dom $\hat{\mu}$. The preceeding part of section 1 consisted of the study of pseudo-differential operators acting on spaces with norms containing exponential weights in t, $\exp(\gamma t)$. Here the symbols have to be holomorphic in the variable that is dual to time. This allowed us to study power law correctness classes of the Cauchy problem for variable coefficients. It is natural to expect, that the addition of exponential weights in the space variables, and the consideration of simultaneous action of PDO's with symbols holomorphic in tubular domains, will allow us to investigate exponential correctness classes.

Let us remark first of all, that if in Propositions 1, 2 of Section 1.2 we replace $\exp(\rho t)$ by the weight $\exp(\rho t + \omega \cdot y)$, and consider the symbol $a(x; \tau, \zeta)$, $\zeta = \eta + i\omega$ for a fixed ω, then all these statements remain valid.

Let now $\mu(t, y)$ be a convex function, and let dom $\hat{\mu} = \{\rho < -\varkappa(\omega)\}$. Let the symbol $a(x; \tau, \zeta)$ be holomorphic in the tubular domain $D(\hat{\mu}) = \mathbb{R}^n + i \operatorname{dom} \hat{\mu}$, and let it satisfy conditions of the form (5) for every $\operatorname{Im}(\tau, \zeta) \in \operatorname{dom} \hat{\mu}$. With each such symbol we associate the PDO ($\xi = \operatorname{Re}(\tau, \zeta)$)

$$a(x, D)u(x) = (2\pi)^{-n/2} \int_{\operatorname{Im}(\tau,\zeta) \in \operatorname{dom} \hat{\mu}} a(x: \tau, \zeta)\hat{u}(\tau, \zeta) \exp(it\tau + iy\cdot\zeta)\, d\xi. \quad (28)$$

Let us consider the spaces $\mathscr{H}^{(\pm\infty)}_{(l),\mu}$ with the system (2.7) of norms. The right hand side of (28) makes sense for every $u \in \mathscr{H}^{(\infty)}_{(l),\mu}$, is independent of the choice of $\operatorname{Im}(\tau, \zeta) \in \operatorname{dom} \hat{\mu}$ and defines a continuous operator that can be extended to $\mathscr{H}^{(-\infty)}_{(l),\mu}$.

Let us associate a class of symbols $\delta(\tau, \zeta)$, that satisfy conditions of the form (9), (9′):

$$|\delta(\tau', \zeta')\delta^{-1}(\tau'', \zeta'')| < c(1 + |\tau' - \tau''| + |\zeta' - \zeta''|)^N,$$

$$(\tau', \zeta'), \quad (\tau'', \zeta'') \in D(\hat{\mu}),$$

$$|\delta^{(\beta)}(\tau, \zeta)\delta^{-1}(\tau, \zeta)| < c_\beta, \quad (\tau, \zeta) \in D(\hat{\mu}),$$

with the domain $D(\hat{\mu})$, and let us use these operators to define the spaces $\mathscr{H}^{\delta}_{(l),\mu}$. Replacing in conditions (10), (11) (τ, η) by $(\tau, \zeta) \in D(\hat{\mu})$, we establish the analogs of Propositions 1, 2 of Section 1.2 for the spaces $\mathscr{H}^{\delta}_{(l),\mu}$. Condition (C) of Section 1.3 carries over trivially to symbols holomorphic in $D(\hat{\mu})$; let us denote the resulting condition by $(C_{\hat{\mu}})$. Let us set $\mu\{\gamma\} = \mu + \gamma t$. Using the argument of Theorem 1.3, we can prove the

Theorem. *Let condition* $(C_{\hat{\mu}})$ *be satisfied. Then for each* l *there exists* $\gamma_0(l)$, *such that for* $\gamma < \gamma_0(l)$ *the mapping*

$$a(x, D)\colon \mathscr{H}^{\delta\lambda}_{(l),\mu\{\gamma\}} \to \mathscr{H}^{\lambda}_{(l),\mu\{\gamma\}}, \quad \delta(\tau, \zeta) = a(x^0, \tau, \zeta), \quad (29)$$

is an isomorphism.

Results relating to the inhomogeneous problem are obtained in a similar way. Let us quote a corollary for differential operators.

Theorem. *Let $P(x; \tau, \zeta)$ be the symbol of a differential operator, such that for each x P is correct in the sense of Petrovskij and $P(x; \tau, \zeta) \neq 0$ for* $\operatorname{Im} \tau < -\chi(\operatorname{Im} \zeta)$, $\chi(\omega) \neq \infty$, $\omega \in \mathbb{R}^{n-1}$. *Let the constant strength condition (condition* B) *be satisfied under these conditions on τ, η. Then the equation $P(x; D_t, D_x)u = f$ is uniquely soluble in $\mathscr{S}_{\nu[\chi]}[a, b), (\mathscr{S}')_{\nu[\chi]}[a, b), \mathscr{S}^{\{-\infty\}}_{\nu[\chi]}[a, b), \mathscr{S}_{-I\nu[\chi]}[a, b), \mathscr{S}^{\{-\infty\}}_{-I\nu[\chi]}[a, b)$.*

If the symbol $P(x; \tau, \eta)$ belongs in any of the classes of § 3, Chapter 2, if the functions $\chi_{P(x)}$ have non-trivial majorants, and if the constant strength condition is satisfied, then the relevant theorem is applicable to P and χ, and the results concerning correctness classes for constant coefficients carry over to the case of variable coefficients.

1) Let the symbol $P(x; \tau, \eta)$ be $2b$-parabolic for each x, and such that $\deg_{2b} P$ does not depend on x, and the estimates are uniform in x. Let $\chi(\omega)$ be the majorant of $\chi_{P(x)}$; it will be a homogeneous function of degree $2b$. Then the conditions of the theorem are satisfied automatically.

2) Let the symbol $P(x; \tau, \eta)$ be N-parabolic for each x, and such that the polygon $N_{P(x)}$ does not depend on x, and let $\chi(\omega)$ be the majorant of $\chi_{P(x)}$. The theorem is applicable to the pair P, χ.

The other classes of exponentially correct symbols of § 3, Chapter 2 can be considered in a similar manner. However, for them the constant strength condition is very restrictive. Thus, in the case of hyperbolic operators, the constant strength condition (with solubility of the symbol with respect to the highest power of τ taken into account) entails constancy of coefficients in the principal part.

Remark. Let the boundary $\partial\Omega$ of the domain Ω be given by the equation $\rho = \varkappa(\omega)$. Let us consider the class of weight functions of the form

$$\mu^2(x) = \int_{\partial\Omega} \exp(2t\varkappa(\theta) + 2\langle y, \theta\rangle)\, d\chi(\theta) \quad \text{for} \quad x \in \Omega,$$

$\mu^2(x) = +\infty$ for $x \notin \Omega$, where $\chi(\theta)$ is some absolutely continuous measure on $\partial\Omega$. For such weights, the square of the norm in H_μ has the form

$$\int_{\partial\Omega} \int_{\mathbb{R}^n} \exp(2t\varkappa(\theta) + 2\langle y, \theta\rangle)|u(x)|^2\, dx\, d\chi. \tag{30}$$

Starting with the "zero" norm (30), it is not hard to construct the scale $H^{\delta}_{(l),\mu}$, for which analogs of propositions of section 1.2 are valid.

Since the norm in $H_{-\mu}$ has the same form (30) (only in the "exp" t and y have to be replaced by $-t$, $-y$), we immediately obtain by duality a calculus in exponentially growing functions.

A drawback of the weights under consideration is their insufficient effectivity: we do not know the value of the weight in different points, and can only find the asymptotic behavior as $|x| \to \infty$ by Laplace's method.

§ 2. Energy Integrals of Differential Operators with Variable Coefficients

As we mentioned at the end of § 1, the parametrix method does not allow us to prove solubility of the Cauchy problem for hyperbolic equations with variable coefficients in the principal part; other methods, based on energy estimates of the solution (see the introduction to this chapter) are better suited for the study of such equations. The present section is devoted to the examination of energy methods and to the description of classes of inhomogeneous differential operators with variable coefficients, for which there are energy estimates (energy integrals) necessary in the proof of the correctness of the Cauchy problem.

2.1. Proof of Solubility of the Homogeneous Cauchy Problem, Based on Energy Estimates in $H^{(s)}_{[\gamma]}$. In § 1, while reducing the Cauchy problem to an auxiliary pseudo-differential equation, we were not especially interested in estimates for the inverse operator. However, from the arguments of § 1, it easily follows that the $H^{(s)}_{[\gamma]}$ norm of an exponentially correct differential operator of constant strength is equivalent (uniformly in γ for $\gamma < \gamma_0$) to the analogous norm of the operator with frozen coefficients. Hence we derive an estimate for "lower order terms" for any $\alpha > 0$

$$\|P^{(\alpha)}(x; D)u\|^{(s)}_{[\gamma]} \leqslant \varepsilon_s(\gamma)\|P(x; D)u\|^{(s)}_{[\gamma]}, \quad \forall u \in H^{(\infty)}_{[\gamma]},$$
$$\varepsilon_s(\gamma) \to 0, \quad \gamma \to -\infty. \tag{31}$$

From conditions (A), (B) also follows an estimate for the formally adjoint operator $P^*(x; D)$:

$$\|\bar{P}^{(\alpha)}(x; D)v\|^{[-s]}_{[-\gamma]} \leqslant \varepsilon^*_s(\gamma)\|P^*(x; D)v\|^{(-s)}_{[-\gamma]}, \quad \forall v \in H^{(\infty)}_{[-\gamma]}.$$
$$\varepsilon^*_s(\gamma) \to 0, \quad \gamma \to -\infty. \tag{31'}$$

Let us clarify the meaning of this estimate. As is well known, the symbol P^* equals

$$P^*(x; \xi) = \bar{P}(x; \xi) + \sum_{\alpha > 0} \frac{1}{\alpha!} \bar{P}^{(\alpha)}_{(\alpha)}(x; \xi).$$

Hence it follows that if conditions (A), (B) are satisfied for $P(x; \tau, \eta)$, then they are satisfied for $P^*(x; -\tau, \eta)$ as well, while an estimate of form (31) holds for the corresponding operator. Substituting $t \to -t$, we obtain (31').

Now we shall show that using weaker estimates than (31), (31') we can prove unique solibility of the homogeneous Cauchy problem in $H^{(s)}_{[\gamma]}$. This statement means that $\forall f \in (H^{(s)}_{[\gamma]})_+$ there exists $u \in (H^{(s)}_{[\gamma]})_+$ that satisfies the equation $Pu = f$ in the sense of distributions:

$$(u, P^*\varphi) = (f, \varphi) \quad \forall \varphi \in D. \tag{32}$$

Existence and uniqueness of (32) in $H^{(s)}_{[\gamma]}$ follow, because of the Hahn-Banach theorem, from the estimates

$$\|u\|_{[\gamma]}^{(s)} \leqslant \text{const}\|P(x;D)u\|_{[\gamma]}^{(s)}, \quad \gamma \leqslant \gamma_0, \quad u \in H_{[\gamma]}^{(\infty)}, \tag{33}$$

$$\|v\|_{[-\gamma]}^{(-s)} \leqslant \text{const}\|P^*(x;D)v\|_{[-\gamma]}^{(-s)}, \quad \gamma \leqslant \gamma_0, \quad v \in H_{[-\gamma]}^{(\infty)}. \tag{33'}$$

The bound (33) guarantees uniqueness of solution of the equation (32) in $H_{[\gamma]}^{(s)}$, and a fortiori also in $H_{[\gamma]+}^{(s)}$. A more difficult question is the one of solubility of equation (32) in a half-space, that is, of existence of a solution $u \in H_{[\gamma]+}^{(s)}$ for the right hand side f in the same space. Estimate (33′) is insufficient to prove this. For the proof of existence it would have been enough to be able to prove the analog of (33′) in the scale of spaces dual to $H_{[\gamma]+}^{(s)}$, that is, in the scale of quotient spaces $H_{[\gamma]\oplus}^{(s)}$. However, a direct derivation of this estimate seems to us difficult, and we propose a different approach. We shall show that if (33′) is replaced by a somewhat stronger estimate in the scale $H_{[-\gamma]}^{(-s)}$ (and not in the scale of quotient spaces), then from this estimate would follow existence not only in $H_{[\gamma]}^{(s)}$, but also in $H_{[\gamma]+}^{(s)}$.

Theorem. *Assume that estimate* (33) *holds for the differential operator* $P(x;D)$, *and that we also have the estimate*

$$\sum_{l=1}^{k} \|\bar{P}^{(l)}(x;D)v\|_{[-\gamma]}^{(-s)} \leqslant \varepsilon(\gamma)\|P^*(x;D)v\|_{[-\gamma]}^{(-s)}$$

$$\forall v \in H_{[-\gamma]}^{(\infty)}, \quad \gamma \leqslant \gamma_1, \quad \varepsilon(\gamma) \to 0, \quad \gamma \to -\infty, \tag{34}$$

where we set

$$P^{(l)}(x;\tau,\eta) = \partial^l P(x;\tau,\eta)/\partial\tau^l, \quad k = \deg_\tau P(x;\tau,\eta).$$

Then for $\forall s \in \mathbb{R}$ *we can find* $\gamma_2(s)$ *such that for* $\gamma < \gamma_2(s)$ $\forall f \in (H_{[\gamma]}^{(s)})_+$ *there exists a unique solution* $u \in (H_{[\gamma]}^{(s)})_+$ *of equation* (32).

Let us indicate the argument of the proof. According to Section 2.2 of Chapter 1, the space $(H_{[\gamma]}^{(s)})_+$ consists of those (and only of those) elements of the intersection $\bigcap_\rho H_{[\rho]}^{(s)}$ having finite norm

$$^+\|f\| = \sup_{\rho \leqslant \gamma} \|f\|_{[\rho]}^{(s)}.$$

Therefore for $f \in (H_{[\gamma]}^{(s)})_+$ equation (32) has in general a whole family of solutions $u_\rho \in H_{[\rho]}^{(s)}$, and, moreover, due to (33)

$$\|u_\rho\|_{[\rho]}^{(s)} \leqslant c\|Pu_\rho\|_{[\rho]}^{(s)} = c\|f\|_{[\rho]}^{(s)} \leqslant c^+\|f\|_{[\gamma]}^{(s)}.$$

The theorem will be proved, if we show that the functions u_ρ do not in fact depend on ρ. For that it is enough to check that

$$u_{\gamma'} = u_{\gamma''}, \quad \gamma' \leqslant \gamma'' \leqslant \gamma, \quad \gamma'' - \gamma' < \delta \quad (\text{small } \delta). \tag{35}$$

Let us denote by $H_{[\gamma',\gamma'']}^{(s)}$, $\gamma' \leqslant \gamma''$, the intersection $\bigcap_{\gamma' \leqslant \rho \leqslant \gamma''} H_{[\rho]}^{(s)}$, equipped with the norm

$$\|f\|_{[\gamma',\gamma'']}^{(s)} = \sup_{\gamma' \leqslant \rho \leqslant \gamma''} \|f\|_{[\rho]}^{(s)}.$$

Condition (35) follows from solubility in the respective space $H^{(s)}_{[\gamma',\gamma'']}$. In fact, let $u_{\gamma',\gamma''}$ be the solution corresponding to the right-hand side $f \in (H^{(s)}_{[\gamma]})_+ \subset H^{(s)}_{[\gamma',\gamma'']}$. By uniqueness in any of $H^{(s)}_{[\rho]}$, we have that $u_{\gamma'} = u_{\gamma',\gamma''} = u_{\gamma''}$.

Solubility in $H^{(s)}_{[\gamma',\gamma'']}$ is equivalent to having an estimate of P^* in the norm of the dual space. This norm is equivalent to the norm

$$\|v\|^{(-s)}_{[-\gamma',-\gamma'']} \overset{\text{def}}{=} \|\chi v\|^{(-s)}_{[-\gamma']} + \|(1-\chi)v\|^{(-s)}_{[-\gamma'']},$$

where $\chi(t) \in C^\infty(\mathbb{R})$, $\chi(t) = 1$ for $t > 1$, $\chi(t) = 0$ for $t < -1$. In view of estimate (34),

$$\sum_{l=1}^{k} \|\bar{p}^{(l)}(\chi v)\|^{(-s)}_{[-\gamma']} \leqslant \text{const}\|P^*(\chi v)\|^{(-s)}_{[-\gamma']} \leqslant \text{const}\|\chi P^* v\|^{(-s)}_{[-\gamma']}$$
$$+ \sum_{l=1}^{k} \|\chi_l P^{*(l)} v\|^{(-s)}_{[-\gamma']}, \quad \chi_l = D_t^l \chi / l!.$$

Let us combine this estimate with an analogous estimate for $(1-\chi)v$ (replacing γ' by γ''). For $\gamma'' - \gamma' < \delta$, the bound (34) allows us to estimate the sum in the right hand side in terms of the one in the left hand side with a constant $\varepsilon < 1$, $\gamma' < \gamma'' < \gamma_1$, and $-\gamma_1$ large enough. We arrive at the estimate

$$\|v\|^{(-s)}_{[-\gamma',-\gamma'']} \leqslant \text{const}\|P^* v\|^{(-s)}_{[-\gamma,-\gamma'']},$$

which concludes the proof of our theorem.

2.2. Sufficient Conditions for Existence of Estimates (33), (34). Deriving energy estimates for hyperbolic equations, Leray [1953] considered the quadratic form

$$-\operatorname{Im}(e^{\gamma t}P(x, D)u, e^{\gamma t}Q(x, D)u), \quad \gamma < 0, \tag{36}$$

in which as the ("separating") operator $Q(x, D)$ we can take the operator with the symbol

$$Q(x; \tau, \eta) = \partial P(x; \tau, \eta)/\partial \tau. \tag{37}$$

The main claim of Leray is that the form (36) can be bounded both from below and from above in terms of $|\gamma|(\|u\|^{(m-1)}_{[\gamma]})^2$, whence we have the estimate

$$|\gamma|\,\|u\|^{(m-1)}_{[\gamma]} \leqslant \text{const}\|P(x, D)u\|_{[\gamma]}.$$

This estimate can be extended to an estimate in the scale:

$$|\gamma|\,\|u\|^{(m-1+s)}_{[\gamma]} \leqslant \text{const}\|Pu\|^{(s)}_{[\gamma]}, \quad \forall s \in \mathbb{R}. \tag{38}$$

The following observation lies at the root of all these estimates: the strict hyperbolicity condition for P is equivalent to the following two-sided bound:

$$c^{-1}|\gamma|(|\tau| + |\xi|)^{2m-2} \leqslant -\operatorname{Im}[P_0(\tau, \eta)\overline{Q_0(\tau, \eta)}] \leqslant c|\gamma|(|\tau| + |\xi|)^{2m-2}, \tag{39}$$

where P_0, Q_0 are the higher order homogeneous parts of P and Q, and m is of ord P. The transition from the algebraic estimate (39) to an estimate of the Leray form (36) is based on the Gårding inequality.

Following Petrovskij, Leray proved solubility of the Cauchy problem by combiling local Cauchy-Kovalevskaya theorems with global estimates (38). Gårding developed a direct (functional) method of proof of the existence theorem. It is based on the fact that the formally adjoint operator P^* is also strictly hyperbolic and, therefore, energy estimates of the form (38) are available for it as well.

In this section we shall consider a generalization of Leray's argument to the case of inhomogeneous operators. Let $P(x; D)$ be a differential operator, such that its symbol $P(x; \tau, \eta)$ is a polynomial that is correct in the sense of Petrovskij at each point $x \in \mathbb{R}^n$. Let the symbol $Q(x; \tau, \eta)$ have the form (37); let us set $\xi = (\operatorname{Re} \tau, \eta)$, $\gamma = \operatorname{Im} \tau$,

$$H_P(x; \gamma, \xi) = -\operatorname{Im}[P(x; \tau; \eta)\overline{Q(x; \tau, \eta)}]. \tag{40}$$

Theorem. *Let the symbol $P(x; \xi)$ satisfy the following conditions:*

(I) *There is γ_0 such that for $\gamma \leqslant \gamma_0$ symbols* (40) *satisfy the constant strength condition:*

$$H_P(x'; \gamma, \xi) \leqslant cH_P(x''; \gamma, \xi), \quad \gamma \leqslant \gamma_0, \quad \xi \in \mathbb{R}^n.$$

(IIa) *For any $x', x'' \in \mathbb{R}^n$* and $\beta > 0$,

$$|P^{(\beta)}(x'; \tau, \eta)| \leqslant \varepsilon(\gamma)|P(x''; \tau, \eta)|, \quad \varepsilon(\gamma) \to 0, \quad \gamma = \operatorname{Im} \tau \to -\infty.$$

(IIb) *For any $\alpha > 0$*

$$|P_{(\alpha)}(x; \tau, \eta)| \leqslant \varepsilon(\gamma)|P(x; \tau, \eta)(1 + |\tau| + |\eta|), \quad \varepsilon(\gamma) \to 0,$$

$$\gamma \to -\infty.$$

Then we have estimates (33), (34), *and, consequently, solubility of the Cauchy problem* (32) *in $(H^{(s)}_{[\gamma]})_+$ for every $s \in \mathbb{R}$.*

Let us first of all discuss the conditions of the theorem. (I) can be interpreted as an analog of Leray's condition for general inhomogeneous symbols. From condition (IIa) it follows that $\forall x \in \mathbb{R}^n$ the symbols $P(x; \tau, \eta)$ will be exponentially correct. Condition (IIa) means in addition that the symbol at a point x'' "restrains" with a large parameter the "lower order" terms at any other point x'.

It can be shown that conditions (I), (IIa), (IIb) hold for exponentially correct symbols of constant strength. Thus, conditions of our theorem are a weakened version of conditions (A), (B) of § 1. As the examples we give in subsequent sections show, this is quite a substantial weakening, which allows us to consider the Cauchy problem for a number of operators with variable coefficients.

The condition of our theorem are invariant with respect to passage to the formally adjoint operator (of course, the direction of time has to be changed). For that reason, we only sketch the proof of a stronger version of estimate (33):

$$\sum_{l=1}^{k} (\gamma - \gamma_0)^l \|P^{(l)}(x; D)u\|^{(s)}_{[\gamma]} \leqslant \text{const } \|P(x; D)\|^{(s)}_{[\gamma]}, \quad \gamma < \gamma_0. \tag{41}$$

First of all we mention a simple algebraic fact.

Lemma. *A polynomial $P(\tau, \eta)$ soluble with respect to the highest power of τ, satisfies the Petrovskij condition if and only if for some γ_0*

$$\sum_{l=1}^{k} (\gamma - \gamma_0)^{2l-1} |P^{(l)}(\tau, \eta)|^2 \leqslant cH_P(\gamma, \xi).$$

The proof of the lemma is based on an explicit formula that expresses H_P in terms of the roots λ_j $(j = 1, \ldots, k)$ of the polynomial P.

$$H_P(\gamma, \xi) = \sum_{j=1}^{k} (-\gamma + \lambda_j(\eta)) \prod_{k \neq j} |\tau - \lambda_k(\eta)|^2. \tag{42}$$

Let the symbol (40) satisfy the conditions of the theorem. Let us take an arbitrary point $x^0 \in \mathbb{R}^n$ and let us set $P(\tau, \eta) = P(x^0; \tau, \eta)$. According to the lemma, for $\gamma \leqslant \gamma_0$ the symbol $H_P(\gamma, \xi)$ will be positive, so that it can be used to define the norm

$$|[u]|_{[\gamma]}^{(s)} = \left(\int H_P(\gamma, \xi)(1 + |\tau|^2 + |\eta|^2)^s |\hat{u}(\tau, \eta)|^2 \, d\xi \right)^{1/2}. \tag{43}$$

From the lemma and from condition (I) we easily derive the estimate

$$\sum_{l=1}^{k} (\gamma - \gamma_0)^l \|P^{(l)}(x; D)u\|_{[\gamma]}^{(s)} \leqslant \text{const}\, |[u]|_{[\gamma]}^{(s)},$$

therefore (41) will follow from the inequality

$$|[u]|_{[\gamma]}^{(s)} \leqslant \text{const}\, \|P(x; D)u\|_{[\gamma]}^{(s)}, \quad \gamma < \gamma_0. \tag{44}$$

To prove (44), let us consider the quadratic form

$$-\operatorname{Im}(P(x; D)u, Q(x; D)u)_{[\gamma]}^{(s)}, \tag{45}$$

where

$$(v, w)_{[\gamma]}^{(s)} = \int e^{2\gamma t} \delta_s^+(D) v \overline{\delta_s^+(D) w} \, dx.$$

If in the form (45) we freeze the coefficients of the operators P and Q, we shall obtain norms equivalent to (43).

Under the conditions of the theorem, the form (45) admits an appropriate estimate: for $\gamma \leqslant \gamma_0(s)$

$$c^{-1}(|[u]|_{[\gamma]}^{(s)})^2 \leqslant -\operatorname{Im}(P(x; D)u, Q(x; D)u)_{[\gamma]}^{(s)} \leqslant c(|[u]|_{[\gamma]}^{(s)})^2$$
$$\forall u \in H_{[\gamma]}^{(\infty)}. \tag{46}$$

Let us clarify the meaning of the estimate (46) in the simpler case $s = 0$. Let us consider the Hermitian form

$$H(w, w) = \operatorname{Re}(H_P(x; \gamma, D)w, w).$$

According to the conditions of the theorem, the symbol $H_P(x; \gamma, \xi)$ is positive, satisfies the constant strength condition, and, in view, of (IIa) for any $\beta > 0$

$$|H_P^{(\beta)}(x;\gamma,\xi)| < \varepsilon_\beta(\gamma) H_P(x;\gamma,\xi), \quad \varepsilon_\beta(\gamma)\to 0, \quad \gamma\to -\infty.$$

With these properties, the form $H(w, w)$ can be estimated both from above and from below by analogous forms with frozen coefficients:

$$c^{-1}(H_P(\gamma, D)w, w) \leqslant H(w, w) \leqslant c(H_P(\gamma, D)w, w). \tag{47}$$

In case of estimate (39) this inequality was established by Gårding, so that (47) can be regarded as an analog of the Gårding inequality for inhomogeneous quadratic forms.

Let us go back to estimate (46). After the substitution $w = \exp(\gamma t)u$ and integration by parts, we can rewrite the form(45) as the sum of two forms:

$$H(w, w) + (C(x;\gamma, D)w, w), \tag{48}$$

where, as a direct computation shows,

$$C(x;\gamma,\xi) = \frac{i}{2}\sum_{\beta>0}\frac{1}{\beta!}[\overline{Q^{(\beta)}(x;\tau,\eta)}P(x;\tau,\eta)$$

$$-\overline{P^{(\beta)}(x;\tau,\eta)}Q(x;\tau,\eta)] - \frac{1}{2}\sum_{\beta>0}\frac{1}{\beta!}H^{(\beta)}_{P(\beta)}(x;\gamma,\xi).$$

From conditions (I), (IIa) we have an estimate for the symbol C:

$$|C(x;\gamma,\xi)| < \varepsilon(\gamma)H_P(x;\gamma,\xi), \quad \varepsilon(\gamma)\to 0, \quad \gamma\to -\infty. \tag{49}$$

Standard techniques of PDO theory allow us to extend the estimate for the symbol to an analogous estimate for the form

$$(C(x;\gamma,D)w, w) < \varepsilon'(\gamma)(H_P(\gamma, D)w, w), \quad \varepsilon'(\gamma)\to 0, \quad \gamma\to -\infty.$$

Comparing this estimate with (47) and returning to the function u, we obtain (46) for $s = 0$. We note that no use of condition (IIb) was made.

In the case $s \neq 0$, the form (45) can also be represented as in (48), but the second summand now will be a pseudo-differential quadratic form, and not a differential one as before. For its symbol we can also obtain an estimate of the type of (49). However, in the proof of this estimate both conditions (IIa) and (IIb) have to be used.

From the lemma and from condition (I) we easily derive that

$$\|Q(x, Q)u\|^{(s)}_{[\gamma]} \leqslant \text{const}\, |[u]|^{(s)}_{[\gamma]}.$$

The desired estimate (44) follows from this inequality and from the left of the inequalities (46).

2.3. The Cauchy Problem in Spaces of Slowly Increasing (Decreasing) and Exponentially Decaying (Growing) Functions. First of all let us remark that under conditions of Theorem 2.2, the same argument delivers an estimate of the form (45) in the space $H^{(s)}_{(l)[\gamma]}$. As a result we obtain the estimate

$$\sum_{l=1}^{k}(\gamma-\gamma_0)^l\|u\|^{(s)}_{(l)[\gamma]} \leqslant \text{const}\,\|P(x;D)u\|^{(s)}_{(l)[\gamma]}.$$

An analogous estimate holds also for P^*. A trivial modification of Theorem 2.1 leads to the following claim.

Theorem. *Let conditions* (I), (IIa), (IIb) *of Theorem* 2.2 *be satisfied. Then* $\forall s, l \in \mathbb{R}$ *we can find a* $\gamma_2(s, l)$ *such that for* $\gamma < \gamma_2(s, l)$ *equation* (32) *has a unique solution* $u \in (H^{(s)}_{(l)[\gamma]})_+$ *for* $\forall f \in (H^{(s)}_{(l)[\gamma]})_+$.

Corollary. *Let* $\Phi = \mathscr{S}, \mathcal{O}, \mathscr{S}', \mathcal{O}', H^{(\infty)}_{(l)}$. *Let conditions of Theorem* 2.2 *be satisfied. Then for* $\forall c > 0$ *we have the space isomorphism*

$$P(x, D): \Phi[0, c) \to \Phi[0, c).$$

Remark. Under the conditions of Theorem 2.2 it is possible to prove energy estimates in the spaces $H^{(s,r)}_{(l)[\gamma]}$, the norms in which are defined by the PDO $(iD_t + \sqrt{1 + |D_y|^2})^s \times (1 + |D_y|^2)^{r/2}$. Then, following the procedure of Section 1.5, it is hard to obtain an isomorphism of the type of (27), that is, to prove the correctness of the inhomogeneous Cauchy problem in $\mathscr{S}$, $\mathcal{O}$, or in $C^{(\infty)}_{(l)}$, $l \in \mathbb{R}$.

Conditions of Theorem 2.2 will still be valid if in them we replace the symbol $P(x; \xi)$ by $P_\omega(x; \xi) = P(x; \xi + i\omega)$, $\omega \in \mathbb{R}^n$. Hence it follows that estimate of the type of (33), (34) will still hold if the weight $\exp(\gamma t)$ is replaced by $\exp(\gamma t + \omega \cdot x)$. With the use of these estimates it is possible to establish solubility of the homogeneous Cauchy problem in the scales $\mathscr{H}^{\delta}_{(l),m}$ of Section 1.6. From that, using the arguments of Chapter 2, it is possible to obtain solubility in spaces of exponentially growing functions. Due to lack of space, we do not go into these details.

Next we shall give concrete applications of results obtained above to classes of exponentially correct operators considered in § 3 of Chapter 2.

2.4. Strictly Hyperbolic Operators with constant coefficients were defined in § 3, Chapter 2. The symbol $P(x; \tau, \eta)$ is called strictly hyperbolic if for each $x \in \mathbb{R}^n$ the polynomial $P_x(\tau, \eta) = P(x; \tau, \eta)$ is strictly hyperbolic, that is, its homogeneous principal part $P_0(x; \tau, \eta)$ has distinct real roots $\lambda_j(x, \eta)$, such that the root simplicity condition is uniform in the parameter x, that is, $\exists \delta > 0$ such that

$$|\lambda_j(x, \eta) - \lambda_l(x, \eta)| > \delta|\eta|, \quad x \in \mathbb{R}^n, \quad j \neq l. \tag{50}$$

This definition can be restated in terms of the symbol H_P (see estimates (39)).

Lemma. *The symbol* $P(x; \xi)$, $m = \deg_\xi P$ *is strictly hyperbolic if and only if* $\exists c > 0$, $\gamma_0 < 0$, so that

$$c^{-1}|\gamma|(|\tau| + |\eta|)^{2m-2} \leqslant H_P(x; \gamma, \xi) \leqslant c|\gamma|(|\tau| + |\eta|)^{2m-2},$$
$$\gamma \leqslant \gamma_0, \quad \xi \in \mathbb{R}^n. \tag{51}$$

Let us clarify this statement. If the symbol P is strictly hyperbolic, then, in view of the equality (42),

$$H_{P_0}(x; \gamma, \xi) = -\gamma \sum_{j=1}^{b} \prod_{l \neq j} [(\sigma - \lambda_l(x, \eta))^2 + \gamma^2].$$

From (50) the inequalities (51) with $\gamma_0 = 0$ follow for the symbol H_{P_0}. Noticing that $H_P - H_{P_0}$ is a polynomial in ξ and γ of degree not higher than $2m - 2$, we obtain (51).

From the left inequality of (51) follows the estimate

$$c|\gamma|(|\tau| + |\eta|)^{m-1} \leqslant |P(x; \tau, \eta)|, \quad \gamma \leqslant \gamma_0, \tag{52}$$

from which the strict hyperbolicity condition (50) is easily derived.

In the case of strictly hyperbolic symbols all conditions of Theorem 2.2 are fulfilled: (I) trivially follows from (52), while (IIa), (IIb) follow from (52). In the case at hand, Theorem 2.2 can be not only refined, but also supplemented by a converse statement. We have the following

Theorem. *The following conditions are equivalent:*

(i) *The symbol $P(x; \xi)$ is strictly hyperbolic.*

(ii) *For $\forall s \in \mathbb{R}$ there is $\gamma(s)$, such that for $\gamma < \gamma(s)$ equation* (32) *has unique solution $u \in (H^{(s)}_{[\gamma]})_+$ for $\forall f \in \mathrm{H}^{(s)}_{[\gamma]})_+$.* For this solution we have the estimate

$$|\gamma| \, \|u\|^{(s+m-1)}_{[\gamma]} \leqslant \text{const} \, \|P(x, D)u\|^{(s)}_{[\gamma]}, \quad \gamma \leqslant \gamma_0, \quad m = \operatorname{ord} P. \tag{53}$$

Let us sketch the proof of (ii) → (i). If $P(x; \xi)$ is a homogeneous symbol independent of x, that is, $P(x; \xi) = P_0(\xi)$, then from the Parseval equality we derive inequality (52), and thus the condition (50).

In the general case we construct, starting with $f \in (H^{(s)}_{[\gamma]})_+$ the sequence $f_\varepsilon = \varepsilon^{-m} f((x - x^0)/\varepsilon) \in (H^{(s)}_{[\gamma/\varepsilon]})_+$. According to (ii), to this sequence there corresponds a sequence of solutions $v_\varepsilon \in (H^{(s+m-1)}_{[\gamma/\varepsilon]})_+$, for which we have estimate (53) with γ replaced by γ/ε. Setting $v_\varepsilon = u_\varepsilon((x - x^0)/\varepsilon)$, we see that $u_\varepsilon \in (H^{(s+m-1)}_{[\gamma]})_+$ and that

$$|\gamma| \, \|u_\varepsilon\|^{(s+m-1)}_{[\gamma]} \leqslant \text{const} \, \|f\|^{(s)}_{[\gamma]}.$$

By this inequality the sequence u_ε is weakly compact in $(H^{(s+m-1)}_{[\gamma]})_+$ and has the weak limit $u \in (H^{(s+m-1)}_{[\gamma]})_+$. Moreover, in the sense of distributions $P_0(x^0, D) = f$, so that we are reduced to the case we already considered above.

2.5. 2*b*-Parabolic Operators. A symbol $P(x; \xi)$ is called 2*b*-parabolic if all the polynomials $P_x(\xi) = P(x; \xi)$ are 2*b*-parabolic with the following (uniform in x) estimate for the roots:

$$\operatorname{Im} \lambda_j(x, \eta) > \delta|\eta|^{2b}, \quad \delta > 0, \quad \eta \in \mathbb{R}^{n-1}, \quad j = 1, \ldots, k. \tag{54}$$

Here $\lambda_j(x, \eta), j = 1, \ldots, k$, are the roots of the principal $(2b, 1, \ldots, 1)$-homogeneous part $P_0(x; \xi)$ of the symbol $P(x; \xi)$. As we mentioned in Section 3.3 of Chapter 2, the number b has to be an integer.

Lemma. *Condition* (4) *of 2b-parabolicity is satisfied if and only if $\exists c > 0$, $\gamma_0 < 0$, such that*

$$\begin{aligned} c^{-1}(|\gamma| + |\eta|^{2b})(|\tau| + |\eta|^{2b})^{2k-2} &\leqslant H_P(x; \gamma, \xi) \\ &\leqslant c(|\gamma| + |\eta|^{2b})(|\tau| + |\eta|^{2b})^{2k-2}, \quad \gamma \leqslant \gamma_0, \quad \xi \in \mathbb{R}^n. \end{aligned} \tag{55}$$

The estimate (55) is derived from (52) with the use of (42), From the left inequality of (55) it follows that

$$c(|\gamma| + |\eta|^{2b})(|\tau| + |\eta|^{2b})^{k-1} \leqslant |P(x; \tau, \eta)|, \quad \operatorname{Im} \tau \leqslant \gamma_0. \tag{55'}$$

Though this is a weaker estimate than

$$c_1(|\tau| + |\eta|^{2b})^k \leqslant |P(x; \tau, \eta)|, \quad \operatorname{Im} \tau \leqslant \gamma_0,$$

which follows from (54), (54) itself already follows from (55′).

Conditions of Theorem 2.2 follow trivially from (55) and (55′). We shall not refine Theorem 2.2 for the case of $2b$-parabolic operators, since we are going to formulate this refinement in a slightly more general setting.

2.6. Pluriparabolic Operators. Let the variables be subdivided, as in Section 3.7, into two sets: $\xi = (\sigma, \eta)$, $\sigma = (\sigma_1, \ldots, \sigma_l) \in \mathbb{R}^l$ $\eta \in \mathbb{R}^{n-l}$. We prescribe weight $2b$ to σ, and weight 1 to η, denote by $P_0(x; \sigma, \eta)$ the principal $(2b, \ldots, 2b, 1, \ldots, 1)$-homogeneous part of the symbol $P(x; \xi)$, and let $\lambda_j(x, \sigma', \eta)$, $j = 1, \ldots, k$, $\sigma' = (\sigma_2, \ldots, \sigma_l)$ be the roots of $P_0(x; \sigma, \eta)$ with respect to the variable σ_1.

A symbol $P(x; \sigma, \eta)$ is called strictly pluriparabolic if the symbol $P(x; \sigma, 0)$ is strictly hyperbolic in the sense of Section 2.4 and $\exists \delta > 0$ such that

$$\operatorname{Im} \lambda_j(x, \sigma', \eta) > \delta |\eta|^{2b}, \quad \forall x \in \mathbb{R}^n, \quad \forall \sigma' \in R^{l-1}, \quad \forall \eta \in \mathbb{R}^{n-l}. \tag{56}$$

Let us set $Q(x; \sigma, \eta) = \partial P(x; \sigma, \eta)/\partial \sigma_1$ and

$$H_P(x; \gamma, \xi) = -\operatorname{Im}[P(x; \sigma_1 + i\gamma, \sigma', \eta)\overline{Q(x; \sigma_1 + i\gamma, \sigma', \eta)}].$$

Lemmas 2.4 and 2.5 are particular cases of the following claim.

Lemma. *The symbol $P(x; \sigma, \eta)$ is strictly pluriparabolic if and only if $\exists c > 0$, $\gamma_0 < 0$, so that*

$$c^{-1} \leqslant \frac{H_P(x; \gamma, \xi)}{(\gamma + |\eta|^{2b})(|\gamma| + |\sigma| + |\eta|^{2b})^{2k-2}} \leqslant c, \quad \gamma \leqslant \gamma_0, \quad \xi \in \mathbb{R}^n. \tag{57}$$

This estimate is derived from the strict hyperbolicity of $P(x; \sigma, 0)$ and (56) using (42). On the other hand, pluriparabolicity will already follow from the weaker estimate

$$c(|\gamma| + |\eta|^{2b})(|\gamma| + |\sigma| + |\eta|^{2b})^{k-1} \leqslant |P(x; \sigma_1 + i\gamma, \sigma', \eta)|, \quad \gamma \leqslant \gamma_0 \tag{57'}$$

All the conditions of Theorem 2.2 follow from inequalities (57), (57′). Let us introduce all the spaces necessary in order to refine this theorem.

Let us denote by $H_{[\gamma]}^{(s, r; 2b)}$ the set of all $u \in H_{[\gamma]}^{(-\infty)}$ having finite norm

$$\|u\|_{[\gamma]}^{(s, r; 2b)} = \left(\int (1 + |\eta|^{2b})^r |(\sigma_1 + i\gamma - i(|\sigma'| + |\eta|^{2b}))^s \hat{u}(\sigma_1 + i\gamma, \sigma', \eta)|^2 \, d\xi \right)^{1/2}.$$

Theorem. *The following conditions are equivalent:*

(i) *The symbol $P(x; \sigma, \eta)$ is strictly pluriparabolic.*

(ii) *For $\forall s \in \mathbb{R}$ there is $\gamma(s)$, such that for $\gamma < \gamma(s)$ equation (32) has unique solution $u \in (H_{[\gamma]}^{(s, 0; 2b)})_+$ for $\forall f \in (H_{[\gamma]}^{(s, 0; 2b)})_+$. For this solution we have the estimate*

$$|\gamma| \, \|u\|_{[\gamma]}^{(s+k-1,0;2b)} + \|u\|_{[\gamma]}^{(s+k-1,2b;2b)} \leqslant \text{const} \, \|P(x,D)u\|_{[\gamma]}^{(s,0;2b)},$$

$$\gamma \leqslant \gamma_0.$$

This theorem is proved using the scheme of Theorem 2.4, and becomes that theorem for $l = n$; for $l = 1$ we obtain the solubility theorem for $2b$-parabolic equations.

2.7. Dominantly Correct Polynomials. Let us consider a symbol $P(x; \xi)$. For each x to this symbol there corresponds a polynomial $P_x(\xi) = P(x; \xi)$ having a Newton polygon $N_{P(x)}$ and a polygon of lower order terms $U_{P(x)}$. The symbol is called dominantly correct if

(i) the lower order polygons $U_{P(x)}$ are independent of x, that is, $U_{P(x)} = \bigcup_x U_{P(x)} = U_P$.

(ii) there exist $\gamma_0 \in \mathbb{R}$ and a function $\varepsilon(\gamma) \to 0$ as $\gamma \to -\infty$, such that

$$|\tau^k \eta^\beta| < \varepsilon(\gamma)|P(x; \tau, \eta)|, \quad \forall x \in \mathbb{R}^n, \quad \operatorname{Im} \tau = \gamma \leqslant \gamma_0, \quad \xi \in \mathbb{R}^n, \quad (k, |\beta|) \in U_P;$$

in other words, the polynomials $P_x(\xi)$ satisfy conditions of Section 3.6 of Chapter 2 uniformly in x.

Dominantly correct polynomials satisfy all conditions of Theorem 2.2. The verification of these conditions is non-trivial. It is based on an equivalent description of dominantly correct polynomials in terms of the symbol H_P. Let us present this description. Let us write

$$H_P(\gamma, \xi) = -\operatorname{Im} P(\tau, \eta)\overline{Q(\tau, \eta)} = \sum h_{jk\beta} \gamma^i \sigma^k \eta^\beta$$

and let us denote by Γ_P the convex polyhedron in $\mathbb{R}^3$ stretched over all triples $(j, k, |\beta|)$ such that $h_{jk\beta} \neq 0$, and their projections on all coordinate planes and axes.

Lemma. *The polynomial $P(\tau, \eta)$ is dominantly correct if and only if the following conditions hold:*

(a) *We have the estimate*

$$\sum_{(j,k,|\beta|) \in \Gamma_P} |\gamma^j \sigma^k \eta^\beta| \leqslant \text{const} \, H_P(\gamma, \xi), \quad \gamma \leqslant \gamma_0.$$

(b) *For every monomial $B = b\tau^k\eta^\beta$ $(k, |\beta|) \in U_P$ and for every $\varepsilon > 0$ there exists $\gamma(\varepsilon, B)$ such that*

$$|H_{P+B}(\gamma, \xi) - H_P(\gamma, \xi)| < \varepsilon H_P(\gamma, \xi), \quad \gamma < \gamma(\varepsilon, B), \quad \xi \in \mathbb{R}^n. \tag{58}$$

(c) *The polyhedron $\Gamma_P \subset \mathbb{R}^3$ can be uniquely reconstructed from the lower order terms polygon $U_P \subset \mathbb{R}^2$.*

The idea of the proof of this lemma is to represent P as a product of a strictly hyperbolic and a $2b_j$-parabolic quasi-polynomials (see Section 3.6 of Chapter 2) and then to estimate the functions (42) for each of the factors.

Remark. The polyhedron Γ_P can be given quite an explicit description. From this description and from inequality (58) we can derive the estimate

$$\sum_{(k,|\beta|)\in U_P} |\tau^k\eta^\beta|^2 < \varepsilon(\gamma)H_P(\gamma,\xi).$$

Hence follows an estimate for an operator with variable coefficients,

$$\sum_{(k,|\beta|)\in U_P} \|D_t^k D_y^\beta u\|_{[\gamma]}^{(s)} \leqslant \varepsilon(\gamma)\|P(x;D)u\|_{[\gamma]}^{(s)}.$$

Thus, in the case of dominantly correct operators, equation (32) $\forall f \in (H_{[\gamma]}^{(s)})_+$ has solution u belonging to $(H_{[\gamma]}^{(s)})_+$ together with all its derivatives $D_t^k D_y^\beta u$, $(k, |\beta|) \in U_P$.

2.8. $(2b+1)$-Hyperbolic Operators. The symbol $P(x;\xi)$ is called strictly q-hyperbolic, where $q = 2b+1$ if the polynomials $P_x(\tau,\eta)$ satisfy the conditions of Section 3.4 of Chapter 3 uniformly in x, that is, its $(q,1,\ldots,1)$-homogeneous principal part $P_0(x;\xi)$ has distinct real roots $\lambda_j(x,\eta)$ and there exist δ_1, δ_2 such that

(i) $$|\lambda_j(x,\eta) - \lambda_k(x,\eta)| > \delta|\eta|^q, \quad k \neq j, \quad x \in \mathbb{R}^n;$$

(ii) if $q > 1$, there is a direction $N \in \mathbb{R}^{n-1}$, such that

$$\langle \operatorname{grad}_\eta \lambda_j(x,\eta), N\rangle > \delta_1|\eta|^{q-1}, \quad j = 1,\ldots,k, \quad x \in \mathbb{R}^n.$$

As we mentioned in Section 3 of Chapter 2, a q-hyperbolic polynomial $P(\tau,\eta)$ is correct in the sense of Petrovskij, as are all its shifts $P(\tau,\eta+i\nu N)$, $\nu > 0$. From here follows the positivity of the symbol

$$H_P(\gamma,\nu,\xi) = -\operatorname{Im}[P(\sigma+i\gamma,\eta+i\nu N)\overline{Q(\sigma+i\gamma,\eta+i\nu N)}] > 0 \tag{59}$$

under the condition

$$\nu > \nu_0, \quad \gamma < -R\nu^q, \quad \nu_0 > 0, \quad R > 0. \tag{60}$$

Moreover, a complete description of strictly q-hyperbolic polynomials and of strictly q-hyperbolic symbols (dependent on the parameter x) can be given in terms of (59).

Lemma. *The symbol $P(x;\xi)$ is strictly q-hyperbolic if and only if there are ν_0, R, c, such that under the conditions* (60) *the following two-sided estimate holds*:

$$c^{-1} \leqslant \frac{H_P(x;\gamma,\nu,\eta)}{[|\gamma| + \nu(\nu+|\eta|)^{q-1}][|\tau| + (\nu+|\eta|)^q]^{2k-2}} \leqslant c. \tag{61}$$

From the left inequality we also have the estimate

$$c\nu(\nu+|\eta|)^{q-1}(|\tau| + \nu^q + |\eta|^q)^{k-1} \leqslant |P(x;\tau,\eta+i\nu N)|. \tag{62}$$

The estimate (61) also gives a clue to the derivation of energy estimates for q-hyperbolic differential operators. Let us consider the scalar product with weight $\exp(\gamma t + \nu\langle y, N\rangle)$:

$$(u,v)_{[\gamma,\nu N]}^{(s|q)} = \int \exp(2\gamma t + 2\nu\langle N,y\rangle)\delta_s(D)u\overline{\delta_s(D)v}\,dx,$$

where the symbol $\delta_s(\tau, \zeta)$ is defined as follows. In the space of variables $y \in \mathbb{R}^{n-1}$ we change by rotation to variables $\{y_1, y'\}$, where the direction of the $\{y_1\}$ axis coincides with N and $y' \in \mathbb{R}^{n-2}$. Let (ζ_1, ζ') be the dual variables. Then we set

$$\delta_s(\tau, \zeta) = (i\tau + \zeta_1^q + |\zeta'|^q)^s.$$

Employing the procedure of Section 2.2, the estimate (60) allows us to obtain upper and lower bounds for the form

$$-\operatorname{Im}(P(x; D)u, Q(x, D)u)^{(s|q)}_{[\gamma, \nu N]}$$

in terms of the analogous form with frozen coefficients. Taking (62) into consideration, we obtain the a priori bound

$$|\gamma| \, \|u\|^{(s+k-1|q)}_{[\gamma, \nu N]} + \nu \sum_{|\beta| \leqslant q-1} \|D_y^\beta u\|^{(s+k-1|q)}_{[\gamma, \nu N]} \leqslant c \, \|P(x; D)u\|^{(s|q)}_{[\gamma, \nu N]}, \tag{63}$$

where we set

$$\|u\|^{(s|q)}_{[\gamma, \nu N]} = [(u, u)^{(s|q)}_{[\gamma, \nu N]}]^{1/2} \tag{64}$$

replacing γ by $-\gamma$ and N by $-N$, we can prove corresponding bounds for the adjoint operator P^*. Following the scheme of Section 2.1, we can establish solubility of the homogeneous Cauchy problem (32) in spaces corresponding to left and right norms of (63).

As the case $q = 1$ is included in the definition of q-hyperbolicity given above, we have estimates (63) (and the corresponding existence theorem) for strictly hyperbolic operators. Here as N we can take any $N \in \mathbb{R}^{n-1}$. Analogous results can also be established for $2b$-parabolic operators.

Results of Sections 2.4, 2.5, and of the present section, can be combined to give the following general theorem.

Theorem. *The following conditions are equivalent for a differential operator $P(x; D)$.*

(I) *For $\forall s \in \mathbb{R}$ there exist constants $\nu_0(s)$, $R(s)$, $c > 0$, such that for $\nu > \nu_0(s)$ and $\gamma < -R(s)\nu^q$, equation (32) has a unique solution $u \in (H^{(s)}_{[\gamma, \nu N]})_+$ (here we denote by $(H^{(s)}_{[\gamma, \nu N]})_+$ the space with norm (64)) for any right hand side $f \in (H^{(s)}_{[\gamma, \nu N]})_+$. For this solution we have the estimate (63).*

(II) *Condition (A_q) below is satisfied.*

(A_q) *q is a positive integer and*

(a) *if $q = 2b + 1$, $b > 1$, then conditions* (i), (ii) *hold, that is, P is a strictly q-hyperbolic symbol, and $N \in V = \bigcap_x V_x$, where V_x is the q-hyperbolicity cone of the polynomial $P(x; \xi)$ (see Section 3.4, Chapter 2);*

(b) *if $q = 1$, then we have* (i), *that is, P is a strictly hyperbolic symbol, and the vector N is arbitrary;*

(c) *if $q = 2b$, then P is a $2b$-parabolic symbol (see sectuib 2.5), and the vector $N \in \mathbb{R}^{n-1}$ is arbitrary.*

In the appendix we present the precise analog of this theorem for the mixed problem.

Appendix
The Mixed Problem

Let a family of differential operators $A(x; D)$, $B_j(x; D)$, $j = 1, \ldots, \mu$, be given in a cylindrical domain $\Omega = G \times \mathbb{R}_t^1$, $G \subset \mathbb{R}_y^{n-1}$. A mixed problem (with homogeneous initial data) is the problem of determination of a function $u(t, y)$ from the conditions

$$A(x; D)u(x) = f(x), \quad x \in \Omega; \tag{1}$$

$$B_j(x; D)u(x) = g_j(x), \quad j = 1, \ldots, \mu, \quad x \in \partial G \times \mathbb{R}_t^l, \tag{2}$$

where ∂G is the boundary of G:

$$u(t, y) = g_j(t, y) = f(t, y) = 0 \quad \text{for} \quad t < 0 \tag{3}$$

As the mixed problem is of interest in a "curved" domain, it is natural to make an isotropy assumption in space variables for equations under consideration. Since the Cauchy problem enters as an ingredient of the mixed problem, it is natural to construct a theory of the problem (1), (2), (3), in the spaces of functions considered in this chapter. However, even this framework is too general for the mixed problem. We shall restrict ourselves to three classes of differential operators: $2b$-parabolic, strictly hyperbolic, and q-hyperbolic (for some details see Volevich and Gindikin [1980]). Well posedness results for the mixed problem can be stated in one unified theorem, which is exactly what we are going to do.

For simplicity, we restrict ourselves to the case when G is the half-space $\langle N, y\rangle \geqslant 0$; the points of the plane $\langle N, y\rangle = 0$ will be denoted by x', and the dual variables by ξ'. In other words, we set $\xi = \lambda N + \xi'$, and shall write $\xi = (\lambda, \xi')$.

Let condition (A_q) of Theorem 2.8 be satisfied. Let us consider the polynomial $A_0(\lambda) = A_0(t, x'; \tau, \xi' + \lambda N)$, where A_0 is the principal $(q, 1, \ldots, 1)$-homogeneous part of A.

Condition (B_q). *The polynomial* $A_0(\lambda)$ *has the form*

$$A_0(\lambda) = a(t, x')\lambda^m + o(\lambda^{m-1}), \quad |a(t, x')| > a > 0$$

and has no real zeroes for $\gamma < 0$.

Let $m_\pm$ be the number of zeroes of $A_0(\lambda)$ in the half-plane $\pm \operatorname{Im} \lambda > 0$ ($m_\pm$ is independent of t, x', γ, σ, ξ' by (B_q)), and let $A_0(\lambda) = A_0^+(\lambda)A_0^-(\lambda)$ be the corresponding factorization.

Condition (C_q). $m_+ = \mu$, *where* μ *is the number of boundary conditions* (2).

Now we only have to impose conditions on the boundary operators. Let us denote by B_{0j} the principal $(q, 1, \ldots, 1)$-homogeneous part of B_j, and let $B_{0j}(\lambda) = B_{0j}(t, x'; \tau, \xi' + \lambda N)$.

Condition (D_q). *The polynomials* $\{B_{0j}(\lambda)\}$, $j = 1, \ldots, \mu$, *are independent* mod $A_0^+(\lambda)$ *uniformly with respect to* $(t, x) \in \mathbb{R}^n$, $\gamma \leqslant 0$, *and* $|\sigma| + |\gamma| + |\xi'|^q = 1$.

Remark. For $q = 2b$, condition (D_q) becomes the Lopatinskij condition for parabolic equations, which was introduced by Zagorskij, and refined and generalized by Agranovich and Vishik [1964], Ejdel'man [1964], and Solonnikov [1965].

For $q = 1$, we have the "uniform" Lopatinskij condition introduced by Agmon. This condition was generalized in the works of Sakamoto and Kreiss.

For $q = 2b + 1$, this condition was introduced in Volevich and Gindikin [1980].

Let us introduce the spaces necessary for the study of the mixed problem. Let $H^{(s,r|q)}_{[\gamma,\nu N]}$ be the space of distributions with the norm

$$\left(\int \sum_{j=0}^{r} |\lambda + i\nu|^{2j}(|\tau|^{1/q} + |\xi'|)^{2(l+s-j)}|\hat{u}(\tau, \xi' + \lambda N + i\nu N)|^2 \, d\sigma \, d\xi'\right)^{1/2}.$$

Let $H^{(s,r|q)}_{[\gamma,\nu N]}(\mathbb{R}^n_N)$ be the space of restrictions to the half-space $\mathbb{R}^n_N = \{(t, y), \langle y, N\rangle > 0\}$, equipped with he induced norm, which we denote by $\| \ \|^{(s,r|q)}_{[\gamma,\nu N]}$. On the plane $\langle N, y\rangle$ we define the space $H^{(s|q)}_{[\gamma]}$ with the norm

$$\{g\}^{(s|q)}_{[\gamma]} = \left(\int (|\tau|^{1/q} + |\xi'|)^{2s}|\hat{g}(\tau, \xi')|^2 \, d\sigma \, d\xi'\right)^{1/2}.$$

Theorem. *For the problem* (1)–(3), *the following conditions are equivalent*:

(i) $\forall s \in \mathbb{R}$, *there are* $\nu(s)$, $R(s)$, $c(s) > 0$, *such that for* $\nu > \nu(s)$, $\gamma < -R(s)\nu^q$ *for* $\forall f \in (H^{(s+1,0|q)}_{[\gamma,\nu N]})_+$ *and all* $g_j \in H^{(m+s-\beta_j|q)}_{[\gamma,\nu N]}$ (β_j is the $(q, 1, \ldots, 1)$*-order of* B_j) *there exists unique solution* $u \in (H^{(s,m|q)}_{[\gamma,\nu N]})_+$ *of the problem* (1)–(3). *Moreover,*

$$\langle N, D\rangle^j u|_{\langle y,N\rangle=0} \in H^{(m+s-jq)}_{[\nu]}$$

and the energy estimate

$$c(s)\left[\nu(\|u\|^{(s,m|q)}_{[\gamma,\nu N]})^2 + \sum_{j=0}^{m-1} (\{\langle N, D\rangle^j u\}^{(m-1+s-\beta_j|q)}_{[\gamma]})^2\right]$$

$$\leqslant \frac{1}{\nu}\left(\|A(x, D)u\|^{(s+1,0|q)}_{[\gamma,\nu N]})^2 + \sum_{j=0}^{\mu} (\{B_j(x, D)u\}^{(m+s-\beta_j|q)}_{[\gamma]}\right)^2$$

holds.

(II) *Conditons* (A_q), (B_q), (C_q), (D_q) *are satisfied.*

The proof of (I) ⇒ (II) is based on energy bounds for the original and for a certain "allied" problems. To prove (II) ⇒ (I), we first derive (I) for a problem with frozen coefficients, after which everything is reduced to the study of ordinary equations with constant coefficients. A detailed discussion of this theorem is to be found in Volevich and Gindikin [1980].

References*

Classical facts concerning the Cauchy problem for hyperbolic and parabolic second order equations can be found in the monographs of Hörmander [1963], Courant [1962], and in the survey paper of Il'in, Kalashnikov, and Olejnik [1962]; these also contain a host of historic material (see also Courant, Friedrichs and Lewy [1928] and Leray [1953]).

Petrovskij's results on the Cauchy problem are presented in the original papers (Petrovskij [1938, 1937]) (see also Schwarz [1950]). Further results on the Cauchy problem for equations with constant coefficients can be found in the monographs of Gel'fand and Shilov [1958], Palamodov [1967], Hörmander [1963], and Hörmander [1983]. On the Cauchy problem for higher order hyperbolic equations and systems see Petrovski [1937], Levi [1907], Gårding [1957], Courant [1962], Hörmander [1963, 1983]. The basic material concerning the Cauchy problem for parabolic equations is contained in the monograph of Ejdel'man [1964].

Theory of distributions is presented in the monographs of Schwarz [1950/51], Leray [1953], Vladimirov [1978], Gel'fand and Shilov [1958], Palamodov [1967], Hörmander [1963, 1983]. Functional spaces and their embedding theorems can be found in the books of Sobolev [1950], Palamodov [1967], Hörmander [1963, 1983], and Taylor [1981].

Proofs of the main results of Chapter 1 can be found in Volevich and Gindikin [1972]; further references are in Volevich and Gindikin [1982].

For energy methods see Volevich [1974], Volevich and Gindikin [1968, 1980], Gindikin [1974].

Agranovich, M.S., Vishik, M.I. [1964] Elliptic problems with a parameter and parabolic problems of general type. Usp. Mat. Nauk *19*, No. 3, 53–161. English transl.: Russ. Math. Surv. 19, No. 3, 53–157 (1964). Zbl.137,296

Boutet de Monvel, L. [1966] Comportement d'un operateur pseudo-differentiel sur une varieté à bord, I, II. J. Anal. Math. 7, 241–253, 255–304. Zbl.161,79

Courant, R. [1962] Partial Differential Equations. New York-London. Zbl.99,295

Courant, R., Friedrichs, K., Lewy, H. [1928] Über die partiellen Differentialgleichungen der mathematischen Physik. Math. Ann. *100*, 32–34

Ejdel'man, S.D. [1964] Parabolic systems. Nauka: Moscow. English transl.: North-Holland Publ. Co. (1969). Zbl.121,319

Gårding, L. [1957] The Cauchy Problem for Hyperbolic Equations. Univ. of Chicago: Chicago. Zbl.91,269

Gel'fand, I.M., Shilov, G.E. [1958] Some Questions of the Theory of Differential Equations (Generalized Functions, vol. 3). Fizmatgiz: Moscow. Zbl.91,111

Gindikin, S.G. [1974] Energy estimates related to the Newton polyhedron. Tr. Mosk. Mat. O-va *31*, 189–236. English translation: Trans. Mosc. Math. Soc. 31, 193–246 (1976). Zbl.311.35012

Hadamard, J. [1932] Le Problème de Cauchy et les Équations aux Dérivées Partielles Linéaires Hyperboliques. Paris. Zbl.6,205

Hörmander, L. [1963] Linear Partial Differential Operators. Springer-Verlag: Berlin, Göttingen, Heidelberg. Zbl.108,93

Hörmander, L. [1983–1985] The Analysis of Linear Partial Differential Operators. I, II, III. Springer-Verlag: Berlin, Heidelberg, New York, Tokyo. Zbl.521.35001, Zbl.521.35002, Zbl.601.35001

Il'in, A.M., Kalashnikov, A.S., Olejnik, O.A. [1962] Linear second order equations of parabolic type. Usp. Mat. Nauk *17*, No. 3, 3–146. English transl.: Russ. Math. Surv. 17, No. 3, 1–146 (1962). Zbl.108,284

Kohn, J.J., Nirenberg, L. [1965] On the algebra of pseudo-differential operators. Commun. Pure Appl. Math. *18*, 269–305. Zbl.166,338

Krein, M.G. [1958] Integral equations on semi-infinite real line with a difference kernel. Usp. Mat. Nauk *13*, No. 5, 3–120. Zbl.88,309

* For the convenience of the reader, references to reviews in Zentralblatt für Mathematik (Zbl.), compiled using the MATH database, have been included as far as possible.

Ladyzhenskaya, O.A. [1950] On the uniqueness of solution to the Cauchy problem for a linear parabolic equation. Mat. Sb., Nov. Ser. *27*(2), 175–184. Zbl.39,107

Ladyzhenskaya, O.A. [1952] Solution of the Cauchy problem for hyperbolic systems by the method of finite differences. Uch. Zap. LGU, Ser. Mat. *23*, 192–246

Ladyzhenskaya, O.A., Solonnikov, V.A., Uraltseva, N.N. [1967] Linear and Quasilinear Equations of Parabolic Type. Nauka: Moscow. English transl.: Providence, Ann. Math. Soc. (1968). Zbl.164,123

Leray, J. [1953] Hyperbolic Differential Equations. Inst. Adv. Studies: Princeton. Zbl.75,98

Levi, E.E. [1907] Sulle equazione lineari totalmente ellitiche alle derivate parzialli. Rend. Circolo Mat. Palermo *24*, 275–317

Palamodov, V.P. [1967] Linear Differential Operators with Constant Coefficients. Nauka: Moscow. English transl.: Springer-Verlag (1970). Zbl.191,434

Paley, R., Wiener, N. [1934] Fourier Transforms in the Complex Domain. Am. Math. Soc.: New York. Zbl.11,16

Petrovski, I.G. (-Petrovskij, I.G.) [1937] Über das Cauchysche Problem für Systeme von partiellen Differentialgleichungen. Mat. Sb. *2*(44), 815–868. Zbl.18,405 (see also Petrovskij, I.G. [1986]. Collected works. Systems of Partial Differential Equations. Algebraic Geometry. pp. 34–97. Nauka: Moscow. Zbl.603.01018)

Petrovskij, I.G. [1938] On the Cauchy problem for systems of linear partial differential equations in spaces of non-analytic functions. Bull. Univ. Etat. Mosk., Ser. Int., Sect. A Mat. Mech. *1*, No. 7, 1–74. Zbl.24,37 (see also Petrovskij, I.G. [1986] Collected works. Systems of Partial Differential Equations. Algebraic Geometry. pp. 98–168. Nauka: Moscow. Zbl.603.01018)

Schwarz, L. [1950] Les équations d'évolution liées au produit de composition. Ann. Inst. Fourier *2*, 19–49. Zbl.42,331

Schwarz, L. [1950/51] Théorie des Distributions, I, II. Paris. Zbl.37,73 and Zbl.42,114

Schwarz, L. [1952] Transformation de Laplace des distributions. Commun. Semin. Math. Univ. Lund (Tome suppl., dedie a Marcel Riesz), 196–206. Zbl.47,349

Sobolev, S.L. [1950] Some Applications of Functional Analysis in Mathematical Physics. LGU: Leningrad. English transl.: Ann. Math. Soc. (1963). Zbl.123,90

Solonnikov, V.A. [1965] On Boundary Value Problems for Linear Parabolic Equations of General Form. Tr. Mat. Inst. Steklova *83*. English transl.: Tr. Steklov Inst. Math. 83 (1967). Zbl.161,84

Taylor, M. [1981] Pseudodifferential operators. Princeton, New Jersey. Zbl.453.47026

Vishik, M.I., Eskin, G.I. [1964] Convolution equations in bounded domains. Usp. Mat. Nauk *20*, No. 3 89–152. English transl.: Russ. Math. Surv. 20, No. 3, 85–151 (1965). Zbl.152,342

Vladimirov, V.S. [1978] Generalized Functions in Mathematical Physics 2nd. ed. Nauka: Moscow. English transl.: Mir: Moscow (1979). Zbl.515.46033, Zbl.403.46036

Volevich, L.R. [1974] Energy methods in the Cauchy problem for differential operators that are correct in the sense of Petrovskij. Tr. Mosk. Mat. O-va *31*, 147–187. English transl.: Trans. Mosc. Math. Soc. 31, 149–191 (1976). Zbl.324.35007

Volevich, L.R., Gindikin, S.G. [1967] Pseudo-differential operators and the Cauchy problem for differential equations with variable coefficients. Funkts. Anal. Prilozh. *1*, No. 4, 8–25. English transl.: Funct. Anal. Appl. 1, 262–277 (1967). Zbl.164,117

Volevich, L.R., Gindikin, S.G. [1968] The Cauchy problem for differential operators with dominant principal part. Funkts. Anal. Prilozh. *2*, No. 3, 22–38. English transl.: Funct. Anal. Appl. 2, 204–218 (1969). Zbl.177,196

Volevich, L.R., Gindikin, S.G. [1972] The Cauchy problem and convolution equations related to it. Usp. Mat. Nauk *27*, No. 4, 65–143. English transl.: Russ. Math. Surv. 27, No. 4, 71–160 (1973). Zbl.244.35016

Volevich, L.R., Gindikin, S.G. [1980] Energy methods in the mixed problem. Usp. Mat. Nauk *35*, No. 5, 53–120. English transl.: Russ. Math. Surv. 35, No. 5, 57–137 (1980). Zbl.455.34045

Volevič, L.R. (-Volevich, L.R.), Gindikin, S.G. [1982] Convolutors in spaces of distributions and related problems for convolution equations. Sel. Math. Sov. *2*, 9–30. Zbl.548.46033

Added in proof. Gindikin, S.G., Volevich, L.R.: Distributions and convolution equations. To appear at Gordon and Breach 1991. – The method of Newton polyhedron in partial differential operators. To appear at Reidel.

II. Qualitative Theory of Second Order Linear Partial Differential Equations

V.A. Kondrat'ev, E.M. Landis

Translated from the Russian
by M. Grinfeld

Contents

Introduction

In the general theory of equations of elliptic and parabolic type much attention is devoted to questions of existence and uniqueness of solutions to various problems. In contrast, in the qualitative theory of these equations it is assumed that the solution of the problem at hand is given a priori, and one considers the question of what conclusions can be made about the properties of solutions.

In the context of harmonic functions and of solutions to the equation of heat conduction, such questions have been studied for a long time. The simplest and the most representative result in this direction is the maximum principle. Questions such as the validity of Harnack's inequality, of Louiville's theorem for harmonic functions, of Holmgren's theorem stating that solutions of the heat equation belong in a certain Gevrey class, also fall in this domain. In the early twenties Wiener obtained his well-known criterion of boundary point regularity. In the early thirties Carleman showed that solutions of elliptic systems with simple characteristics in the plane have the same uniqueness properties as analytic functions.

All these questions can be considered as belonging in the qualitative theory. However, it was only in the mid-fifties that the qualitative theory came into being as an independent field. Its emergence was partially connected with the fact that after the appearance of Petrovskij's works on classification of systems, the general theory of equations of various types started developing rapidly. This theory was mostly concerned with questions of existence and uniqueness, and thus a need appeared to find an appelation for the branch of theory dealing with other questions.

The papers of Heinz [1955], Cordes [1956], and Landis [1956b] dealing with questions of the connection between uniqueness of solutions to equations not of hyperbolic type and their vanish rates, as well as theorems of the Phragmen-Lindelöf type (Landis [1956a], Lax [1957]) appeared in the early fifties. These works had numerous extensions. At that time the name "qualitative theory of elliptic and parabolic equations" (in analogy with the qualitative theory of ordinary differential equations) came into use.

In this survey we want to give an up to date description of the state of this field. We shall mostly present the work of the last 10–15 years. Moreover, we shall deal not only with the results obtained, but also with some of the methods used to obtain them. Having in mind the space limitations of this survey, we consciously refrained from covering too wide a range of works in this field, as just a complete listing of these works would have turned the present survey into not more than a bibliographic handbook. Such a handbook might be of some use, but in the case of key theorems we wanted not only to present their statement, but also to explain as far as possible the methods used in the proof. Moreover, we wanted to present in a sufficiently detailed manner some recent work that we consider important and that had not appeared in monographs.

The study of higher order equations, naturally, has the character of generalizing results pertaining to second order equations. However, not all the results obtained for second order equations can be extended to cover the case of equations of higher order, and not all the methods used to study second order equations can be applied to higher order ones. Therefore we restrict ourselves to second order equations, even though higher order ones are doubtless no less interesting and are encountered in applications.

The work comprises two chapters, the first of which deals with elliptic equations, while the second deals with parabolic equations.

Each of the chapters is subdivided into three paragraphs: equations in non-divergence form, equations in divergence form, and equations with smooth coefficients.

In physical problems, second order elliptic (parabolic) equations appear in such a form, that their principal part is in divergence form, div(A grad u). Coefficients of the matrix A do not have to be smooth functions. Lately the requirements of mechanics, and from inside mathematics, of probability theory, have led to the need for considering matrices A with coefficients that are non-smooth, for example, just measurable. In these cases solutions are of course to be understood in a generalized sense: they are to belong to a natural energy space. Thus § 2, Chapter 1 and § 2, Chapter 2 are devoted to the study of qualitative properties of solutions of such equations.

In § 1, Chapter 1 we consider elliptic equations in non-divergence form, under no restriction on the coefficients apart from the uniform ellipticity condition. At first glance it would appear that such equations constitute an exotic object of inquiry. However, the significance of theorems proved in that paragraph lies in the following: let the coefficients of the equation at hand be smooth (C^1 for example), and we have to find estimates on solutions that do not depend on derivatives of the coefficients, but only on the ellipticity constant of the operator. This allows us to use the method of Leray-Schauder to prove solubility of nonlinear equations. The same can be said about parabolic equations (§ 1, Chapter 2).

We use consecutive numeration of formulae in each of the chapters. For theorems, a dual system is used: chapter number followed by the number of the theorem in the chapter.

We tried to make the various chapters and even paragraphs independent of each other. This led to the same equation being identified sometimes by different numbers in different places.

We use well established notation for the classical Sobolev spaces and for spaces of continuously differentiable functions, as well as for the functions, k-th derivatives of which satisfy the Hölder condition. Let us only note that in our work the space $W_2^{(l)}$ is denoted by W^l.

Chapter 1
Second Order Elliptic Equations

§ 1. Equations in Non-divergence Form with Measurable Coefficients

In this section we shall consider operators of the form

$$L = \sum_{i,k=1}^{n} a_{ik}(x)\frac{\partial^2}{\partial x_i\,\partial x_k} + \sum_{i=1}^{n} b_i(x)\frac{\partial}{\partial x_i} + c(x), \tag{1}$$

where $x = (x_1, \ldots, x_n)$, $a_{ik}(x) = a_{ki}(x)$, and all the coefficients of the operator L are assumed to be measurable functions. The operator L is called *elliptic* at a point $x \in \mathbb{R}^n$ if the matrix $\|a_{ik}(x)\|$, $i, k = 1, \ldots, n$, is positive definite. The operator L is called elliptic in a domain $G \subset \mathbb{R}^n$, if it is elliptic at every point of the domain. L is called uniformly elliptic in G if there exist constants $\lambda > 0$, $M > 0$, such that

$$\lambda^{-1}|\xi|^2 \leqslant \sum_{i,k=1}^{n} a_{ik}(x)\xi_i\xi_k \leqslant \lambda|\xi|^2, \quad \xi \in \mathbb{R}^n, \quad x \in G; \tag{2}$$

$c(x) \leqslant M$, $|b_i(x)| \leqslant M$, $|a_{ik}(x)| \leqslant M$, $i, k = 1, \ldots, n$, $x \in G$. A function $u(x) \in C^2(G)$ satisfying the equation

$$Lu = f \tag{3}$$

is called a solution of this equation. Let us note that sometimes solutions $u(x) \in W_p^2(G)$, which satisfy (3) almost everywhere, are considered.

We call a function $u(x) \in C^2(G)$ which satisfies the inequality $Lu \leqslant 0$ ($Lu \geqslant 0$) superelliptic (subelliptic).

Note that if $b_i \equiv 0$, $i = 1, \ldots, n$, $c \equiv 0$, $v(x)$ is a subelliptic function and the function $f(t)$, defined for $\inf v \leqslant t \leqslant \sup v$, satisfies the conditions $f'(t) \geqslant 0$ and $f''(t) \geqslant 0$, then the function $w(x) = f(v(x))$ is also subelliptic.

An important role is played by subelliptic functions of the form

$$v(x) = \frac{1}{|x - x_0|^s},$$

where s is a large enough positive constant. The function $v(x)$ is necessarily subelliptic if

$$s \geqslant n(\lambda^2 + M) - 2, \quad c(x) \equiv 0.$$

1.1. A Maximum Principle. The following statements are well known, and their proofs are completely elementary.

Theorem 1.1. *If for a uniformly elliptic operator the coefficient* $c(x) \leqslant 0$, *and* $u(x)$ *is a solution in a bounded domain* G, *then*

$$\sup_G u \leqslant \max\left(0, \sup_{x_0 \in \partial G} \overline{\lim_{x \to x_0}} u(x)\right),$$

$$\inf_G u \geqslant \min\left(0, \inf_{x_0 \in \partial G} \underline{\lim_{x \to x_0}} u(x)\right).$$

Theorem 1.2. *If for a uniformly elliptic operator the coefficient $c(x) \leqslant 0$, and $u(x)$ is a subelliptic (superelliptic) function in a bounded domain G, then*

$$\sup_G u \leqslant \max\left(0, \sup_{x_0 \in \partial G} \overline{\lim_{x \to x_0}} u(x)\right),$$

$$\inf_G u \geqslant \min\left(0, \inf_{x_0 \in \partial G} \underline{\lim_{x \to x_0}} u(x)\right).$$

In particular, if $u(x)$ is continuous in $\bar{G}$ then $\max_{\bar{G}} u = \max_{\partial G} u$ ($\min_{\bar{G}} u = \min_{\partial G} u$).

Theorem 1.3. *If for a uniformly elliptic operator the coefficient $c(x) \leqslant 0$, and $u(x)$ is its subsolution in G which reaches a negative minimum at a point $x_0 \in G$, then*

$$u(x) \equiv \text{const}.$$

If there are no restrictions on the sign of the coefficient $c(x)$ then, naturally, there is no maximum principle. However, for domains of small diameter or, in greater generality, for domains contained in a sufficiently narrow layer, the following weakened version of the maximum principle is true.

Theorem 1.4. *For every $\varepsilon > 0$ there is a $\delta > 0$ which depends on ε, λ, and M, such that if the domain G is contained in the layer $|x_1| < \delta$, and if $u(x)$ is a subelliptic (superelliptic) function for a uniformly elliptic operator L, then*

$$\sup_G u \leqslant \max\left(0, (1+\varepsilon) \sup_{x_0 \in \partial G} \overline{\lim_{x \to x_0}} u(x)\right),$$

$$\inf_G u \geqslant \min\left(0, (1+\varepsilon) \inf_{x_0 \in \partial G} \underline{\lim_{x \to x_0}} u(x)\right).$$

Indeed, let us set $u = v(1 - Kx_1^2)$. Choosing K sufficiently large, we see that if δ is small enough, $1 - Kx_1^2 > 0$ in G and v satisfies the inequality $L'v \geqslant 0$ ($L'v \leqslant 0$), where L' is an operator of the same form as L, but with a non-positive coefficient of v. Therefore, the maximum principle applies to v. Taking δ small enough so that $1 - Kx_1^2 > \dfrac{1}{1+\varepsilon}$, we obtain the statement of the theorem.

Let us now consider a solution of equation (3) and let us try to estimate $\max_{\bar{G}} |u(x)|$ in terms of f. It is not hard to obtain an estimate of $\max_{\bar{G}} |u(x)|$ in terms of $\sup_G |f|$. However, such an estimate is too rough. A considerably more refined one is given by a theorem due to Aleksandrov (Aleksandrov [1966]).

Theorem 1.5. (A.D. Aleksandrov). *Let $c(x) \leqslant 0$ in the operator L, and let $u(x)$ be a solution of equation* (3) *in a bounded domain G, which is continuous in $\bar{G}$ and satisfies the condition $u(x)|_{\partial G} = 0$. Then*

$$\max_{\bar{G}} |u(x)| \leqslant c \|f\|_{L^n(G)}, \tag{4}$$

where C is a constant which depends on λ, n, M, and on the diameter of the domain G.

We give here a proof of this theorem in the case $b_i(x) \equiv c(x) \equiv 0$. In the general case the proof is in principle the same, but requires a number of technical modifications.

Let $\max_{\bar{G}} |u(x)| = |u(x_0)| = M$, $x_0 \in G$, with R being the diameter of G. Without loss of generality, we can assume that $u(x_0) < 0$. Let us denote by E the set of points $x \in G$ such that $u(x) \leqslant 0$. For each $p \in \mathbb{R}^n$ with

$$|p| < M/R, \tag{5}$$

let us consider the n-dimensional *hyperplane* $z = (p, x) + C$ in the $(n+1)$-dimensional space $(x_1, \ldots, x_n, z)$, which *supports* from below the graph of the function $z = u(x)$. In view of condition (5), the set H of points $x' \in E$, for which $u(x') = (p, x') + C$, is non-empty. In each of these points the second differential is non-negative definite. Let us consider on this set the gradient mapping $x \to \operatorname{grad} u(x) = p$. The image of H under this mapping contains the ball $|p| < M/R$. The Jacobian of this map is given by $\det \|u_{x_i x_k}(x)\|$ $i, k = 1, \ldots, n$. Taking into account the fact that the *Hessian* $\|u_{x_i x_k}(x)\|$ $i, k = 1, \ldots, n$ is non-negative definite, we obtain

$$\int_H \det \|u_{x_i x_k}(x)\| \, dx \geqslant \Omega_n (M/R)^n,$$

where Ω_n is the volume of the unit ball.

We have

$$\begin{aligned} \int_H \det \|u_{x_i x_k}(x)\| \, dx &= \int_H \frac{\det \|u_{x_i x_k}(x)\| \cdot \det \|a_{ik}(x)\|}{\det \|a_{ik}(x)\|} \, dx \\ &= \int_H \frac{\det(\|u_{x_i x_k}(x)\| \cdot \|a_{ik}(x)\|)}{\det \|a_{ik}(x)\|} \, dx \\ &\leqslant \lambda^n \int_H \det(\|a_{ik}\| \cdot \|u_{x_i x_k}\|) \, dx = \lambda^n \int_H \prod_{i=1}^n \mu_i(x) \, dx, \end{aligned} \tag{6}$$

where $\mu_1(x), \ldots, \mu_n(x)$ are the eigenvalues of the matrix $\|a_{ik}\| \cdot \|u_{x_i x_k}\|$. As the matrix $\|a_{ik}\|$ is positive definite, and the matrix $\|u_{x_i x_k}\|$ is non-negative definite, we obtain that $\mu_i(x) \geqslant 0$, $i = 1, \ldots, n$. (In fact, by positive definiteness of $\|a_{ik}\|$, both matrices can simultaneously be reduced to diagonal form.) From non-negativity of $\mu_i(x)$ it follows that

$$\left(\prod_{i=1}^{n} \mu_i(x)\right)^{1/n} \leqslant 1/n \sum_{i=1}^{n} \mu_i(x) = 1/n \operatorname{Sp}(\|a_{ik}(x)\| \cdot \|u_{x_i x_k}(x)\|)$$

$$= 1/n \sum_{i,k=1} a_{ik}(x)_{x_i} u_{x_i x_k} = f(x)/n. \tag{7}$$

From this inequality and from the inequality (6) we obtain that

$$\lambda^n(1/n)^n \int_G |f(x)|^n \, dx \geqslant (\lambda/n)^n \int_H |f|^n \, dx \geqslant \Omega_n(M/R)^n \tag{8}$$

or

$$M \leqslant \lambda R/n(\Omega_n)^{1/n} \|fn\|_{L^n} \tag{9}$$

which had to be shown. □

1.2. The Two-dimensional Case. The case of two variables occupies a special place in the theory of elliptic equations. In this case, on one hand a number of theorems are, naturally, simplified, while, on the other hand, some phenomena specific to this case, occur.

In the case $n = 2$ the maximum principle by itself allows one to obtain theorems on the qualitative behaviour of functions. We shall say that a continuous function $z = f(x, y)$ has a graph of non-positive generalized curvature if for every function $z = ax + by + c$ neither of the sets $\{(x, y): f(x, y) > ax + by + c\}$ and $\{(x, y): f(x, y) < ax + by + c\}$ contains connected components belonging to the domain of definition of the function (that is, a "cap" cannot be cut by any plane from the graph of the function).

The following theorem was obtained by Adel'son-Vel'skij [1945].

Theorem 1.6. *If $z = f(x, y)$ is a continuous function with a graph of non-positive generalised curvature defined on all of $\mathbb{R}^2$, then either the graph of $z = f(x, y)$ is a cylinder or $\overline{\lim}_{r\to\infty}\left(\max_{x^2+y^2=r^2} |f(x, y)|/r\right) > 0$.*

In particular, the two-sided Liouville theorem holds for a non-cylindrical function $f(x, y)$ with a graph of non-positive generalised curvature, that is, if $|f(x, y)| \leqslant M$ on $\mathbb{R}^2$ then $f(x, y) \equiv$ const. The one-sided Liouville theorem is false as shown by Bernstein: the functions $u = e^{x-y^2}$ is a solution of the equation $2(1 + 2y^2)u_{xx} + 4yu_{xy} + u_{yy} = 0$, and it is non-negative.

The following formulation (due to Gerasimov, see Landis [1971]) of a Phragmen-Lindelöf type theorem holds for functions of two variables with a graph of non-positive generalised curvature.

Theorem 1.7. *Let $f(x, y)$ be a continuous function with a graph of non-positive generalised curvature, defined in the half-plane $y \geqslant 0$ and such that $f(x, 0) < 0$. Then either $f(x, y) \leqslant 0$ for $y > 0$ or*

$$\overline{\lim_{r\to\infty}}\left(\max_{x^2+y^2=r^2} |f(x, y)|/r\right) > 0. \tag{10}$$

These theorems cannot be extended to functions of a larger number of variables. A counterexample to Theorem 1.6 has been constructed for $n \geqslant 3$ by Hopf [1929].

A slight modification of this example provides also a counterexample to Theorem 1.7 for $n \geqslant 3$.

In the proofs of Theorems 1.6 and 1.7 topological properties of the plane are extensively exploited.

In the planar case of equations in non-divergence form with discontinuous coefficients and satisfying the uniform ellipticity condition (2), Nirenberg obtained an internal estimate of the Hölder norm of first derivatives of the solution. Nirenberg's method is presented in detail in the monograph of Ladyzhenskaya and Uraltseva [1973].

In the case of many variables this result is false. More than that, Safonov [1987] has shown that for any $\alpha > 0$ a sequence of equations $L_k u = 0$ of form (1) can be constructed in a domain $\Omega \subset \mathbb{R}^n$, satisfying condition (2) with the same constant λ, and having solutions $u_k(x)$ for which

$$\sup_{\substack{x', x'' \in \Omega \\ x' \neq x''}} \frac{|u(x') - (u(x'')|}{|x' - x''|^\alpha} \geqslant C_k \sup_\Omega |u_k|,$$

and, moreover, $C_k \to \infty$.

1.3. The Inward Directional Derivative Lemma

Lemma 1. *Let Q be an open ball in $\mathbb{R}^n$ of radius R let $c(x) \equiv 0$, and let $u(x)$ be a continuous function on $\bar{Q}$, which is subelliptic (superelliptic) in Q. Let us assume that $u(x)$ reaches its maximum (minimum) at a point $x_0 \in \partial Q$ and that moreover in the interior of Q the function is strictly less (larger) than $u(x_0)$. If a vector l based at x_0 is directed inward into Q, then the lower (upper)* derived number *of u at x_0 is negative (positive).*

The proof is completely elementary. One uses for sufficiently large s the barrier function $u(x_0) \pm \varepsilon\left(\frac{I}{|x - x'|^s} - \frac{I}{R^s}\right)$ in the layer $Q \backslash Q'$ where Q' is a ball concentric with Q and x' is the center of the ball Q.

Immediate corollaries of this lemma are the *strong maximum* principle and the uniqueness theorem for solutions of the Neumann problem in the following version.

Theorem 1.8. *Let $G \subset \mathbb{R}^n$ be a bounded domain such that for each boundary point x_0 a ball Q^{x_0} tangent to ∂G at x_0 can be inscribed into G, and let $u(x)$ be a solution of equation* (3) *which is continuous in $\bar{G}$ and satisfies the conditions $\partial u/\partial v|_{\partial G} = 0$, where the vector v is directed into Q^{x_0}. If $c(x) \equiv f(x) \equiv 0$, then* $u \equiv$ const.

The condition requiring every boundary point to be a tangency point for an inscribed ball is a restrictive one. A number of authors made efforts directed at proving Lemma 1 of § 1.3 under less stringent restrictions on the boundary of the domain. The following problem was considered: let the origin belong to ∂G, and let the "paraboloid"

$$\Pi = \left\{ x \in \mathbb{R}^n, 0 < x_n < a, x_n > \phi\left[\left(\sum_{i=1}^{n-1} x_i^2\right)^{1/2}\right]\right\}$$

lie in G; under what conditions on $\phi(t)$ at 0 is the statement of the Lemma of § 1.3 valid? In Kamynin and Khimchenko [1972] it is shown that for $\phi(t)$ one can take the function $\phi(t) = t\,h(t)$ where $h(t)$ is a continuous monotone decreasing function such that

$$\int_0^a \frac{h(t)}{t}\,dt < \infty, \quad h(0) = 0. \tag{11}$$

It turns out that condition (11) cannot be improved. It is shown in Kamynin and Khimchenko [1972] that if

$$\int_0^a \frac{h(t)}{t}\,dt = \infty, \quad h(0) = 0,$$

there exists in Π a harmonic function which is continuous in $\bar{\Pi}$, positive in Π, zero at 0 and such that

$$\partial u/\partial x_n|_{x=0} = 0.$$

Thus, uniqueness of solutions of the Neumann problem for equation (3) holds for a bounded domain, each of boundary points of which can be touched from inside the domain by the vertex of a paraboloid congruent to Π with $\phi(t) = t\,h(t)$, where $h(t)$ satisfies condition (11).

The question of uniqueness of solutions of the Neumann problem in the classical formulation remained unresolved for a long time. This formulation is: given that G is a bounded domain with a C^1 boundary, and that $u(x)$ is a function harmonic in G, continuous in $\bar{G}$ and having at each point of boundary zero normal derivative, does it follow that $u \equiv \text{const}$. The solution of this problem, known as the Lavrientiev-Keldysh problem, has been obtained by Nadirashvili [1983] in a considerably more general setting. We quote here the statement and the proof of Nadirashvili's lemma from which uniqueness of solutions to the Neumann problem will follow. The essence of this lemma is that although at an extremum point the derivative of a solution of equation (3) with $c(x) \leqslant 0$, $f \equiv 0$, in a direction which is not tangent to the boundary, can be zero, in any neighborhood of such a point there exists a point at which this derivative is non-zero.

We shall say that a bounded domain $G \subset \mathbb{R}^n$ has a boundary satisfying the *cone condition* if there are numbers $H > 0$, α_1, $0 < \alpha_1 < \pi/2$ and a continuous vector field on ∂G such that for every point $x \in \partial G$ a cone of height H with vertex at x, opening angle 2α, and with the axis directed along $v(x)$ is contained in its entirety (except for the vertex) in G. In particular, the cone condition is satisfied by a domain with a *Lipschitz boundary*.

Lemma 2 (N.S. Nadirashivili [1983]). *Let $G \subset \mathbb{R}^n$ be a bounded domain satisfying the cone condition. Let us consider a superelliptic (subelliptic) function $u(x)$ which is continuous in $\bar{G}$ and non-constant. If $x' \in G$ and $u(x) \geqslant u(x')$ $(u(x) \leqslant u(x'))$ for all $x \in G$, then in any neighborhood $\Omega \in \partial G$ of the point x' there is a point $x'' \in \partial G$ such that*

$$\left.\frac{\underline{\partial u}}{\partial v}\right|_{x=x''} > 0 \quad \left(\left.\frac{\overline{\partial u}}{\partial v}\right|_{x=x''} < 0\right),$$

where $\frac{\underline{\partial u}}{\partial v}\left(\frac{\overline{\partial u}}{\partial v}\right)$ is the lower (upper) derived number of the function u in the direction of the vector v.

The proof is based on the following elementary lemma from function theory.

Lemma 3. *Let $f(x)$, $g(x)$ be functions on the interval $[0, \beta]$ such that $f(0) = 0$, $f(x) > 0$ for all $x \in (0, \beta)$, $f(x)$ is lower semicontinuous on $[0, \beta]$ and the lower right derived number of f at 0 is zero; $g(x)$ is in $C^1[0, \beta]$, $g(0) = 0$, $g'(x) > \delta > 0$ for all $x \in [0, \beta]$. Then there exists a continuous function $h(x)$ defined on $(0, g(\beta))$, that is twice differentiable on that interval, such that $h(x) \geqslant 0$, $h'(x) > 0$, $h''(x) \geqslant 0$ for all $x \in (0, g(\beta))$, and such that $h(g(x)) \leqslant f(x)$ on $(0, \beta)$. Furthermore, $h(g(x_m)) = f(x_m)$ where $x_m \in (0, \beta)$ is a sequence such that $x_m \to 0$ as $m \to \infty$.*

Proof of Lemma 2. We break the proof into a number of steps.

a) Let u be a superelliptic function, $u(x') = 0$. Let us set $E = \{x \in \bar{G}: u(x) = 0\}$. By the maximum principle, $E \subset \partial G$, $u(x) > 0$ for $x \in G$. Let us show that there is a ball $Q_{r_0}^{x_0}$, $Q_{r_0}^{x_0} \cap E \neq \varnothing$, and such that there exists $x_1 \in \Omega \cap E \cap S_{r_0}^{x_0}$ satisfying the condition $(x_0 - x_1, v(x_1)) > 0$.

By assumption there exists a cone of height H with vertex at x' and axis directed along $v(x')$, which is completely, apart from the vertex, contained in G. Let us denote it by K. We can assume (making, if needed, a linear transformation, and choosing a smaller H) that the opening angle of the cone is larger than $3\pi/4$.

Let us choose $r' > 0$ small enough so that $\partial G \cap G_{r'}^{x'} \subset \Omega$. In view of the continuity of the vector field v, there exists r'', $0 < r'' < r'$ such that for all $x \in \partial G \cap Q_{r''}^{x'}$ the angle between $v(x)$ and $v(x')$ is less than $\pi/8$. Let us select any ball $Q \subset K \cap Q_{r''}^{x'}$ and let us translate it along the axis of the cone K so that it touches the set $E \cap Q_{r''}^{x'}$. The resulting ball clearly has the desired property.

b) Let us set $r_2 = \min(r_0/2, 1)$. In the ball $Q_{x_0}^{r_0}$ consider a radius connecting the center x_0 and the point x_1, and let us take on that radius a point x_2 which lies at a distance r_2 from x_1. Then the (open) ball $Q_{r_2}^{x_2}$ is disjoint from E, while the sphere $S_{r_2}^{x_2}$ intersects E in only one point x_1. Moreover, $(x_2 - x_1, v(x_1)) > 0$.

c) If there exists an $\varepsilon > 0$ such that $Q_{r_2}^{x_2} \cap Q_\varepsilon^{x_1}$ contains no points of ∂G, then $Q_{r_2}^{x_2} \cap Q_\varepsilon^{x_1} \subset G$, and then the boundary point x_1 can be touched from inside the domain by a small enough ball, so that, by Lemma 1, $\left.\frac{\underline{\partial u}}{\partial v}\right|_{x=x_1} > 0$. Let us assume now that for any $\varepsilon > 0$ the set $Q_{r_2}^{x_2} \cap Q_\varepsilon^{x_1} \cap \partial G$ is non-empty.

d) Let us set $A(\xi) = S^{x_2}_{r_2-\xi} \cap \partial G$. Let us define on the interval $[0, r_2/2]$ the function $f(\xi)$ by setting

$$f(r_2/2) = \min_{S^{x_2}_{r_2/2} \cap G} u(x); \quad f(\xi) = \min_{x \in A(\xi)} u(x), \quad \xi \in [0, r_2/2),$$

if $A(\xi)$ is non-empty. In all the other points we set $f = \max u$. As the set of ξ for which $A(\xi) = \varnothing$ is open, $f(\xi)$ is lower semicontinuous.

e) Let us assume that $f_r'(0) = 0$. For s sufficiently large the function $|x - y|^{-s}$, as a function of x, is subelliptic for $0 < |x - y| < 1$. Let us fix such an s. Set $g(\xi) = \dfrac{1}{(\xi - r_2)^s} - \dfrac{1}{r_2^s}$. The functions f and g satisfy the conditions of Lemma 3. Let us find the corresponding function h and the sequence $\xi_m \to 0$, $\xi_m > 0$, $h(g(\xi_m)) = f(\xi_m)$. Then there exists a sequence z_m, $z_m \in Q^{x_2}_{r_2} \cap \partial G$, $|z_m - x_1| \to 0$, $h(|z_m - x_2|^{-s} - r_2^{-s}) = u(z_m)$. The function $h(|x - x_2|^{-s} - r_2^{-s})$ is also subelliptic, and by the maximum principle $h(|x - x_2|^{-s} - r_2^{-s}) \leqslant u(x)$ in $G \cap (Q^{x_2}_{r_2} \setminus Q^{x_2}_{r_2/2})$. In a neighborhood of x_1 we have $\dfrac{\partial}{\partial v} h(|x - x|_2^{-s} - r_2^{-s}) > 0$. Therefore $\left.\dfrac{\partial u}{\partial v}\right|_{x=x_m} > 0$ for all m starting with some m_0.

f) $f_r'(0) = \delta > 0$. In this case we can choose an $\varepsilon > 0$ such that $\varepsilon(|x - x_2|^{-s} - r_2^{-s}) \leqslant u$ in $G \cap (Q^{x_2}_{r_2} \setminus Q^{x_2}_{r_2/2})$. As $\dfrac{\partial}{\partial v}(|x - x_2|^{-s} - r_2^{-s})|_{x=x_1} > 0$, $\left.\dfrac{\partial u}{\partial v}\right|_{x=x_1} > 0$. Lemma 2 is proved. □

1.4. A Growth Lemma. In this section we consider a technical lemma known as the "growth lemma" which is useful in applications. Let us assume that an open set D is contained in the ball $Q^{x_0}_{4R}$, $0 < R \leqslant 1/4$, such that its intersection with $Q^{x_0}_R$ is non-empty. Let a solution of the equation $Lu = 0$ be defined in D and let moreover $c(x) \leqslant 0$. If u is continuous in $\bar{D}$, positive in D and $u = 0$ on that portion of ∂D which is strictly contained in $Q^{x_0}_{4R}$, then by the strong maximum principle $\max_{\bar{D}} u > \max_{\overline{D \cap Q^{x_0}_R}} u$. Let us try to estimate

$$\max_{\bar{D}} u \Big/ \max_{\overline{D \cap Q^{x_0}_R}} u = h.$$

To do that some restriction must be placed on the structure of the domain under consideration. Let $H = Q^{x_0}_R \setminus D$. Then the larger is the ratio of the measure of the set H to the measure of $Q^{x_0}_R$, the larger is h.

Next we state the *growth lemma*. Let us use the following notation: $D \subset Q^{x_0}_{4R}$, $0 < R \leqslant 1/4$ is an open set, $H = Q^{x_0}_R \setminus D$, $\Gamma = \partial D \cap Q^{x_0}_{4R}$, and $u(x)$ is the solution of the equation $Lu = 0$ which is positive in D and zero on Γ. Then

$$\max_{\bar{D}} u > \left(1 + \xi\left(\frac{\text{meas}\, H}{R^n}\right)\right) \cdot \max_{\overline{D \cap Q^{x_0}_R}} u \tag{12}$$

where ξ is a positive function, which depends on the ellipticity constants of the operator.

This lemma was proved by Krylov and Safonov [1979, 1980]. Here we present the main steps of the proof, restricting ourselves to the case $b_i(x) \equiv c(x) \equiv 0$. The presence of lower order terms (and they are present in works quoted above) introduces non-essential complications in the proofs.

a) Let H contain the ball $Q_\rho^{x'}$, where ρ is of the same order of magnitude as $(\text{meas } H)^{1/n}$.

Let us consider the function $1/|x - x'|^s$. The number s can be chosen so that this function is a subsolution of the equation $Lu = 0$. Let $M = \max_D u$. Let us set

$$V = M\left(1 - \frac{\rho^s}{|x - x'|^s} + \frac{\rho^s}{(3R)^s}\right)$$

We see that $V > u$ in D. Therefore $u|_{D \cap Q_Q^{x_0}} < M\left(1 - \frac{\rho^s}{(2R)^s} + \frac{\rho^s}{(3R)^s}\right)$, which proves the lemma in this case: here $\xi = \left(\frac{\rho}{R}\right)^s \left(\frac{1}{2^s} - \frac{1}{3^s}\right)$.

b) Let us now consider the case of D of small measure. Namely, we shall show that there exists a $\delta > 0$ depending on λ and n, such that if $\text{meas } D/\text{meas } Q_R^{x_0} < \delta$, then $\max_{\bar{D}} u > 2 \max_{\overline{D \cap Q_R^{x_0}}} u$.

We shall assume that the functions $a_{ik}(x)$ are continuously differentiable. It will turn out that the number δ will depend on λ and n only, and not on the magnitude of derivatives of the coefficients of the equation. A simple limiting procedure allows us then to remove all restrictions on the smoothness of the coefficients.

We can assume that meas $\Gamma = 0$. In the opposite case, we could consider instead of D the set $D' = \{x \in Q_{4R}^{x_0}, u(x) > \varepsilon\}$, $\varepsilon > 0$ and instead of $u(x)$ the function $u'(x) = u(x) - \varepsilon$ for ε arbitrarily small. If meas $\Gamma = 0$, there exists an open set D_0 such that $\bar{D} \subset D_0$ and $\text{meas } D_0/\text{meas } Q_R^{x_0} < \delta$. Let us construct a C^∞ function $f(x)$ such that

$$f(x) = \begin{cases} -\dfrac{n\lambda}{8R^2} & \text{if } x \in D, \\ 0 & \text{if } x \in Q_R^{x_0} \setminus D_0, \end{cases}$$

and $0 \geqslant f(x) \geqslant -\frac{n\lambda}{8R^2}$ for $x \in (D_0 \cap Q_R^{x_0}) \setminus D$.

Let us solve in the ball $Q_{4R}^{x_0}$ the Dirichlet problem,

$$Lv = f, \quad v|_{S_{4R}^{x_0}} = 0.$$

Smoothness of coefficients was needed for existence of solutions to this problem (Miranda [1970]). By Theorem 1.5,

$$|v| \leqslant C\delta^{1/n}.$$

Let $\max_{\bar{D}} u = M$. Let us set $\omega(x) = M\left(\frac{(x - x_0)^2}{(4R)^2} + v(x) + 1/4\right)$. It is easy to see that $L\omega(x) \leqslant M\left(\dfrac{n\lambda}{8R^2} + f(x)\right) \leqslant 0$ in D. The statement of the lemma now follows from the maximum principle applied to the function $\omega - u$ in D.

c) Let us move on now to the general case: let meas H be an arbitrary number less than meas $Q^{x_0}_{R/4}$.

By a similarity transformation the equation under consideration is transformed into an equation with the same ellipticity constant λ, and therefore it is sufficient to consider the case $R = 1/4$.

Let the number h be such that $\text{meas}(H \cap Q^{x_0}_{1/4-h}) = (\text{meas } H)/2$, $v_1 = 1 - u/\max_{\bar{D}} u$. We have to prove that there exists a $\sigma = \sigma(\text{meas } H)$, such that $\inf_{\overline{Q^{x_0}_{1/4} \cap D}} v_1 \geqslant \sigma$. Let $H = H_1$. Let us set $Q^{x_0}_{1/4-h} \cap H_1 = H_1^0$ so that meas $H_1^0 =$ (meas $H_1)/2$. Let x be a point of density of the set H_1^0 and let r_x be the largest number not exceeding h such that $\text{meas}(Q^x_{r_x} \setminus H_1) \leqslant \delta r_x^n$, where δ is the constant from part b).

Two cases can occur: 1) $r_x < h$ and 2) $r_x = h$. In case 2), by part b) we conclude that $v_1|_{Q^x_h} > 1/2$, and then by part a) $v_1|_{D \cap Q^{x_0}_{1/4}} > \sigma$ where σ depends on λ and h, that is on λ and meas H. The lemma is proved. □

Let case 1) hold for all points $x \in H_1^0$. Then $\text{meas}(Q^x_{r_x} \setminus H_1) = \delta r_x^n$. All of the set H_1^0 is covered by balls $Q^x_{r_x}$, $x \in H_1^0$ for which this equality holds. Let us select, using the Banach procedure, either a finite or a countable number of pairwise disjoint balls, the sum of measures of which is larger than meas $H_1^0/5^n =$ meas $H_1/(2 \cdot 5^n)$. Let these balls be $Q_1, \ldots, Q_s, \ldots$ By part b)

$$v_1|_{Q_i \setminus H_1} > 1/2.$$

Let us set $v_2 = 2v_1$, $D_2 = \{x \in D: v_2 < 1\}$ and $H_2 = Q^{x_0}_{1/4} \setminus D_2$ so that meas $H_2 >$ meas $H_1(1 + \delta/(2 \cdot 5^n)) =$ meas $H(1 + \delta/(2 \cdot 5^n))$.

We now repeat the same argument substituting v_2, D_2, and H_2, respectively, for v_1, D_1, and H_1. We shall find $v_3 = 2v_2 = 2^2 u$, $D_3 = \{x \in D: v_3 < 1\}$ and $H_3 = Q^{x_0}_{1/4} \setminus D_3$ such that

$$\text{meas } H_3 > \text{meas } H(1 + \delta/(2 \cdot 5^n))^2,$$

if case 1) holds, or, if case 2) holds, the lemma will be proved. Continuing this process, we shall obtain v_k, D_k, H_k such that $v_k = 2^k v_1$, $D_k = \{x \in D: v_k < 1\}$, $H_k = Q^{x_0}_{1/4} \setminus D_k$ and meas $H_k >$ meas $H(1 + \delta(2 \cdot 5^n))^{k-1}$. But $H_k \subset Q^{x_0}_{1/4}$, and therefore this process must terminate after a finite number of steps k_0, which depends on δ, meas H, and h, that is, on meas H, λ, and n. On the other hand it will not terminate before $\text{meas}(Q^{x_0}_{1/4} \setminus H_k) = 0$. Thus $v_{k_0}|_{Q^{x_0}_{1/4} \cap D} > 1$, or $v_1|_{Q^{x_0}_{1/4} \cap D} > 1/2^{k_0}$, that is, $\sigma = 1/2^{k_0}$, and the lemma is proved.

From the growth lemma we can deduce a-priori estimates of the Hölder norm and the Harnack inequality (for equations (2) in non-divergence form with arbitrary coefficients they were first obtained by Krylov and Safonov [1979, 1980] in papers quoted above).

Theorem 1.9 (Hölder norm estimate theorem). *If $u(x)$ is a solution of the equation $Lu = 0$ in a domain $G \subset \mathbb{R}^n$, then for any subdomain G' compactly* *embedded in G, there exist* $\alpha > 0$ *depending on* λ, μ, *and* n, *and* $C > 0$ *depending on the distance of* G' *from* ∂G, *such that*

$$\frac{|u(x) - u(y)|}{|x - y|^\alpha} \leqslant C \sup_G u \tag{13}$$

for all $x, y \in G'$, $x \neq y$.

This theorem is a direct consequence of the following oscillation theorem.

Oscillation Theorem. *Let a solution of the equation* $Lu = 0$ *be defined on* $Q_R^{x_0}$, $R < R_0$. *Then*

$$\operatorname{osc}_{Q_R} u > (1 + \eta) \operatorname{osc}_{Q_{R/4}} u \tag{14}$$

where $\eta > 0$ *depends on* λ, M, n, *and* R_0. *If* $b_i \equiv c \equiv 0$, *the restriction* $R < R_0$ *can be removed and* η *depends only on* λ *and* n.

Proof. We shall assume that $b_i \equiv c \equiv 0$. Let us set $\sup_{Q_{R/4}^{x_0}} u - \inf_{Q_{R/4}^{x_0}} u = d$, and $v = \inf_{Q_{R/4}^{x_0}} u - d/2$. Let furthermore $E^+ = \{x \in Q_R^{x_0}: v \geqslant 0\}$ and $E^- = \{x \in Q_R^{x_0}: v \leqslant 0\}$. At least one of the sets E^+ or E^- satisfies the condition $\operatorname{meas} E^{+(-)} \geqslant (\operatorname{meas} Q_{R/4}^{x_0})/2$. Let it be E^-. Then, by the growth lemma $\sup_{Q_R} v > (1 + \xi)d/2$, or $\operatorname{osc}_{Q_R} u > (1 + \xi/2)d$. □

As we already stated in § 1.2, Safonov (Safonov [1987]) showed that the constant α in Theorem 1.9 can turn out to be arbitrarily small.

Theorem 1.10. (the Harnack inequality). *Let a positive solution of the equation* $Lu = 0$ *be defined in* $Q_R^{x_0}$, $R < R_0$. *There exists a constant* $C > 0$, *depending on* λ, M, n, *and* R_0 *such that*

$$\sup_{Q_{R/16}} u \Big/ \inf_{Q_{R/16}} u < C.$$

If $b_i \equiv c \equiv 0$, *the restriction* $R < R_0$ *is not needed.*

We detail here the procedure for obtaining the Harnack inequality from the *growth lemma*; it is similar to the procedure used in Landis [1971] for the analogous proof in the particular case of an equation with a small spread of eigenvalues of the matrix $\|a_{ij}(x)\|$.

We shall assume that $b_i(x) \equiv c(x) \equiv 0$. In this case we can use similarity transformations on $\mathbb{R}^n$ and also change the solution by adding constants. We shall assume that $R = 16$ and for convenience let the center of the ball be at the origin O.

Thus, $u|_{Q_{16}^0} > 0$. By multiplying the solution by a constant we can get $\sup_{Q_1} u = 2$. The problem at hand is then to show that there exists a constant $\sigma > 0$ depending on λ and n such that $\inf_{Q_1} u > \sigma$. We break the proof into steps:

a) The growth lemma can be modified in the following way: for any $A > 1$ there exists a $\varepsilon_0 > 0$ such that if $u(x)$ is a positive solution of the equation in $D \subset Q_R^{x_0}$, $D \cap Q_{R/4}^{x_0} \neq \varnothing$, $u|_{\partial D \cap Q_R^{x_0}} = 0$, and $\operatorname{meas} D < \varepsilon_0 \operatorname{meas} Q_R^{x_0}$, then $\sup_D u > A \sup_{D \cap Q_{R/4}^{x_0}} u$.

In fact, let δ be the constant from part b) of the proof of the growth lemma, and let N be the smallest natural number such that $2^N > A$. Let us subdivide $Q_R^{x_0} \setminus Q_{R/4}^{x_0}$ into N layers of thickness $3R/4N$ by concentric spheres S_i. Let us consider the point $x_i \in S_i$ such that $u(x_i) = \max_{D \cap S_i} u$. Applying part b) of the proof of the growth lemma to the balls $Q_{3R/4N}^{x_i}$, $Q_{3R/16N}^{x_0}$, and to $D \cap Q_{3R/4N}$, we obtain the desired statement.

In what follows, given λ and n, A will be chosen as above. We shall assume henceforth that it is given and shall find the corresponding ε_0. We shall use the notation $G = \{x \in Q_4^0 : u(x) > 1\}$.

b) Let meas $G \geqslant \varepsilon_0 \Omega_n$, where Ω_n is the volume of the unit ball. Let us set in the growth lemma $\xi = \xi((\text{meas } Q_{4-\varepsilon_0}^0)/4^n)$, and $v = 1 - u + \xi$. Let us denote $D = \{x \in Q_{16}^0 : v(x) > 0\}$. Then $\text{meas}(Q_4^0 \setminus D) \geqslant \text{meas } Q_{4-\varepsilon_0}^0$. If at some point $x_0 \in Q_4^0, v(x_0) \leqslant \xi$, then $\sup_{D \cap Q_{16}^0} v \geqslant 1$. Thus by the growth lemma, $\sup_{D \cap Q_{16}^0} v \geqslant 1 + \xi$, that is, $\inf_{D \cap Q_{16}^0} u \leqslant 0$, and we have a contradiction.

Therefore, as σ we can take ξ and in this case our inequality is proved. Let us from now on assume that meas $G < \varepsilon_0$.

c) Let us set $u_1 = u - 1$. Then there is a point x_1 on the sphere S_1^0 such that $u_1(x_1) \geqslant 1$. Let us denote by G_1 the component of the set $\{x \in Q_4^0 : u_1(x) > 0\}$ which contains x_1. For ρ small enough, $\text{meas}(G_1 \cap Q_\rho^1) > \varepsilon_0 \text{ meas } Q_0^{x_1}$. For $\rho = 1$, meas $(G_1 \cap Q_1^{x_0}) < \varepsilon_0$ meas $Q_1^{x_0}$. Therefore by continuity of the measure $\text{meas}(G_1 \cap Q_\rho^{x_1})$ in ρ, there exists ρ_1 such that $\text{meas}(G_1 \cap Q_{\rho_1}^{x_1}) = \varepsilon_0 \text{ meas } Q_{\rho_1}^{x_0}$. Let us denote $G_1 \cap Q_{4\rho_1}^{x_1}$ by D_1. Then $\sup_{D_1} u > A$. Let us set $B = A/2$. We have, firstly, $u|_{D_1} > 1 = B^0$, meas $D_1 > \varepsilon_0$ meas $Q_{\rho_1}^{x_1}$, and, secondly, there is a point x_2 on the sphere $S_{1+4\rho_1}^0$, such that $u_1(x_2) > 2B$. Let us set $u_2 = u_1 - B$. Denote by G_2 the component of the set $\{x \in Q_4^0 : u(x) > 0\}$ which contains x_2. Repeating the foregoing argument, we can find ρ_2 such that $\text{meas}(G_2 \cap Q_{\rho_2}^{x_2}) = \varepsilon_0 \text{ meas } Q_{\rho_2}^{x_2}$. If we set $D_2 = G_2 \cap Q_{4\rho_2}^{x_2}$, then

$$u|_{D_2} > B^1, \quad \text{meas } D_2 > \varepsilon_0 \text{ meas } Q_{\rho_2}^{x_2}.$$

Let us continue this process. Then we shall obtain the sequences $\rho_1, \ldots, \rho_i, \ldots$, $x_1, \ldots, x_i, \ldots, D_1, \ldots, D_i, \ldots$ such that

$$x_i \in S_{1+4\rho_1+\cdots+4\rho_i}^0, \quad u|_{D_i} > B^{i-1}, \quad \text{meas } D_i > \varepsilon_0 Q_{\rho_i}^{x_i}.$$

Let k be the smallest index such that $4\rho_i + \cdots + 4\rho_k < 1$. (such a k exists as $u(x_i)$ growth without bound with i). Therefore there exists $i_0 \leqslant k$ such that $\rho_i > 1/4 \cdot 1/2^{i_0}$. Thus

$$D_{i_0} \subset Q_3^0, \quad \text{meas } D_{i_0} > \varepsilon_0 \Omega_n 1/(4 \cdot 2^{n i_0}), \quad u|_{D_{i_0}} > B^{i_0 - 1}.$$

d) Using the growth lemma we can find the constant $\xi_0 = \xi(1 - 1/4^n)$. Let $B = 1/\xi_0$. Therefore $A = 2/\xi_0$, and thus ε_0 depends only on λ and n.

Let us set $\xi = \xi(1 - \varepsilon_0)$, $v = u(1 - \xi_1)$. Using the argument of part a) we find that $u|_{Q_{\rho_{i_0}}^{x_{i_0}}} > \xi_1 B^{i_0 - 1} = \xi_1 (1/\xi_0)^{i_0 - 1}$. Hence, applying the same argument to $v = u(1 - \xi(4/\xi_0)^{i_0 - 1})$, we find that

$$u|_{Q_{4i_0}^{x_{i_0}}} > \xi_0 \xi_1 B^{i_0 - 2} = \xi_0 \xi_1 \left(\frac{1}{\xi_0}\right)^{i_0 - 2}.$$

Applying the same argument, consecutively, to the balls $Q^{x_{i_0}}_{4^2\rho_{i_0}}$, $Q^{x_{i_0}}_{4^3\rho_{i_0}}$, etc., we find that

$$u|_{Q^{x_{i_0}}_{4^l\rho_{i_0}}} > \xi_0^l \cdot \xi_1 \left(\frac{1}{\xi_0}\right)^{i_0-l-1}.$$

There exists an l such that $Q^{x_{i_0}}_{4^l\rho_{i_0}} \supseteq Q^{x_{i_0}}_{4^l \cdot 1/4 \cdot (1/2)^{i_0}} \supset Q^0_1$,

$$l < i_0/2 + 3.$$

Therefore

$$(1/\xi_0)^{i_0-1} \cdot \xi_1 \cdot \xi_0^{i_0/2+3} \leqslant 1,$$

or

$$i_0 < (\ln 1/\xi_1/\ln 1/\xi_0) + 4.$$

Therefore i_0 depends on λ and n, and thus meas $D_{i_0} > \varepsilon_1$, that is, meas $G > \varepsilon_1$, where $\varepsilon_1 > 0$ depends on λ and n, so that the argument of part a) allows us to conclude that there exists $\sigma_1 > 0$ depending on λ and n such that $\inf_{Q_1} v > \sigma_1$. The *Harnack inequality* is proved. □

A proof of the growth lemma, and, consequently, an estimate of the *Hölder norm* and the Harnack inequality, for solutions of the equation $Lu = 0$ was later presented by Novruzov (Novruzov [1983a]) employing different methods. He uses barrier functions constructed with the help of potentials with kernels $|x|^{-s}$ where $s > 0$ and small enough. Instead of the original equations a new one is considered

$$Lu + \frac{\partial^2 u}{\partial x_{n+1}^2} = 0 \tag{15}$$

and $u(x_1, \ldots, x_n, x_{n+1}) \equiv u(x_1, \ldots, x_n)$. The function $|x|^{-s}$ for $s > 0$ and small enough is a supersolution of the new equation.

From Harnack's inequality we can deduce Liouville's theorem.

Theorem 1.11 (Liouville's theorem). *Let $u(x)$ be a solution of the equation $Lu = 0$ in $\mathbb{R}^n$ and let $b_i(x) \equiv 0, i = 1, \ldots, n, c(x) \equiv 0$. If $u(x)$ is bounded either from above or from below, then $u \equiv$* const.

Without loss of generality, we can assume that $u \geqslant 0$ and $\inf_{\mathbb{R}^n} u = 0$. In view of the homogeneity of the equation, Harnack's inequality provides us with the estimate $\sup_{Q_R^{x_0}} u/\inf_{Q_R^{x_0}} u < C$ where C is independent of R, x_0. Let us consider the sequence of points $x_m \in \mathbb{R}^n$ such that $u(x_m) \to 0$ as $m \to \infty$, and a sequence of balls $Q^{x_0}_{R_m} \ni x_m$. Then $u(x_0) \leqslant \sup_{Q^{x_0}_{R_m}} u \leqslant Cu(x_m) \to 0$ as $m \to \infty$; that is, $u(x_0) = 0$, which means that $u \equiv 0$.

Liouville's theorem admits the following generalization.

Theorem 1.12. *There exists a positive number α depending on λ and n such that if $u(x)$ is a solution of the equation*

$$\sum_{i,k=1}^{n} a_{ik}(x) \frac{\partial^2 u}{\partial x_i \partial x_k} = 0 \tag{16}$$

in $\mathbb{R}^n$ *and*

$$u(x) > -C|x|^{\alpha} \tag{17}$$

for any positive C *and* $|x|$ *large enough, then* $u \equiv \text{const}$.

Proof. Let us set $R_k = 2^k R_0$, $k = 0, 1, \ldots, R_0 > 0$. Let $M_k = u(0) - \inf_{Q^0_{R_k}} u$, $k = 0, 1, \ldots$ From Harnack's inequality it follows that

$$M_k \leqslant \frac{1}{1+\xi} M_{k+1}$$

where $\xi > 0$ depends only on λ and n. Therefore

$$u(0) - \inf_{Q^0_{R_0}} u \leqslant \frac{1}{(1+\xi)^k}(u(0) + CR_0^{\alpha} \cdot 2^{\alpha k})$$

for sufficiently large k. If $\alpha < \ln(1+\xi)/\ln 2$, then $u(0) - \inf_{Q^0_{R_0}} u = 0$, and $u \geqslant u(0)$ in $Q^0_{R_0}$. As R_0 was arbitrary, we see that $u \equiv \text{const}$. □

The question whether in Theorem 1.2 condition (17) can be changed to $u(x) > \phi(x)$ with $\phi(x) < 0$, $\phi(x) = O(|x|)$, is of interest.

We conclude this section with the following proposition.

Proposition (Landis and Nadirashvili [1985]). *Let* K *be a cone in* $\mathbb{R}^n$ *with vertex at the origin and with a Lipschitz boundary. Let the equation*

$$Lu = \sum_{i,j=1}^{n} a_{ij}(x) \frac{\partial^2 u}{\partial x_i \partial x_j} = 0$$

with coefficients satisfying the Hölder condition, be defined in K. *Then there exists a unique (up to multiplication by constants) positive solution of this equation which is zero on* ∂K.

In what follows we denote by Q_R the open ball of radius R in $\mathbb{R}^n$ with center in the origin, and by S_k—its boundary. We break the proof of this proposition into steps.

1. Let $K_m = Q_{2^m}$ and let $u_m(x)$ be the solution in K_m of the Dirichlet problem $u|_{\partial K \cap Q_{2^m}} = 0$, $u|_{K \cap S_{2^m}} = M_m = \text{const.}$, where M_m is chosen so that at some fixed point $x_0 \in K$, $u_m(x_0) = 1$. From the Harnack inequality and the growth lemma it follows that u_m are uniformly bounded on every K_l, $l \leqslant m$. Therefore from the application of the Hölder condition to the coefficients of the equation, and from the fact that all boundary points are regular, it follows that we can choose a subsequence which converges to a positive (in view of the condition $u_m(x_0) = 1$) solution which is zero on ∂K.

2. We shall show that there is a constant $\sigma > 0$ such that if $u(x)$ is a solution which is positive in K_4 and zero on $\partial K \cap Q_4$ such that $\sup_{K_2} u = 1$, then, given that $v(x)$ is a solution in K_2 which is zero on $\partial K \cap Q_2$, and such that $\sup_{K_2} |v| < \sigma$, it follows that $u + v > 0$ in K_1.

Let $\varepsilon > 0$, $G_\varepsilon = \{x \in K_4: \rho(x, \partial K \cap (Q_{1+\varepsilon} \backslash Q_{1-\varepsilon})) > 0\}$ and $0 < r < 1/2$. Let us use the notation $D^r = K_{1+r} \backslash K_{1-r}$, $\Gamma = \partial K \cap (Q_{1+\varepsilon} \backslash Q_{1-\varepsilon})$, $\gamma_{\varepsilon,r} = \partial G_\varepsilon \cap D^r$, $d_{\varepsilon,r} = G_\varepsilon \cap D^r$, $H_{\varepsilon,r} = D^r \backslash d_{\varepsilon,r}$, $\Sigma_\varepsilon = (G_\varepsilon \cap S^{1+\varepsilon}) \cup (G_\varepsilon \cap D^r)$.

Let q be a solution of the equation $Lq = 0$ in D^r, $q|_\Gamma = 0$, $q > 0$ in $H_{\varepsilon,r}$, $q \geqslant -1$ in D^r. Let us denote $q^+ = \max(0, q)$, $q^- = \min(0, q)$. Let us consider functions s, ω, z which solve the following Dirichlet problems: $Ls = 0$ in D^r, $s|_{\partial D^r} = q^+$, $L\omega = 0$ in $d_{\varepsilon,r}$, $(q - \omega)|_{\partial d_{\varepsilon,r} \backslash \Sigma_\varepsilon} = 0$, $\omega|_{\Sigma_\varepsilon} = q^+|_{\Sigma_\varepsilon}$, $Lz = 0$ in $d_{\varepsilon,r}$, $z|_{\partial d_{\varepsilon,r} \backslash \Sigma_\varepsilon} = 0$, $z|_{\Sigma_\varepsilon} = q^-|_{\Sigma_\varepsilon}$.

Thus $q = \omega + z$ in $d_{\varepsilon,r}$. Furthermore $s > 0$ in D^r, $s \geqslant 0$ in $H_{\varepsilon,r}$. Therefore by Harnack's inequality $s > C$ in $H_{\varepsilon/2,r-\varepsilon}$ where $C > 0$ depends on the ellipticity constant and K. We have that $s = \omega$ on $\partial d_{\varepsilon,r} \backslash \gamma_{\varepsilon,r}$, $s \leqslant 2\omega$ on $\gamma_{\varepsilon,r}$. Hence, by the maximum principle we obtain that $\omega > C/2$ in $H_{\varepsilon/2,r-\varepsilon}$. By the growth lemma there exists a constant $N > 0$ depending on the ellipticity constant and on K, such that $|z| < C/4$ in $d_{\varepsilon,r-N\varepsilon}$. Therefore $q > C/4$ in $H_{\varepsilon/2,r-N\varepsilon}$. Hence it follows that if q' is a solution of the equation $Lq' = 0$ in D^r, $q'|_\Gamma = 0$ and $\min_{H_{\varepsilon,r}} q' \geqslant \min_{D^r} q'$, then $\min_{H_{\varepsilon/2,r-N\varepsilon}} q' \geqslant \min_{D^{r-N\varepsilon}} q'$.

Let us set $r_1 = 1/2$, $\varepsilon_1 = 1/(4N)$. We define r_i, ε_i by induction, setting $r_i = r_{i-1} - \varepsilon_{i-1} N$, $\varepsilon_i = \varepsilon_{i-1}/2$. Let us examine the solution u. By Harnack's inequality, we have $u > C'$ in $H_{\varepsilon_1, r_1/2}$ for some C'. Let us set $\sigma = C'/2$ and write $q = u + v$. We have that $Lq = 0$ in $D^{1/2}$, $q|_\Gamma = 0$, and $\min_{H_{\varepsilon_1,r_1}} q \geqslant \min_{D^{r_1}} q$. From the argument above $q \geqslant 0$ in H_{ε_i,r_i} for any $i \geqslant 0$. As $K \cap S' \subset \bigcup_{i=1}^\infty H_{\varepsilon_i,r_i}$, $q \geqslant 0$ on $K \cap S_1$. From the fact that $\sup(u + v) > 0$ and from the strong maximum principle it follows that $u + v > 0$ in K_1.

3. Let u and v be positive solutions in K which are zero on ∂K. Let us set $K_m = K \cap Q_{2^m}$. Let $N > 1$ be a natural number and let u^*, v^* be, respectively, the restrictions of u and v to K_N. We set $u^*/\|u^*\|_{C(K_n)} = U$, $v^*/\|v^*\|_{C(K_N)} = V$, and denote by u_m and v_m the restrictions of U and V, respectively, to K_m, $m = 0, 1, \ldots, N$. Then, from part 2, and taking into account the fact that under a similarity transformation the equation is transformed into an equation of the same form, we find that there exists a number σ, $0 < \sigma < 1$ such that $u_m - \sigma v_m|_{K_m} > 0$, $m = 0, 1, \ldots, N-1$. Let a_m be equal to the lowest upper bound of α such that $u_m - \alpha v_m > 0$ in K_{m-1}, $m = 1, 2, \ldots, N$. If at a point of K_{m-1} $u_m - a_m v_m = 0$, then by the strong maximum principle $u_m - a_m v_m \equiv 0$ in K_{m-1}, and, in particular, for some a, $u \equiv av$ in K_0. Let us assume that $u_m - a_m v_m|_{K_{m-1}} > 0$ for $m = 1, \ldots, N$. As $\sigma u_m - v_m|_{K_{m-1}} < 0$, $\sigma \leqslant a_m \leqslant 1/\sigma$ for all $m = 1, \ldots, N$, and, therefore $a_{m_1}/a_{m_2} \leqslant 1/(\sigma^2)$, $m_1, m_2 = 1, \ldots, N$. Moreover,

$$\frac{u_{m-1} - a_m v_{m-1}}{\|u_m - a_m v_m\|_{C(K_m)}} - \sigma \frac{u_m}{\|u_m\|_{C(K_m)}} < 0 \quad \text{in } K_{m-1}.$$

We denote $\dfrac{\|u_m - a_m v_m\|_{C(K_m)}}{\|u\|_{C(K_m)}} = \delta_m$, so that $u_{m-1} - a_m v_{m-1} - \sigma\delta_m v_{m-1}$ in K_{m-1}, or

$u_{m-1} - \dfrac{a_m}{1 - \sigma\delta_m} v_m > 0$ in K_{m-1}, and thus in K_{m-2} as well, so that $a_{m-1} > \dfrac{a_m}{1 - \sigma\delta_m}$.

Since $-\dfrac{1}{\sigma}\dfrac{u_m}{\|u_m\|_{C(K_m)}} + \dfrac{u_{m-1} - a_m v_{m-1}}{\|u_m - a_m v_m\|_{C(K_m)}} < 0$ in K_{m-1}, and $a_{m-1} > a_m$, we have that a fortiriori $-\dfrac{1}{\sigma}\dfrac{u_{m-1}}{\|u_m\|_{C(K_m)}} + \dfrac{u_{m-1} - a_{m-1} v_{m-1}}{\|u_m - a_m v_m\|_{C(K_m)}} < 0$ in K_{m-1}, from which it follows that $\delta_{m-1} < \sigma^2 \delta_m$.

We shall show that $\|u_1 - a_1 v_1\|_{C(K_1)} \leqslant \dfrac{1 - \sigma^{2/(N-1)}}{\sigma} \|u\|_{C(K_1)}$. Let us assume that $\|u_1 - a_1 v_1\|_{C(K_1)} > \dfrac{1 - \sigma^{2/(N-1)}}{\sigma} \|u\|_{C(K_1)}$. Then $\delta_1 > \dfrac{1 - \sigma^{2/(N-1)}}{\sigma}$, and as $\delta_{m+1} > \delta_m$, $\delta_m > \dfrac{1 - \sigma^{2/(N-1)}}{\sigma}$.

We have shown above that $a_{m-1} > \dfrac{a_m}{1 - \sigma\delta_m}$, and therefore $a_{m-1} > \dfrac{a_m}{\sigma^{2/(N-1)}}$, and thus $a_1 > a_N/\sigma^2$. But earlier it was shown that $a_{m_1}/a_{m_2} \leqslant 1/\sigma^2$ for all m_1, m_2, so that $a_1 \leqslant a_N/\sigma^2$, which is a contradiction.

This means that $\|u_1 - a_1 v_1\|_{C(K_1)} \leqslant \dfrac{1 - \sigma^{2/(N-1)}}{\sigma} \|u\|_{C(K_1)}$. Since N was arbitrary, it follows from this that for some a $u \equiv av$ in K_0. Thus in all the cases $u \equiv av$ in K_0. As we remarked earlier, under a similarity transformation the solution is transformed into a solution of an equation with the same properties. Therefore $u \equiv av$ in K, and our proposition is proved. □

Corollary. *If $u(x)$ is a solution of the equation $Lu = 0$ in the upper half-space $x_n > 0$, which is zero for $x_n = 0$, then $u(x)$ is a linear function.*

1.5. Phragmen-Lindelöf Type Theorems. A number of theorems dealing with the behaviour of solutions of elliptic equations in unbounded domains, which generalize the classical Phragmen-Lindelöf theorem, have been proved in Landis [1963, 1971, 1979]. The Phragmen-Lindelöf theorem says that if $u(x)$ is a harmonic function in the strip $\Pi = ((x, y)\colon 0 < y < \pi, -\infty < x < \infty)$, which is continuous in $\bar{\Pi}$ and non-positive on the boundary of Π, then either $u \leqslant 0$ in Π or

$$\varliminf_{t\to\infty} \max_{|x|=t} u(x, y)^{-t} e > 0.$$

In the works quoted above certain restrictions were imposed on the coefficients. These restrictions were used only in the proof of the growth lemma, while the main results were derived from the growth lemma without making direct use of the restrictions on the coefficients. Since at present the growth lemma in terms of the measure (see Section 1.4) is proved without any restriction on the coefficients of a uniformly elliptic operator, we have corresponding results for equations with coefficients satisfying the inequalities (2).

We shall say that an unbounded domain $G \subset \mathbb{R}^n$ is a cylinder type domain with parameters $R > 0$ and ε_0, $0 < \varepsilon_0 < 1$, if meas $(Q_R^x \setminus G) > \varepsilon_0 R^n$, $\forall x \in \mathbb{R}^n$.

Theorem 1.13. *If $G \subset \mathbb{R}^n$ is a cylinder type domain with parameters R and ε_0, and $u(x)$ is a solution of the equation $Lu = 0$ in G which is continuous in $\bar{G}$ and non-positive on ∂G, then either $u \leqslant 0$ in G, or $\underline{\lim}_{r\to\infty} M(r)e^{-Cr/R} > 0$, where $M(r) = \max_{|x|=r} u(x)$, and $C > 0$ depends on ε_0, λ, M, and n.*

We shall say that an unbounded domain $G \subset \mathbb{R}^n$ is a cone type domain with parameter e_0, if for sufficiently large integers N meas $(Q^0_{2^N}\setminus G) > \varepsilon_0$ meas $Q^0_{2^N}$.

Theorem 1.14. *If $u(x)$ is a solution of equation (16) in a cone type domain G with parameter ε_0, which is continuous in $\bar{G}$ and non-positive on ∂G, then either $u \leqslant 0$ in G, or*

$$\underline{\lim}_{r\to\infty} M(r)r^{-C(\varepsilon_0)} > 0,$$

where $C(\varepsilon_0) >$ depends on ε_0, λ, and n.

We quote another theorem of the same type which extends the *Hadamard theorem* about the connection between the growth of an *entire function* and the density of its zeros, to the case of solutions of elliptic equations.

Theorem 1.15. *Let $u(x)$ be a solution of equation* (16) *in $\mathbb{R}^n$. Let there be N regions in which $u(x)$ is of one sign. Then*

$$\underline{\lim}_{r\to\infty} M(r)r^{cN^{1/(n-1)}} > 0,$$

where $C =$ const > 0 depends on λ and n.

We also state a different version of this theorem. Consider $u(x)$, a solution of equation (16) in $\mathbb{R}^n$. Assume that for N sufficiently large, the number of regions in which $u(x)$ is of one sign in the ball Q^0_N, is large than N. Then $\underline{\lim}_{t\to\infty} M(r)e^{-Cr} > 0$, where $C =$ const > 0 depends on λ, M, and n.

These theorems follows quite easily from the growth lemma (see Landis [1963], where they are proven under the assumption that the coefficients of the equation are smooth functions).

Theorem 1.14 can be modified as follows:

Theorem 1.16. *Let $G' \subset \mathbb{R}^n$ be a cone type domain with parameter ε_0, and for R arbitrary, let there be defined a solution of equation* (16) *in $G = G'\setminus Q^0_N$, which is continuous up to $\Gamma = \partial G\setminus Q^0_N$ and zero on Γ. Then there exists $C(\varepsilon_0) =$ const. such that either $\underline{\lim}_{r\to\infty} M(r)r^{-C(\varepsilon_0)} > 0$, or $\underline{\lim}_{r\to\infty} M(r)r^{C(\varepsilon_0)} = 0$. The constant C depends on λ and n as well as on ε_0.*

A very simple and effective generalization of the Phragmen-Lindelöf theorem can be derived from the simplest version of the growth lemma, in which the set H contains a ball of commesurate with R radius (see p. 99 of the work).

Theorem 1.17. *Let $\phi(t)$, $0 \leqslant t < \infty$ be a positive continuously differentiable function with $|\phi'(t)| <$ const. For convenience we can take $|\phi'(t)| < 1/16$ (the general case is reduced to this one by a linear transformation). Let us set*

$$\Omega_\varphi = \{x \in \mathbb{R}^n : |x_n| < \varphi(\rho)\},$$

where $\rho = \left(\sum_{i=1}^{n-1} x_i^2\right)^{1/2}$. *Then the following claim holds:*

Let Ω_ϕ *contain an unbounded domain G. Let us consider in G a solution u(x) of equation* (1), *the coefficients of which satisfy condition* (2) *and* $c \leqslant 0$ *if* ϕ *is bounded, and* $b_i \equiv c \equiv 0$ *if* ϕ *is unbounded. Let* $\overline{\lim}_{\substack{x \to \partial G \\ x \in G}} u(x) \leqslant 0$. *Then either* $u \leqslant 0$ *everywhere in G, or*

$$\underline{\lim}_{r \to \infty} (M(r)/e^{\alpha \int_0^r dt/\varphi(t)}) > 0, \qquad M(r) = \max_{|x|=r} |u(x)|,$$

where $\alpha > 0$ *is a constant depending on* λ *and M in inequality* (2), *and on n, the dimension of the space.*

Proof. Let there exist a point $x' \in G$, such that $|u(x')| = a > 0$, $u(x') = a$. Let us set

$$r_0 = |x'|, \quad r_1 = r_{i-1} + \varphi(r_{i-1}).$$

Let

$$M_i = \max_{|x|=r_i} u(x) = u(x^{(i)}), \quad |x^{(i)}| = r_i, \quad i = 0, 1, \ldots .$$

Let us consider the balls $Q^{x^{(i)}}_{8\phi(r_i)}$ and $Q^{x^{(i)}}_{2\phi(r_i)}$ with center in $x^{(i)}$ and with radii, respectively, $8\phi(r_i)$ and $2\phi(r_i)$. In $Q^{x^{(i)}}_{8\phi(r_i)}$ let us consider the set of points $x \in G$ in which $u(x) > 0$. Applying the growth lemma in the elementary formulation given above, to this set, we find that there exists $\xi > 0$ depending on λ, M, and n, such that

$$M_{i+1} > (1 + \xi)M_i, \quad i = 0, 1, \ldots .$$

Consequently, $M_i > a(1 + \xi)^i = ae^{\beta i}$, where $\beta > 0$ depends on λ, M, and n. On the other hand, in view of the condition $|\phi'(t)| < 1/16$, we have that

$$\frac{1}{8} \int_{r_i}^{r_{i+1}} \frac{dt}{\varphi(t)} < 1.$$

Therefore

$$M_i > ae^{\alpha \int_{r_0}^{r_{i+1}} dt/\varphi(t)},$$

where $\alpha > 0$ depends on λ, M, and n. Using the maximum principle, we obtain an estimate for r lying between r_i and r_{i+1}.

1.6. *s*-Capacity. Let $s > 0$, let E be a Borel set in $\mathbb{R}^n$. Let us consider on E all measures defined on some σ-algebra of subsets of E which contains all Borel sets. We call a measure μ admissible if

$$\int_E \frac{d\mu(y)}{|x - y|^s} \leqslant 1$$

outside of $\bar{E}$. Let us put $c_s(E) = \sup \mu(E)$, where the supremum is taken over all admissible measures. The number $c_s(E)$ is called the *s-capacity* of the set E.

The following properties of *s-capacity* can be easily proved (the proofs are to be found, for example, in Landis [1971]).

a) If $E_1 \subset E_2$, $c_s(E_1) \leqslant c_s(E_2)$.

b) Under the similarity transformation $x = kx'$, s-capacity of a set is multiplied by k^s.

c) s-capacity of the ball $Q_R^{x_0}$ is not less that R^s.

d) Let $s > 1$, let $Z_{\rho,h}$ be the cylinder

$$Z_{\rho,h} = \left\{x\colon \sum_{i=1}^{n-1} x_i^2 < \rho^2, 0 < x_n < h\right\}.$$

Then $c_s(Z_{\rho,h}) \geqslant Ch\rho^{s-1}$ where $c = \text{const} > 0$ depends on s.

e) If $s < n$ and $E \subset Q_1^{x_0}$, then $c_s(E) \geqslant K \operatorname{meas} E$, where $K = \text{const} > 0$ depends on s and n.

Let us set

$$e = \sup_{x \in G} \frac{\sum_{i=1}^{n} a_{ii}(x)}{\lambda_{\min}(x)},$$

where G is a domain in $\mathbb{R}^n$ and $\lambda_{\min}(x)$ is the smallest eigenvalue of the matrix $\|a_{ik}(x)\|$. Let us call e the ellipticity constant of the operator L (of form (1) with $c \leqslant 0$).

Let $s \geqslant e - 2$. If $G \subset Q_R^{x_0}$ where R is sufficiently small, then $1/|x - x_0|^s$ is a subelliptic function (see Landis [1971]). If $b_i \equiv c \equiv 0$, this function is subelliptic for all R.

We remark that e is always larger or equal to n, and if $e = n$, the equation becomes an equation, the principal part of which is a Laplacian multiplied by a bounded function with a positive lower bound.

Let $x_0 \in \partial G$ and let L be the operator (1). We call x_0 a regular boundary point if the following conditions hold. For every pair $\varepsilon_1 > 0$, $\varepsilon_2 > 0$ there exists $\delta > 0$ such that for any domain $G' \subset G$ from the facts that $Lu = 0$ and that

$$\varlimsup_{x \to \partial G' \cap Q_{\varepsilon_1}^{x_0}} u(x) < 0, \quad x \in G' \cap Q_{\varepsilon_1}^{x_0},$$

it follows that

$$u|_{G' \cap Q_\delta^{x_0}} < \varepsilon_2,$$

In the case of sufficiently smooth equation coefficients this definition of regularity coincides with the usual definition of regularity due to Wiener (see § 2.7).

Let $D \subset R^n$ be an open set with boundary Γ, $x_0 \in \Gamma$ and let $e \geqslant n$ be some number. We call the point x_0 an e-regular boundary point if the following conditions hold. For every pair $\varepsilon_1 > 0$, $\varepsilon_2 > 0$ there exists $\delta > 0$ such that for any domain $D' \subset D$ with boundary Γ', uniformly elliptic operator L defined on D' with ellipticity constant $e' \leqslant e$, and for any function $u(x)$ subelliptic with respect to L and bounded from above, from the fact that

$$\overline{\lim_{\substack{x \to \Gamma' \cap Q^{x_0}_{\varepsilon_1} \\ x \in D'}}} u(x) \leqslant 0,$$

it follows that

$$u|_{D' \cap Q^{x_0}_{\delta}} < \varepsilon_2.$$

Theorem 1.18. *Let x_0 belong to the boundary Γ of domain D. Let $e \geqslant n$ be an arbitrary number. Set $s = e - 2$. Let us denote*

$$c_s(Q^{x_0}_{4^{-m}} \setminus D) = \gamma_m.$$

Then in order that the point x_0 is e-regular, it is sufficient that the series

$$\sum_{m=1}^{\infty} 4^{ms} \gamma_m \tag{18}$$

diverges.

Thus, if a point of the boundary ∂G of a bounded domain $G \subset \mathbb{R}^n$ is e-regular, then it is regular for the Dirichlet problem for every equation $Lu = 0$, such that the ellipticity constant of the operator L does not exceed e.

This allows us to obtain sufficient conditions for regularity.

We quote here an example of e-regularity conditions for a boundary point in the case $e \geqslant n$.

Theorem 1.19. *Let $x_0 \in \partial G$, $n \geqslant 4$, $e \geqslant n$, and let the domain G be such that an orthogonal system of coordinates $y_1, \ldots, y_n$ with origin at x_0 can be introduced in such a way that for some $h > 0$ the set of points*

$$B = \{y \colon \rho < y_n / |\ln y_n|^{1/(e-3)}, \quad 0 < y_n < h\},$$

where $\rho = (\sum_{i=1}^{n-1} y_i^2)^{1/2}$ belongs to the complement of the domain G. Then the point x_0 is e-regular.

1.7. Removable and Non-essential Sets. Let $\Omega \subset \mathbb{R}^n$ be a domain, and let $K \subset \Omega$ be compact. A compact set K is called removable for a harmonic function $u(x)$ defined and bounded on $\Omega \setminus K$, if this function can be extended to K in such a way that the extension is harmonic in Q.

This definition can be modified in the following way.

Let $D \subset \Omega$ be a neighbourhood of K with smooth boundary. Let us solve the Dirichlet problem: $\Delta v = 0$ in D, $v|_{\partial D} = u|_{\partial D}$. Let us set $\omega = u - v$. We have obtained a harmonic function ω which is defined and bounded in $D \setminus K$ and zero on ∂D. The set K is removable if and only if $\omega \equiv 0$.

We shall apply the definition above to solutions of equations $Lu = 0$.

Let the operator L be defined and uniformly elliptic in a domain Ω with a smooth boundary, and let $K \subset \Omega$ be a compact set. We shall say that K is a non-essential set if from the fact that $u(x)$ is a bounded solution of the equation $Lu = 0$ in $\Omega \setminus K$ which is zero on $\partial\Omega$, it follows that $u \equiv 0$.

It is known that a necessary and sufficient condition for a set to be non-essential for Laplace's equation is the condition

$$c_{n-2}(K) = 0. \tag{19}$$

In particular, an isolated point is always non-essential for a harmonic function. In the case of a uniformly elliptic equation with discontinuous coefficients, this is not necessarily so (even if the coefficients have a single discontinuity point, the point to which K can be reduced). This was first observed by Gilbarg and Serrin [1954/56], who constructed an example of an equation in the punctured ball $Q_1^0\backslash(0)$ for which a solution is given by the function

$$u(x) = 1 - |x|^{\varepsilon}, \quad 0 < \varepsilon < 1. \tag{20}$$

Conditions on the operator which make such a phenomenon possible in $\mathbb{R}^n$, have been investigated by Miller [1966].

Remark. The function (20) belongs to the space W^2 for $n > 3$. Hence we see that the generalized solution in W^2 (in the sense in which it is defined in the introduction to this section) does not necessarily satisfy the maximum principle.

To state sufficient conditions under which a set K is non-essential for the equation $Lu = 0$, let us introduce the number

$$\bar{e} = \sup_{x \in \Omega} \frac{\sum_{i=1}^{n} a_{ii}(x)}{\lambda_{\max}(x)},$$

where $\lambda_{\max}(x)$ is the largest eigenvalue of the matrix $\|a_{ik}(x)\|$. We shall call the number $\bar{e}$ the super ellipticity constant of the operator L.

Theorem 1.20. *In order that a set K to be non-essential it is sufficient that*

$$c_{\bar{e}-2}(K) = 0.$$

The proof of this theorem follows easily from the fact that the function $U(x) = 1/|x-y|^{\bar{e}-z}$ is superelliptic.

This theorem gives a sufficient condition for non-essentiality, which, however, is not necessary. A set of non-zero $(\bar{e}-2)$-capacity can be non-essential if it is spread well enough in the space. The following theorem clarifies the situation.

Theorem 1.21 (Landis and Bagotskaya [1983]). *Let K be contained in a smooth l-dimensional manifold M_l, $2 \leqslant l \leqslant n$. For any number s, $0 < s < n-2$, there exist a constant $\delta > 0$ and a compact set K, such that $c_s(K) > 0$, and such that for any operator L, the superellipticity constant of which satisfies the inequality*

$$|\bar{e} - (s+2)| < \delta,$$

the set K is non-essential.

For any s, $l = 2 < s < m$, a compact set $K \subset M$, and a number $\delta > 0$, there exists an operator L in Ω, the superellipticity constant of which satisfies the inequality

$$|\bar{e}-(s+2)|<\delta$$

and for which the set K is essential.

§2. Equations in Divergence Form

In this section we shall consider operators of the form

$$L=\sum_{i,k=1}^{n}\frac{\partial}{\partial x_i}\left(a_{ik}(x)\frac{\partial}{\partial x_k}\right)+\sum_{i=1}^{n}b_i(x)\frac{\partial}{\partial x_i}+c(x), \tag{21}$$

where $x=(x_1,\dots,x_n)$, $a_{ik}(x)\equiv a_{ki}(x)$ are measurable and bounded functions. The operator L is called *elliptic* in a domain $G\subset\mathbb{R}^n$, if at every point of the domain the matrix $\|a_{ik}(x)\|$ is positive definite. L is called *uniformly elliptic* in G if there exist constants $\lambda>0$, $M>0$, such that

$$\lambda^{-1}|\xi|^2\leqslant\sum_{i,k=1}^{n}a_{ik}(x)\xi_i\xi_k\leqslant\lambda|\xi|^2,$$

$$|b_i(x)|\leqslant M,\quad i=1,\dots,n,\quad c(x)\leqslant M;\quad |a_{ik}(x)|\leqslant M, \tag{22}$$

$$i,k=1,\dots,n,\quad x\in\bar{G};\quad \xi\in\mathbb{R}^n.$$

A function $u(x)\in W^1(G)$ is called a *generalized solution* of the equation

$$Lu=f+\sum_{i=1}^{n}\frac{\partial f_i}{\partial x_i},\quad f\in L_2(G),\quad f_i\in L_2(G),\quad i=1,\dots,n, \tag{23}$$

if for every $v(x)\in\mathring{W}^1(G)$ we have the equality

$$\begin{aligned}E(u,v)\equiv{}&\int_G\sum_{i,k=1}a_{ik}(x)\frac{\partial v}{\partial x_i}\frac{\partial u}{\partial x_k}\,dx-\int_G\sum_{i=1}^{n}b_i(x)v(x)\frac{\partial u}{\partial x_i}\,dx\\&-\int_G c(x)v(x)u(x)\,dx+\int_G fv\,dx-\int\sum_{i=1}^{n}f_i\frac{\partial v}{\partial x_i}\,dx=0.\end{aligned} \tag{24}$$

We call a function $u(x)\in W^1(G)$ superelliptic (subelliptic) if for every $v(x)\in\mathring{W}^1(G)$, $v(x)\geqslant 0$, the following inequality holds:

$$\begin{aligned}&\int_G\sum_{i,k=1}^{n}a_{ik}(x)\frac{\partial v}{\partial x_i}\frac{\partial u}{\partial x_k}\,dx-\int_G\sum_{i=1}^{n}b_i(x)v(x)\frac{\partial u}{\partial x_i}\,dx\\&-\int_G c(x)v(x)u(x)\,dx\geqslant 0\quad(\leqslant 0).\end{aligned} \tag{25}$$

We denote by L_0 the operator L in the case $b_i\equiv 0$, $i=1,\dots,n$, $c\equiv 0$.

2.1. The Maximum Principle. We shall state now a maximum principle due to Stampacchia [1965], which holds for solutions of equations (21). We call a function $v(x) \in W^1(G)$ non-negative on $E \subset \bar{G}$ if there is a sequence v_m of $C^1(G)$ functions such that 1) $v_m \geqslant 0$ on E, 2) $v_m \to v$ in $\mathring{W}^1(G, E)$.

A function $v(x) \in W^1(G)$ satisfies the inequality $v(x) \leqslant k$, $x \in \partial G$, $k = \text{const}$, if the function $k - v(x)$ is non-negative on ∂G.

Stampacchia's maximum principle takes the following form:

Theorem 1.22. *Let $u(x) \in W^1(G)$ be a subelliptic function, let the operator L be uniformly elliptic and such that, moreover, $c(x) \leqslant 0$, $x \in G$. Then if $u(x) \leqslant k$ on ∂G,* $\operatorname{ess\,sup}_G u \leqslant \max(0, k)$.

Let $u(x) \in \mathring{W}^1(G)$ be a generalized solution of (23). Let us set ourselves the problem of estimating the solution in terms of the right-hand side. In the case $c(x) \leqslant 0$, $f \equiv 0$, such an estimate has been obtained in by Chicco [1971]:

$$\|u(x)\|_{L_2(G)} \leqslant C \sum_{i=1}^{n} \|f_i\|_{L_2(G)},$$

where C depends on G, n, λ and M of inequality (22). For the same equation an estimate of $|u(x)|$ in terms of the right-hand side has been obtained by Stampacchia [1958]:

$$\max|u| \leqslant C[\text{meas } G]^{1/p-1/n} \sum_{i=1}^{n} \|f_i\|_{L_p}, \quad p > n,$$

where C dpends on p, n, λ, and M. The best possible value of the constant C has been found by Weinberger [1962]. The same type of estimates for the solutions were obtained by Maz'ya [1969].

Next we state a theorem due to Ladyzhenskaya and Uraltseva [1973]. This theorem gives an estimate of $\max|u(x)|$ up to the boundary of the domain.

Theorem 1.23. *Let $u(x)$ be a generalized solution (23) with $f_i \in L_q(G)$, $i = 1, \ldots, n$, $f \in L_{q/2}(G)$, $q > n$. Then for any $G' \subset\subset G$, the number* $\operatorname{ess\,max}|u(x)|$ *is finite and bounded from above by a constant depending only on λ, m, $\|f_i\|_{L_q}$, $i = 1, \ldots, n$, $\|f\|_{L_{q/2}(G)}$, $\|u\|_{L_2(G)}$, and on the distance between G' and ∂G. If moreover* $\operatorname{ess\,max}|u(x)| < \infty$ *for any part S_1 of the boundary ∂G, then if G_1 is a subdomain of G positively separated from $\partial G \backslash S_1$,* $\operatorname{ess\,max}_{G_1}|u(x)|$ *is finite and bounded from above by a constant depending on λ, m, $\|f_i\|_{L_q}$, $i = 1, \ldots, n$, $\|f\|_{L_{q/2}}$, q, $\|u\|_{L_2(G)}$, and the distance between G_1 and $\partial G \backslash S_1$.*

2.2. Continuity of Generalized Solutions. One of the most important results concerning the properties of the generalized solution of the equation

$$\sum_{i,j=1}^{n} \frac{\partial}{\partial x_i}\left(a_i(x)\frac{\partial u}{\partial x_j}\right) = 0 \tag{26}$$

is the following theorem which was first proved by Nash [1958] and De Giorgi [1957].

Theorem 1.24. *For any domain G' such that $\bar{G}' \subset G$, there exist constants $K = K(\lambda, G', G)$, $\alpha = \alpha(\lambda, G', G)$ such that*

$$\|u(x') - u(x'')| \leqslant K\|u\|_{L_2(G)}|x' - x''|^{\alpha}, \tag{27}$$

where $x', x'' \in G'$.

A simpler proof of this theorem has been given by Moser [1960] (see also Landis [1971]). Moser considered the case when $b_i \equiv c \equiv 0$. Without this restriction theorem 2.2 has been proved by Ladyzhenskaya and Uraltseva [1973].

We present here Moser's method for obtaining the estimate (27). Though this proof was obtained a long time ago, its basic idea is still widely in use.

We consider the equation

$$Lu \equiv \sum_{i,j=1}^{n} \frac{\partial}{\partial x_i}\left(a_{ij}(x)\frac{\partial u}{\partial x_j}\right) = 0, \tag{28}$$

where $a_{ij} = a_{ji}$ and

$$\lambda^{-1}|\xi|^2 \leqslant \Sigma a_{ij}\xi_i\xi_j \leqslant \lambda|\xi|^2 \quad \forall \xi \in \mathbb{R}^n. \tag{29}$$

Let us denote by Q_R the ball of radius R with center at the origin. It is sufficient to prove the following fact. Let $u(x)$ be a generalized solution in W^1 defined in the ball Q_3. Then

$$\operatorname*{osc}_{Q_3} u \geqslant (1 + p) \operatorname*{osc}_{Q_1} u, \tag{30}$$

where $p > 0$ depends on λ and on the dimension n of the space.

Below, in Section 2.4, we show that the generalized W^1 solution of equation (28) in Q_3 can be approximated in the norm of that space by classical solutions of equations (28) with infinitely differentiable coefficients, and such that all the equations satisfy inequalities (29) with the same constant λ. Therefore we shall consider only such smooth solutions of equations with smooth coefficients.

We divide the proof into a number of steps:

1°. Let u be a solution of equation (28), let $f(t)$ be a function in $C^2(\inf_{Q_3} u, \sup_{Q_3} u)$ satisfying the conditions $f(t) > 0$, $f'(t) \geqslant 0$, $f''(t) \geqslant 0$. If $\varphi(t)$, $t > 0$ is a twice differentiable function such that $\varphi' \geqslant 0$, $\varphi'' \geqslant 0$, then

$$L\varphi(f(u)) \geqslant 0. \tag{31}$$

This inequality is verified by direct substitution. Let us put $f(u(x)) = z(x)$.

2°. Let $1 < R_1 < R_2 < 3$. We shall show that then

$$\int_{Q_{R_1}} |\nabla z|^2\, dx \leqslant \frac{C_1}{(R_2 - R_1)^2} \int_{Q_{R_2}} z^2\, dx, \tag{32}$$

where C_1 depends on λ and the dimension n of the space (in what follows, we denote constants depending on λ and n by C_i).

Let $\eta \in C^2(Q_3)$ be a cut-off function in Q_3 such that $0 \leqslant \eta \leqslant 1$, $\eta = 1$ in Q_{R_1}, $\eta \equiv 0$ outside of Q_{R_2}. We have that $\int_{Q_{R_2}} \eta^2 z L z\, dx \geqslant 0$. Integration by parts gives us

$$\int_{Q_{R_3}} \eta^2 \sum_{ij} a_{ij} z_{x_i} z_{x_j}\, dx + \int_{Q_{R_2}} 2\eta z \sum_{ij} a_{ij} z_{x_i} \eta_{x_j}\, dx \leqslant 0.$$

Using (29) and the inequality

$$2ab \leqslant \varepsilon a^2 + \frac{1}{\varepsilon} b^2 \quad (\varepsilon > 0), \tag{33}$$

we obtain (32).

3°. We shall use the following Sobolev embedding theorem: for $g \in W_2^1$ in the ball Q_R, $1 \leqslant R \leqslant 3$, for $p = 2n/(n-2)$, we have $\|g\|_{L^p} \leqslant C_2 \|g\|_{W_2^1}$. Let us set $p_1 = p/2$.

4°. Let $\beta > 1$. In view of (31), we can apply the result of part 2° to the function z^β, so that

$$\int_{Q_{R_1}} |\nabla(z^\beta)|^2 \, dx \leqslant \frac{C_1}{(R_2 - R_1)^2} \int_{Q_{R_2}} z^{2\beta} \, dx. \tag{34}$$

Let $\beta = p_1^k$ where k is a non-negative integer. Then, using for $k = 0$ the result of part 2° and for $k > 0$ inequality (34), we obtain

$$\int_{Q_{R_1}} |\nabla(z^{p_1^k})|^2 \, dx \leqslant \frac{C_1}{(R_2 - R_1)^2} \int_{Q_{R_2}} z^{2p_1^k} \, dx. \tag{35}$$

setting in the embedding theorem $g = z^{p_1^k}$, we find that

$$\left(\int_{Q_{R_1}} (z^2)^{p_1^{k+1}} \right)^{1/p} \leqslant C_2 \left(\int_{Q_{R_1}} |\nabla(z^{p_1})|^2 \, dx + \int_{Q_{R_1}} (z^{p_1^k})^2 \right)^{1/2}.$$

Using (35), we obtain

$$\int_{Q_{R_1}} (z^2)^{p_1^{k+1}} \, dx \leqslant \left(\frac{C_3}{(R_2 - R_1)^2} \right)^p \left(\int_{Q_{R_2}} (z^2)^{p_1^{k_1}} \right)^{p_1} dx. \tag{36}$$

Let us set $R^{(l)} = 1 + \frac{1}{2^l}$, $l = k, k-1, \ldots, 0$. We substitute l for k in (36) and set $R_2 = R^{(l)}$, $R_1 = R^{(l+1)}$. Letting now l take all the values from k to 0, we get by induction

$$\int_{Q_{R^{(k+1)}}} (z^2)^{p_1^{k+1}} \, dx \leqslant C_3^p \cdot 2^{2(k+1)p_1} \left(\int_{Q_{R^{(k)}}} (z^2)^{p_1^k} \, dx \right)^{p_1}$$

$$\int_{Q_{R^{(k+1)}}} (z^2)^{p_1^{k+1}} \, dx \leqslant C_3^{p^3} \cdot 2^{2((k+1)p_1 + kp_1^2)} \left(\int_{Q_{R^{(k-1)}}} (z^2)^{p_1^{k-1}} \, dx \right)^{p_1^2}$$

$$\cdots\cdots\cdots\cdots\cdots\cdots\cdots\cdots\cdots\cdots\cdots\cdots$$

$$\int_{Q_{R^{(k+1)}}} (z^2)^{p_1^{k+1}} \, dx \leqslant C_3^{p_1^{k+1}} \cdot 2^{2((k+1)p_1 + kp_1^2 + \cdots + p_1^{k+1})} \left(\int_{Q_R^{(0)}} z^2 \, dx \right)^{p_1^{k+1}},$$

that is,

$$\left(\int_{Q_1} (z^2)^{p_1^{k+1}} \, dx \right)^{1/p_1^{k+1}} \leqslant (C_3^{(2p_1)^{k+1}} \cdot 2^{2(p_1^{k+2}/(p_1 - 1))})^{1/p_1^{k+1}} \int_{Q_2} z^2 \, dx.$$

Letting k go to ∞, we find that

$$\sup_{Q_1} |z| \leqslant C_4 \left(\int_{Q_2} z^2 \, dx \right)^{1/2}. \tag{37}$$

5°. Multiplying u by a constant and adding another constant, we can get $\sup_{Q_3} u = 1$, $\inf_{Q_3} u = 0$, so that by the strong maximum principle $0 < u(x) < 1$ in Q_3 (we remind the reader that Q_3 is an open ball. Let $E = \{x \in Q_2 | u(x) \leqslant 1/2\}$. We can assume that meas $E \geqslant$ (meas Q_2)/2, otherwise we could consider the function $1 - u$. Let us set $f(t) = \ln \dfrac{1}{1-t}$, $0 < t < 1$. The function $f(t)$ satisfies the requirements of part 1°. Let $z(x) = f(u(x)) = \ln \dfrac{1}{1 - u(x)}$. We set $z_1(x) = z(x) - \ln 2$, then E is the set of points of Q_2 where $z_1 \leqslant 0$. Therefore, by the generalized Friedrichs' inequality (see, for example (Kondrat'ev [1967]), $\int_{Q_2 \setminus E} z_1^2 \, dx \leqslant C_4 \int_{Q_2 \setminus E} |\nabla z_1|^2 \, dx$, and since $z|_E < \ln 2$, we have that

$$\int_{Q_2} z^2 \, dx \leqslant C_5 \int_{Q_2} |\nabla z|^2 \, dx + C_5. \tag{38}$$

6°. We remark that for our choice of f, $f''/f'^2 = 1$ and $f' > 1/2$. Let η be a cut-off function in Q_3: $\eta_1 \in C^2$, $0 \leqslant \eta_1 \leqslant 1$, $\eta_1 \equiv 1$ in Q_2, and $\eta_1 \equiv 0$ in a neighborhood of ∂Q_3. Then $\int_{Q_3} \eta_1 f'(u) Lz \, dx \geqslant 0$, or

$$\int_{Q_3} \eta_1^2 f''(u) \sum_{ij} a_{ij} u_{x_i} z_{x_j} \, dx + \int_{Q_3} 2\eta_1 f'(u) \sum a_{ij} z_{x_i} \eta_{1x_j} \, dx \leqslant 0.$$

Taking into account that $f'(u) u_{x_i} = z_{x_i}$, and that $f''/(f')^2 = 1$, $f' > 1/2$, we obtain that

$$\int_{Q_3} \eta_1^2 \sum_{ij} a_{ij} z_{x_i} z_{x_j} \, dx + \int_{Q_3} 2\eta_1 \sum_{ij} a_{ij} z_{x_i} \eta_{x_j} \, dx \leqslant 0.$$

Using (29) and (33) we find that $\int_{Q_2} |\nabla z|^2 \, dx < C_6$. Together with (38) this means that $\int_{Q_2} z^2 \, dx < C_7$, and therefore, by (37), $\sup_{Q_1} z < C_8$, which concludes the proof of this theorem. □

Remark. In the proof of the theorem we did not use the fact that meas $E \geqslant$ (meas Q_2)/2. We could assume that meas $E \geqslant \alpha$ meas Q_2, where $\alpha > 0$ is some constant. This means that Moser's proof also provides us with a proof of the growth lemma in the following formulation:

Let there exist a solution of equation (28) in the ball Q_R, and let $E^- = \{x \in Q_{R/3} | u(x) \leqslant 0\}$. Let meas $E^- \geqslant \alpha$ meas $Q_{R/3}$. Let us assume that $u(x) > 0$ at some point $x \in Q_{R/3}$. Then

$$\sup_{Q_R} u \geqslant (1 + \xi) \sup_{Q_{R/3}} u, \tag{39}$$

where $\xi > 0$ depends on λ, n, and α.

This version of the growth lemma is sufficient in order to obtain the Harnack inequality by following the general construction (see Landis [1971] or Safonov [1980]).

Moser [1961] proved the Harnack inequality by appealing to the complicated John-Nirenberg lemma which uses the Euclidian structure of the space. The method of proof of the Harnack inequality we propose here does not use this lemma and therefore is applicable in more general cases, such as, for example, the Laplace-Beltrami equation on manifolds (if a number of isoperimetric conditions are satisfied).

It is possible to find examples of equation (28) which have generalized solutions that do not satisfy the Hölder condition for any given exponent $\alpha > 0$. Let $n \geqslant 3$ and $\alpha_0 > 0$. Let us consider a cone K with vertex at the origin, and let $(r, \theta_1, \ldots, \theta_{n-1})$ be the spherical system of coordinates. Let us consider the function $r^\alpha \Phi(\theta)$, where $\Phi(\theta)$ is an eigenfunction of the Beltrami operator in K with zero Dirichlet data on $Q_1^0 \cap K$, which corresponds to the eigenvalue $\mu = \alpha(\alpha + n - 1)$. If $Q_1^0 \backslash Q_1^0 \cap K$ is of small $(n-1)$-dimensional measure, then μ, and thus α, will also be small. Let us choose the cone in such a way that $\alpha < \alpha_0$. We can make a change of coordinates $y = g(x)$, $g(0) = 0$, such that the functions $g_i(x)$ satisfy the Lipschitz condition, which maps ∂K onto the hyperplane $y_n = 0$. After the transformation $y = g(x)$, Laplace's equation has the form

$$\sum_{i,j=1}^{n} \frac{\partial}{\partial y_i} A_{ij}(y) \frac{\partial u}{\partial y_j} = 0 \tag{40}$$

and is uniformly elliptic in $\mathbb{R}^n$, while the function $u = r^\alpha \Phi(\theta)$ turns out to be its solution which is zero on $y_n = 0$. Let us define u on all of $\mathbb{R}^n$, by using in the domain $y_n < 0$ its odd extension, and let us extend $A_{ij}(y)$ to this domain by even extension. As a result we have that equation (40) has for $|y| < 1$ a generalized solution that satisfies the Hölder condition with the given exponent α, and does not satisfy it with any larger exponent. Similar examples can be constructed for $n = 2$.

The following claim is only valid for a domain, the boundary of which satisfies certain smoothness conditions. Let G be a ball. Then if $f_i \in L_p$ $(i = 1, \ldots, n)$ and $u \in W_p^1$, $p > n$, then the solution of equation (3) is such that $u - f \in W^1$ is Hölder continuous in $\bar{G}$.

There exist other methods for proving De Giorgi's theorem. One of these (Landis [1967]) is based on the following theorem from function theory (see Landis [1971]). Let $0 < R_1 < R_2$, and let an open set D be contained in the annulus $Q = \{x \in \mathbb{R}^n \colon R_1 < |x| < R_2\}$. We shall say that a piecewise smooth $(n-1)$-dimensional surface Σ separates the spheres $S_{R_1} = \{|x| = R_1\}$ and $S_{R_1} = \{|x| = R_2\}$ in D if any continuous curve connecting S_{R_1} and S_{R_2}, all the points of which, apart from the ends, lie in D, intersects Σ.

Theorem 1.25 (integral mean value theorem). *Let there be defined in D a symmetric positive definite matrix* $A = \|a_{ik}(x)\|$, *coefficients of which are continuously differentiable, and eigenvalues of which are bounded by* λ^{-1} *and* λ $(\lambda \geqslant 1)$.

Then there exists a constant $C > 0$ depending on λ, such that for any twice differentiable function $f(x)$ defined in D, there exists a piecewise-smooth surface Σ separating S_{R_1} and S_{R_2} in D, and such that

$$\int_\Sigma \left|\frac{\partial f}{\partial \nu}\right| ds \leqslant C \frac{\operatorname{osc}_D f \cdot \operatorname{meas} D}{(R_2 - R_1)^2}, \tag{41}$$

where $\frac{\partial}{\partial \nu} = \sum_{i,k=1}^n a_{ik}(x) \cos(O\hat{x}, n) \frac{\partial}{\partial x_k}$ — is the derivative co-normal to Σ.

2.3. The Harnack Inequality. The following theorem holds for a generalized solution of the equation $Lu = 0$.

Theorem 1.26 (the Harnack inequality). *Let $R_0 > 0$, let $u(x)$ be a positive solution of equation* (26) *in the ball $Q_{2R}^{x_0}$, $2R < R_0$. Then*

$$\sup_{Q_R^{x_0}} u(x) / \inf_{Q_R^{x_0}} u(x) < C, \tag{42}$$

where the constant $C > 0$ depends on λ and M in conditions (22).

This theorem was first proved by Moser [1964] in the case $b_i \equiv c \equiv 0$. Later Stampacchia [1965] established its validity for a generalized solution of the full equation $Lu = 0$. In Stampacchia [1965] this theorem is even proved for a generalized (in the sense of an integral identity) solution (belonging to W^1) of the more general equation

$$\sum_{i,j=1}^n \frac{\partial}{\partial x_i}\left(a_{ij}\frac{\partial u}{\partial x_j} + d_i u\right) + \sum_{i=1}^n b_i \frac{\partial u}{\partial x_i} + cu = 0, \tag{43}$$

where a_{ij} satisfy conditions (22), $b_i \in L_n$, $d_i \in L_p$, $c \in L_{p/2}$, $p > n$.

In the case of equation (26), $L_0 u = 0$, the constant C in Harnack's inequality (42) is independent of R_0.

The Harnack inequality is equivalent to the following two inequalities

$$\frac{\sup_{Q_{2R}} u - \sup_{Q_R} u}{\sup_{Q_R} u - \inf_{Q_R} u} > A \quad \text{and} \quad \frac{\inf_{Q_R} u - \inf_{Q_{2R}} u}{\sup_{Q_R} u - \inf_{Q_R} u} > A, \tag{44}$$

where u is a generalized solution in the ball Q_{2R}.

In the case of u being a solution of equation (23), we impose on R the restriction $R < R_0$, and the constant $A > 0$ then depends on λ, M, and R_0. In the case of u solving equation (43), R is arbitrary, and $A > 0$ depends only on λ and M.

A corollary of the Harnack inequality is the following one-sided Liouville theorem.

Theorem 1.27 (Liouville's theorem). *If a generalized solution of equation* (26) *in $\mathbb{R}^n$ is bounded either from below or from above, then it is constant.*

We shall prove now a stronger claim.

Theorem 1.27′. *There exists an $\alpha > 0$ depending on λ and M in condition* (22) *such that if $u(x)$ is a generalized solution of equation* (26) in $\mathbb{R}^n$, *and for sufficiently large R either*

$$\sup_{|x|<R} u < R^\alpha \tag{45}$$

or

$$\inf_{|x|<R} u > -R^\alpha, \tag{45'}$$

then $u \equiv$ const.

Proof. It is sufficient to consider the case (45). Let $R_0 > 0$ be an arbitrary number. Let us put $R_k = R_0 \cdot 2^k$, $k = 1, 2, \ldots$ Let us assume that

$$\sup_{|x|<R_0} u(x) - \inf_{|x|<R_0} u(x) = a.$$

From inequality (44) it follows that

$$\sup_{|x|<R_k} u(x) \geqslant a(1 + A)^k = a \cdot 2^{\ln(1+A)/(\ln 2)} > aR_k^a$$

for α depending on A (and therefore on λ and M) and sufficiently small, and for k sufficiently large. Hence it follows that for such α, $a \equiv 0$, and therefore $u \equiv$ const in the ball $Q^0_{R_0}$. From the fact that R_0 was arbitrary, it follows that $u \equiv$ const in $\mathbb{R}^n$. □

Note that for every $\alpha > 0$ we could construct an example of equation (26) in $\mathbb{R}^n$, the solution of which satisfies

$$|u(x)| < |x|^\alpha$$

for all x. This construction is similar to the construction of the example on page 117.

We state some other elementary consequences of the Harnack inequality.

Theorem 1.28. *Let G be the semi-infinite cylinder,*

$$G = \left\{x \in \mathbb{R}^n \colon x_1 > 0, \sum_{i=2}^n x_i^2 < h^2 < R_0^2\right\}$$

and let there be defined in G a solution u of equation (43), *such that $u|_{\partial G} = 0$. Let us set*

$$M(t) = \max_{x_1=t} u(x).$$

$$\left(\sum_{i=2}^n x_i^2\right)^{1/2} \leqslant \frac{h}{2}$$

Then there exists a constant $B > 0$ depending on λ, M of condition (22), *and on R_0, such that for t large enough $e^{-(B/h)t} < M(t) < e^{(B/h)t}$. If instead of* (43), *the simpler equation $L_0 u = 0$ is considered, then condition $h^2 < R_0^2$ is not required, and consequently, B will depend only on λ and M.*

Theorem 1.29. *Let G be the cone*

$$G = \left\{x \in \mathbb{R}^n : x_1 > 0, \left(\sum_{i=2}^{n} x_i^2\right)^{1/2} < bx_1, b > 0\right\},$$

and let there be defined in G a solution u of equation (43), $u|_{\partial G} = 0$. *If we set*

$$M(r) = \sup_{|x|=r} u(x),$$

$$\left(\sum_{i=2}^{n} x_i^2\right)^{1/2} < b^{r/2}$$

then for sufficiently large r

$$r^{-\alpha/b} < M(r) < r^{\alpha/b},$$

where $\alpha > 0$ is a constant depending on λ and M.

2.4. On the Approximation of Solutions by Solutions of Equations with Infinitely Differentiable Coefficients

Theorem 1.30. *Let $G \subset \mathbb{R}^n$ be a bounded domain, and let u(x) be a generalized solution of equation* (23) *in G. Assume that there exists a sequence of equations*

$$\sum_{i,j=1}^{n} \frac{\partial}{\partial x_i}\left(a_{ij}^{(m)} \frac{\partial u_m}{\partial x_j}\right) + \sum_{i=1}^{n} b_i^{(m)}(x) \frac{\partial u_m}{\partial x_i} + c^{(m)} u_m = 0, \qquad (43^m)$$

the coefficients $a_{ik}^{(m)}$, $b_i^{(m)}$, $c^{(m)}$ of which satisfy condition (22) *with the same constants λ and M as do the coefficients of a_{ik}, b_i, c of the original equation* (43), $c \leqslant 0$. *Let $a_{ik}^{(m)}$, $b_i^{(m)}$, $c^{(m)}$ converge, respectively, to a_{ik}, b_i, c almost everywhere as $m \to \infty$.*

Let u_m be a solution of equation (43^m) *satisfying the condition*

$$u_m - u \in \mathring{W}^1(G).$$

Then

$$u_m \to u \quad \text{in } W^1(G). \qquad (46)$$

Corollary 1. *In view of the a-priori estimate of the* Hölder norm, *for every subdomain $G' \subset\subset G$, the sequence u_m is precompact in $C(G')$ and therefore*

$$u_m \rightrightarrows u \quad \text{on every } G' \subset\subset G. \qquad (47)$$

Remark. For every equation (43) there exists a sequence of equations (43^m) the coefficients of which satisfy (22) with the same constants λ and M, are infinitely differentiable and converge to the respective coefficients of the given equation almost everywhere. The sequences may be obtained, for example, by averaging.

The generalized solution of an equation with infinitely differentiable coefficients is an infinitely differentiable function which satisfies the equation in the classical sense. Therefore, once we have obtained estimates depending only on the constants λ and M of inequalities (22) for classical solutions of equations with smooth coefficients, by (47) these estimates can automatically be applied to generalized solutions of equation (43). For example, it was enough to prove the a-priori estimate of the Hölder norm or the Harnack inequality for classical solutions of equations with smooth coefficients.

In what follows we shall sometimes use this argument: in some cases we shall obtain estimates directly for the generalized solution, while in others we shall obtain them for the classical solution of an equation with smooth coefficients.

We now prove Theorem 1.30 in the particular case of $b_i \equiv c \equiv 0$, that is, for the equation $L_0 u = 0$. Its proof in the general case requires a greater effort and can be found, for example, in Stampacchia [1958].

Thus, let $a_{ik}^{(m)} \to a_{ik}$ almost everywhere and $\forall \varphi \in \overset{\circ}{W}{}^1(G)$

$$\int_G \sum_{i,k=1}^n a_{ik} u_{x_i} \varphi_{x_k}\, dx = 0,$$

$$\int_G \sum_{i,k=1}^n a_{ik}^{(m)} u_{m x_i} \varphi_{x_k}\, dx = 0, \quad u_m - u \in \overset{\circ}{W}{}^1.$$

Subtracting the first equality from the second, adding and subtracting $\int_G \sum_{i,k=1}^n a_{ik} u_{m x_i} \varphi_x\, dx$ and setting $\varphi = u_m - u$, we see that

$$\int_G \sum_{i,k=1}^n a_{ik}(u_m - u)_{x_i}(u_m - u)_{x_k}\, dx$$
$$\leqslant \sqrt{\sum_{i,k=1}^n \int_G (a_{ik}^{(m)} - a_{ik})^2 u_{x_i}^2\, dx \int_G \operatorname{grad}^2 u_m\, dx},$$

and therefore by inequality (22) and by the Friedrichs inequality (see for example Landis [1971]), $\|u_m - u\|_{W^1} \to 0$ as $m \to \infty$. □

2.5. Green's Function. In this and in the two subsequent sections we shall consider for simplicity the case $n > 2$.

Generalized solutions of equation (23) have a number of properties analogous to the properties of the solutions of Laplace's equation. This is connected with the fact that equation (23) has a fundamental solution with the same singularity as the fundamental solution of Laplace's equation.

We call a function $G(x, y)$ a Green's function for operator (21) in a bounded domain Ω if is satisfies the relation.

$$\int_\Omega G\left(\sum_{i,j=1}^n \frac{\partial}{\partial x_i} a_{ij} \frac{\partial \psi}{\partial x_j} - \sum_{i=1}^n \frac{\partial}{\partial x_i} b_i \psi + c\psi\right) dx = \psi(y)$$

for every $\psi \in \overset{\circ}{W}{}^1(\Omega) \cap C^0(\bar{\Omega})$ such that

$$L^*\psi = \sum_{i,j=1}^n \frac{\partial}{\partial x_i} a_{ij} \frac{\partial \psi}{\partial x_j} - \sum_{i=1}^n \frac{\partial}{\partial x_i} b_i \psi + c\psi \in C^0(\bar{\Omega}).$$

We call $G(x, y)$ a Green's function for operator L in $\mathbb{R}^n$ if

$$\forall \varphi \in \overset{\circ}{W}{}^1(\mathbb{R}^n) \cap C^0(\mathbb{R}^n), \quad L^*\varphi \in C^0(\mathbb{R}^n),$$

and for every fixed $y \in \mathbb{R}^n$ we have the equality

$$\int_{\mathbb{R}^n} GL^*(\varphi)\,dx = \varphi(y)$$

The following theorem was proved by Stampacchia [1958].

Theorem 1.31. *There exists a Green's function $G(x, y)$ for operator (21) in a bounded domain G, and for $x, y \in G' \subset\subset G$ the following inequality holds:*

$$\frac{C_1}{|x-y|^{n-2}} \leqslant G(x, y) \leqslant \frac{C_2}{|x-y|^{n-2}}, \tag{48}$$

where $C_1 > 0$ and $C_2 > 0$ depend on constants λ and M of conditions (22) and on the distance from G' to ∂G.

This theorem generalizes the following theorem which was obtained earlier by Littman, Stampacchia, and Weinberger [1963]:

Theorem 1.31′. *There exists a Green's function $G(x, y)$ for the operator L_0 in $\mathbb{R}^n$ such that conditions (48) are satisfied with constants $C_1 > 0$ and $C_2 > 0$ depending on λ and M of conditions (22).*

We supply below two methods of proof of the latter theorem.

First method of proof. Note that it suffices to consider the case of G being the unit ball and G' being a concentric ball of radius one half, with y coinciding with the center of the ball, and $a_{ik} \in C^\infty$.

Let us consider in the annular layer $\Omega_s = \{x\colon 2^{-s} < |x| < 1\}$, $s > 2$, a solution v_s of equation (26), which is continuous in $\bar{\Omega}_s$, zero on S_1^0 and equal on $S_{2^{-s}}^0$ to a positive constant, such that for any surface separating S_1^0 and $S_{2^{-s}}^0$ in Ω_s, $\int_S \frac{\partial u}{\partial v}\,ds = 1$. (By Green's formula all such integrals are equal.)

Proposition 2.1. *There exist constants C_1 and C_2, $0 < C_1 < C_2$, depending on λ and n such that*

$$C_1|x|^{2-n} < v_s(x) < C_2|x|^{2-n}, \quad 2^{s-1} < |x| < 2^{-1}. \tag{49}$$

Proof. 1) Let $1 < k < s$. Let us use the notation $r_k = 2^{-k}$ and $M_k = \max_{|x|=r_k} v_s(x)$. Let us apply Theorem 1.25 (the integral mean value theorem) to the layer $\Omega^k = \{x\colon r_{k+1} < |x| < r_k\}$. Thus there exists a surface Σ contained in the layer which separates the boundary spheres, and such that

$$1 = \int_\Sigma \frac{\partial u}{\partial v}\,ds \leqslant C\frac{\operatorname{osc}_{\Omega_k} v_s \operatorname{meas}\Omega^k}{r_{k+1}^2} < C^* M_k r_k^{n-2}.$$

Hence $M_k < C^* r_k^{2-n}$ and, by the Harnack inequality, $v_s(x) > C_1|x|^{2-n}$, $x \in \Omega^k$.

2) Let us fix r, $2^{1-s} < r < 2^{-1}$, and let $m = \min_{|x|=r} v_s(x)$. We set $S_l = r \cdot 2^{l/\sqrt{n-2}}$, $l = 0, 1, \ldots$, and let us denote by v_l the volume of the ball of radius S_l. We set $H_l = \left\{x \in \Omega_s\colon u(x) > \frac{m}{2^l\sqrt{n-2}}\right\}$, $l = 0, 1, \ldots$.

By assumption, meas $H_0 > v_0$. On the other hand, for l sufficiently large meas $H_l < v$. Therefore there exists l_0 such that meas $H_{l_0} \geqslant v_{l_0}$ and meas $H_{l_0+1} < v_{l_0+1}$. Let us denote by E_t the level set $E_t = \{x \in \Omega_s: v_s(x) = t\}$. (By Sard's theorem for almost all t the set E_t is a smooth $(n-r)$-dimensional surface). We have that

$$\int_{m/2^{(l_0+1)}\sqrt{n-2}}^{m/2^{l_0}\sqrt{n-2}} \left(\int_{E_t} ds \Big/ \left|\frac{\partial u}{\partial n}\right|\right) dt < \text{meas } H_{l_0+1} < v_{l_0+1}.$$

Thus there exists E_{t_0}, $m/2^{(l_0+1)\sqrt{n-2}} < t_0 < m/2^{l_0\sqrt{n-2}}$, ($E_{t_0}$ is a smooth surface) $\left|\frac{\partial u}{\partial n}\right|_{|E_{t_0}} = |\text{grad } u|_{|E_{t_0}} \neq 0$, such that

$$\int_{E_{t_0}} ds \Big/ \left|\frac{\partial u}{\partial n}\right| < \frac{(2^{(l_0+1)\sqrt{n-2}} - 2^{(l_0\sqrt{n-2})})v_{l_0+1}}{m} < C' \frac{2^{l_0\sqrt{n-2}} v_{l_0}}{m} \tag{50}$$

with $C' > 0$ depending on n. As E_{t_0} contains H_{t_0}, by the isoperimetric inequality $\text{meas}_{n-1} E_{t_0} = \int_{E_{t_0}} 1 \cdot ds \geqslant (\text{meas } H_1)^{(n-1)/n} \geqslant v_{l_0}^{(n-1)/n}$.

By the Cauchy-Buniakovsky inequality,

$$\left(\int_{E_{t_0}} 1 \cdot ds\right)^2 = \left(\int_{E_{t_0}} \left(\sqrt{\left|\frac{\partial u}{\partial n}\right|} \Big/ \sqrt{\left|\frac{\partial u}{\partial n}\right|}\right) ds\right)^2$$
$$\leqslant \int_{E_{t_0}} \left|\frac{\partial u}{\partial n}\right| ds \cdot \int_{E_{t_0}} ds \Big/ \left|\frac{\partial u}{\partial n}\right|$$

or

$$\int_{E_{t_0}} \left|\frac{\partial u}{\partial n}\right| ds \cdot \int_{E_{t_0}} ds \Big/ \left|\frac{\partial u}{\partial n}\right| \geqslant v_{l_0}^{(2n-2)/n}. \tag{51}$$

Combining (50) with (51), we find that

$$\int_{E_{t_0}} \left|\frac{\partial u}{\partial n}\right| ds \geqslant C'' m v_{l_0}^{(n-2)/n} / 2^{l_0\sqrt{n-2}} = C''' \cdot r^{n-2}$$

where C'' and C''' depend on n, and as on the level surface $\left|\frac{\partial u}{\partial v}\right| > C^{\text{IV}} \left|\frac{\partial u}{\partial n}\right|$ (C^{IV} depends on λ and n), $m < C^{\text{V},z-u}$, and by the Harnack inequality $u(x) < C_2|x|^{2-n}$, $2^{1-s}|x| < 2^{-1}$. This concludes the proof of Proposition 2.1. □

Let us show that there exists a Green's function $G(x, 0)$ in the unit ball with a singularity in the center of the ball, which satisfies the estimate

$$C_1|x|^{2-n} < G(x, 0) < C_2|x|^{2-n}. \tag{52}$$

In order to prove this statement, it is enough to consider v_s as $s \to \infty$. The family $\{v_s\}$ is precompact in C^2 on each of $\bar{\Omega}_{(t)}$ for t fixed (due to inequality (49) and to interior Schauder estimates (Miranda [1970])). Here $\Omega_{(t)} = \{x: 1/t < |x| < t\}$. The limit function is in fact the desired function $G(x, 0)$.

A different method of obtaining an estimate for Green's function for the operator L_0 was given in the work of Littman, Stampacchia, and Weinberger

[1963] (historically, this method has precedence). The method uses the concept of capacity of a set relative to an operator. As above, we shall assume that the coefficients $a_{ij}(x)$ of the operator are C^∞ functions (in what follows, a passage to the limit is effected). We want to estimate, the Green's function $G(x, 0)$ for the unit ball Q_1, moreover, $y = 0$.

Let $E \subset Q_1$ be a compact set with a smooth boundary. Let us put

$$\operatorname{cap}_{L_0} E = \inf \int_{Q_1} \sum_{ij} a_{ij}(x)\varphi_{x_i}\varphi_{x_j}\, dx, \tag{53}$$

where the infimum is taken over all $\varphi \in C_0^\infty(Q_1)$ such that $\varphi|_E \geqslant 1$. From this definition it follows that if $E_1 \subset E_2$, $\operatorname{cap}_{L_0} E_1 \leqslant \operatorname{cap}_{L_0} E_2$. Let us solve the following Dirichlet problem in $Q_1 \backslash E$: $L_0\varphi_0 = 0$, $\varphi_0|_{\partial Q_1} = 0$, $\varphi_0|_{\partial E} = 1$, and extend its solution by 1 to all interior points of E. Then from the variational principle it follows that $\operatorname{cap}_{L_0} E = \int_{Q_1} \sum a_{ij}\varphi_{0x_i}\varphi_{0x_j}\, dx$. Let us use Green's formula in $Q_1 \backslash E$:

$$0 = \int_{Q_1} \varphi_0 L_0 \varphi_0\, dx = \int_{\partial E} \frac{\partial \varphi_0}{\partial \nu}\, dx - \int_{Q_1} \sum_{ij} a_{ij}\varphi_{0x_i}\varphi_{0x_j}\, dx.$$

Hence $\operatorname{cap}_{L_0} E = \int_{\partial E} \frac{\partial \varphi_0}{\partial \nu}\, ds$, where $\frac{\partial}{\partial \nu}$ is the differentiation in the direction of co-normal interior to ∂E.

Since we assumed that the coefficients of the operator L_0 are smooth, the Green's function $G(x, y)$ exists; all we want is an estimate of $G(x, 0)$ depending on λ, the dimension n of the space, and the distance to the boundary of the ball Q_1.

Let $a > 0$, and $I_a = \{x \in Q_1 \backslash \{0\} | G(x, 0) \geqslant a\}$. Using Sard's theorem, and if necessary, changing a by an arbitrarily small amount, we can assume that ∂I_a is a smooth surface. We have that $G(x, 0)|_{\partial I_a} = a$. Furthermore, since for any surface S in Q_1 encircling the center of the ball, $\int_S \frac{\partial G(x, 0)}{\partial \nu}\, ds = 1$, $\int_{\partial I_a} \frac{1}{a} \frac{\partial G(x, 0)}{\partial \nu}\, ds = \frac{1}{a}$, and therefore $\operatorname{cap}_{L_0} I_a = 1/a$.

Let Q_γ be a ball with center at 0, obtained from Q_1 by uniform shrinking with coefficient γ $(0 < \gamma_0 < \gamma < 1)$, and let $a = \min_{\partial Q_\gamma} G(x, 0)$. By the maximum principle $\bar{Q}_\gamma \subset I_a$. By the monotonicity of the capacity

$$\operatorname{cap}_{L_0} Q_\gamma \leqslant \operatorname{cap}_{L_0} I_a = 1/a = 1/\min_{\partial Q_\gamma} G(x, 0).$$

Similarly, if $b = \max_{x \in \partial Q_\gamma} G(x, 0)$, then

$$\operatorname{cap}_{L_0} \bar{Q}_\gamma \geqslant \operatorname{cap}_{L_0} I_b = 1/b = 1/\max_{\partial Q_\gamma} G(x, 0).$$

Thus

$$\min_{\partial Q_\gamma} G(x, 0) \leqslant 1/\operatorname{cap}_{L_0} \bar{Q}_\gamma \leqslant \max_{\partial Q_\gamma} G(x, 0).$$

As $G(x, 0)$ is a positive solution of the equation $L_0 u = 0$ in $Q_1 \backslash \{0\}$, by Harnack's inequality there exists $C \geqslant 1$ depending on λ, γ_0, and n such that

$$\max_{x \in \partial Q_\gamma} G(x, 0) / \min_{x \in \partial Q_\gamma} G(x, 0) < C.$$

Therefore on ∂Q_γ

$$C^{-1}(\operatorname{cap}_{L_0} \bar{Q}_\gamma) \leqslant G(x, 0) \leqslant C(\operatorname{cap}_{L_0} \bar{Q}_\gamma)^{-1}. \tag{54}$$

Since the ellipticity constant is invariant with respect to similarity transformations, we have that for $0 < \gamma < \gamma_0 < 1$, C depends on λ, γ_0, and n only.

Let us now consider the operator Δ, the ellipticity constant of which equals $1 \leqslant \lambda$. Let $G_0(x, 0) = \dfrac{A}{|x|^{n-2}}$ be its Green's function (A depends on n). According to (54)

$$C^{-1} \operatorname{cap}_\Delta \bar{Q}_\gamma \leqslant G_0(x, 0) \leqslant C(\operatorname{cap}_\Delta \bar{Q}_\gamma)^{-1}. \tag{55}$$

Furthermore, $\lambda^{-1} \operatorname{cap}_\Delta E \leqslant \operatorname{cap}_{L_0} E \leqslant \lambda \operatorname{cap}_\Delta E$. Together with (54), this gives

$$\lambda^{-2} C^{-2} \frac{A}{|x|^{n-2}} \leqslant G(x, 0) \leqslant \lambda^2 C^2 \frac{A}{|x|^{n-2}}.$$

for $x \in Q_{\gamma_0}$ which is exactly what was to be shown.

2.6. A Growth Lemma. A Version of a Phragmen-Lindelöf Type Theorem. In this section, as above, we shall consider the case $n > 2$. We shall call the capacity of a Borel set E, generated by the potential $\dfrac{1}{r^{n-2}}$, the *Wiener capacity*, and denote it by cap E.

Lemma 2.1 (growth lemma). *Let us consider $R_0 > 0$ and let R be a number such that $0 < R < \dfrac{R_0}{4}$. Let an open set D be contained in the ball Q_{4R}. We set $\Gamma = \partial D \cap Q_{4R}^{x_0}$ and let $D \cap Q_{4R}^{x_0} \neq \varnothing$. Let us consider in D a solution u of equation* (41) *which is continuous in $\bar{D}$, positive in D and zero on Γ. Then*

$$\sup_D u > \left(1 + \xi \frac{\operatorname{cap}(Q_R^{x_0} \backslash D)}{R^{n-2}}\right) \sup_{D \cap Q_R^{x_0}} u, \tag{56}$$

where ξ depends on the constants in inequality (22) *and on R_0.*

This version of the *growth lemma* which incorporates capacity was first proved by Maz'ya [1963].

Proof. By Theorem 2.10, the *Green's function* $G(x, y)$ is defined in the ball $Q_{R_0}^{x_0}$. It satisfies for $x, y \in Q_{R_0}^{x_0}$, and thus for $(x, y) \in Q_{4R}^{x_0}$, the conditions

$$\frac{C_1}{|x - y|^{n-2}} \leqslant G(x, y) \leqslant \frac{C_2}{|x - y|^{n-2}}, \tag{57}$$

where $C_1 > 0$, $C_2 \geqslant C_1$, depend on the constants of condition (22) and on R_0.

Let us set $E = Q_R^{x_0} \backslash D$ and let μ_0 be the equilibrium measure for the potential $\dfrac{1}{|x-y|^{n-2}}$ defined on that set, so that

$$\int_E \frac{d\mu_0}{|x-y|^{n-2}} \leqslant 1 \quad \text{outside of } E$$

and cap $E = \mu_0(E)$.

Let

$$V = \left(\sup_D u \left(1 - \frac{1}{C_2}\right) \int_E G(x, y)\, d\mu(y) + \frac{1}{3^{n-2}} \operatorname{cap} E\right).$$

Then

$$\varlimsup_{\substack{x \to \partial D \\ x \in D}} V > \lim_{\substack{(x,y) \to \partial D \\ (x,y) \in D}} u|_\Gamma$$

and, by the maximum principle, $u < V$.

Therefore,

$$\sup_{D \cap Q_R^{x_0}} u \leqslant \frac{C_1}{C_2}\left(\frac{1}{2^{n-2}} - \frac{1}{3^{n-2}}\right) \frac{1}{R^{n-2}} \sup_D u$$

and thus $\xi = \dfrac{C_1}{C_2}\left(\dfrac{1}{2^{n-2}} - \dfrac{1}{3^{n-2}}\right)$. □

Remark. If instead of equation (43) we consider equation (26), then, according to Theorem 1.31, the *Green's function* $G(x, y)$ is defined on $\mathbb{R}^n$ and satisfies (48) with constants depending on λ only (that is, R_0 is not needed, and the ball Q_R can be of arbitrary radius).

Definition. Let R and γ be two positive numbers. We shall say that a domain G (which in general is unbounded) has an internal diameter not exceeding R to order capacity γ if for every point $x_0 \in \mathbb{R}^n$ we have the inequality

$$\operatorname{cap}(Q_R^{x_0} \backslash G) \geqslant \gamma R^{n-2}. \tag{58}$$

Definition. Let $G \subset \mathbb{R}^n$ be an unbounded domain. We shall say that $u \in \mathring{W}^1_{\text{loc}}$ if for any $\varphi \in C_0^\infty(\mathbb{R}^n)$ $u\varphi \in \mathring{W}^1(G \cap D)$ where $D = \{x \in \mathbb{R}^n | \varphi(x) \neq 0\}$.

Theorem 1.32 (a Phragmen-Lindelöf type theorem). *Let R_0 and γ be fixed positive numbers, $0 < R < R_0$, and let $G \subset \mathbb{R}^n$ be an unbounded domain of interior diameter not exceeding R to order capacity γ. Let $u \in \mathring{W}^1_{\text{loc}}$ be a solution of equation (43) in G. Let us set*

$$M(r) = \sup_{\substack{|x| = R \\ x \in G}} |u(x)|.$$

Then if $u \not\equiv 0$, for r large enough

$$M(r) > e^{(\alpha/R)r}, \tag{59}$$

where $\alpha > 0$ depends on R_0, γ, and also on the constants λ and M of inequality (22).

Proof. Let $u \not\equiv 0$. There exists a point $x^0 \in G$ such that $u(x^0) \neq 0$. We can assume that $u(x^0) = a > 0$ (otherwise we could switch the sign of u). Let us consider the balls $Q_{2R}^{x_0}$ and $Q_R^{x_0}$. From (58) and from Lemma 2.1 (the growth lemma)

$$\sup_{Q_R^{x_0} \cap G} u > (1 + \xi\gamma)a.$$

Therefore, there exists a point $x^1 \in G$ such that $|x^1 - x^0| \leqslant R$ and $u(x^1) > (1 + \xi\gamma)a$. Considering the balls $Q_{2R}^{x_1}$ and $Q_R^{x_0}$, we shall find a point $x^2 \in G$ with $|x^2 - x^1| \leqslant R$ and $u(x^2) > (1 + \xi\gamma)^2 a$. Continuing in the same manner, we obtain a sequence $x^0, x^1, \ldots$ such that $|x^i - x^{i-1}| \leqslant R$ and $u(x^i) > (1 + \xi\gamma)^i a$. Therefore, by the maximum principle for generalized solutions, we get

$$M(r) > (1 + \xi\gamma)^{((r - |x_0|)/R) - 1} a = a(1 + \xi\gamma)^{(-|x_0|/R) - 1} \cdot e^{(\ln(1 + \xi\gamma)r)/R},$$

and if $\alpha < \ln(1 + \xi\gamma)$, then for r large enough, $M(r) > e^{(\alpha/R)r}$. □

This theorem is more subtle than the Phragmen-Lindelöf type theorem we obtained in Section 1 (Theorem 1.13).

Next we give an example of a domain with interior diameter less than R to order capacity γ. Let us consider the integer lattice in $\mathbb{R}^n$. Let us place at every point of the lattice the center of an $(n - 1)$-dimensional disc of radius not less than r, $0 < r < 1$. Every disc can be oriented in an arbitrary way. Let us denote by E the union of these (closed) discs. Then our domain is $G = \mathbb{R}^z \setminus E$.

Remark. If instead of equation (43) we consider the simpler equation (26), then in Theorem 1.32 R does not have to be bounded by R_0, thus α will depend only on γ, λ, and M.

Let $E \subset \mathbb{R}^n$ be a bounded Borel set. The capacity cap E can be introduced by a different, though equivalent, method. Let $\mathfrak{M}$ be the set of functions $f \in C_0^\infty(\mathbb{R}^n)$ which take the value one in some neighborhood of E. Then

$$\operatorname{cap} E = \inf_{f \in \mathfrak{M}} \frac{1}{\omega_n} \int_{R^n} |\operatorname{grad} f|^2 \, dx, \tag{60}$$

where ω_n is the area of the $(n - 1)$-dimensional surface of the unit sphere in $\mathbb{R}^n$.

We give another example of the use of the Lemma 2.1 (the growth lemma). For that we use the following lower bound for the Wiener capacity of a set in terms of its n-dimesional Lebesgue measure ($n > 2$). Let $E \subset Q_R^{x_0}$. Then

$$\frac{\operatorname{cap} E}{R^{n-2}} \geqslant \frac{C \operatorname{meas} E}{R^n}$$

where C is a constant depending only on the dimension of the space (see for example Landkof [1966]).

Then the following claim can be derived from Lemma 2.1:

Let $D \subset Q_{2R}^{x_0}$ $(0 < R < R_0)$ be an open set having non-empty intersection with $Q_R^{x_0}$, and let $\Gamma = \partial D \cap Q_{2R}^{x_0}$. If $u(x)$ is a generalized solution of equation (43) in D which is continuous in $\bar{D}$, positive in D and zero on Γ, then

$$\sup_D u > \left(1 + \xi \frac{\operatorname{meas} E}{R^n}\right) \sup u, \quad E = Q_R^{x_0} \backslash D,$$

where $\xi > 0$ is a constant depending on λ and μ of inequality (22) and on R_0. Using this statement and the maximum principle, we can subsequently deduce the following fact.

Lemma 2.1′. *For any $A > 0$ there exists $\varepsilon_0 > 0$ such that if $D \subset Q_R^{x_0}$, $0 < R < 1$, is an open set containing x_0, $\Gamma = \partial D \cap Q_R^{x_0}$, and $u(x)$ is a solution of equation* (43) *in D which is continuous in $\bar{D}$, positive in D and zero on Γ, and*

$$\operatorname{meas} D < \varepsilon_0 R^n,$$

then

$$\max_{\bar{D}} u > Au(x_0).$$

Let us prove now the following proposition.

Let a generalized solution of equation (43) *be defined in a domain $G \subset \mathbb{R}^n$, $G' \subset\subset G$, and $\rho(G', \partial G) = r$, $r < 1$. Then*

$$\max_{\bar{G}'} |u| \leqslant C \|u\|_{L^1(G)},$$

where $C > 0$ depends on λ, M, and r.

Proof. We set in Lemma 2.1′ $R = \dfrac{r}{2}$, $A = 2^n$ and find the corresponding ε_0. Let $\max_{\bar{G}} |u| = 2M = |u(x_0)|$. We can assume that $u(x_0) = 2M$ (if $u(x_0) = -2M$ we can change u to $-u$). Let us set $u_1 = u - M$ and $D_1 = \{x \in Q_{r/2}^{x_0} \cap G \colon u_1 > 0\}$, so that $u|_{D_1} > M$. If

$$\operatorname{meas} D_1 \geqslant \varepsilon_0 \left(\frac{r}{4}\right)^n,$$

then $\int_G u\, dx > M$, and putting $C = \dfrac{2 \cdot 4^n}{\varepsilon_0 r^n}$, we get the desired inequality. Let

$$\operatorname{meas} D_1 < \varepsilon_0 \left(\frac{r}{4}\right)^n.$$

Then there exists ρ_1, $0 < \rho_1 < \dfrac{r}{4}$ such that

$$\operatorname{meas} D_1 \cap Q_{\rho_1}^{x_0} = \varepsilon_0 \rho_1^n.$$

Applying Lemma 2.1′ to the ball $Q_{\rho_1}^{x_0}$, the set $D \cap Q_{\rho_1}^{x_0}$, and the function u, we find a point $x_1 \in G$, $x_0 - x_1| \leqslant \rho_1$ such that

$$u(x_1) > 2 \cdot 2^n \cdot M.$$

Next, put $u_2 = u - 2^2 M$, and $D_2 = \{x \in Q^{x_1}_{r/2}\colon u_2 > 0\}$. If meas $D_2 \geqslant \varepsilon^0 \left(\frac{r}{4}\right)^n$, then the proposition is proved with the same value of C as above.

Let

$$\text{meas } D_2 < \varepsilon_0 \left(\frac{r}{4}\right)^n,$$

Then there exists ρ_2, $0 < \rho_2 < \frac{r}{4}$ such that

$$\text{meas } D_2 \cap Q^{x_1}_{\rho_2} = \varepsilon_0 \rho_2^n.$$

Applying Lemma 2.1′ to the ball $Q^{x_1}_{\rho_2}$, the set $D_2 \cap Q^{x_1}_{\rho_2}$, and the function u_2, we find a point $x_2 \in G$, $|x_2 - x_1| < \rho_2$, such that

$$u(x_2) > 2 \cdot 2n \cdot M.$$

Continuing this process, we obtain a sequence $\rho_1, \ldots, \rho_k, \ldots$. Let ρ_k be the first number such that $\rho_1 + \cdots + \rho_k > \frac{r}{4}$. Such a number exists since $u(x_k) \to \infty$ for $k \to \infty$, while the function u is bounded in the ball $Q^{x_0}_{r/2}$. Therefore there exists a number i_0, $1 \leqslant i_0 \leqslant k$, such that

$$\rho_{i_0} > \frac{r}{4} \cdot \frac{1}{2^{i_0}}.$$

On the set $D_{i_0} u > 2^0 \cdot M$, and meas $D_{i_0} \geqslant \omega_n \left(\frac{r}{4}\right)^n \frac{1}{2\, i_0 n}$. Thus $\int_G u\, dx > \omega_n \left(\frac{r}{4}\right)^n$, where ω_n is the volume of the unit ball, and we obtain the inequality

$$\max_{G'} |u| \leqslant C\omega_n \int_G |u|\, dx,$$

where G is the constant characterized above. □

Kondrat'ev [1967] introduced the concept of capacity of order k connected with elliptic equations of order $2k$, which generalizes definition (60). Subsequently Maz'ya [1965] used a different (but equivalent for bounded sets) definition. Let E be a bounded Borel set in $\mathbb{R}^n$,

$$\text{cap } E = \inf_{f \in \mathfrak{M}} \int_{R^n} \sum_{|\alpha|=k} (D^\alpha f)^2\, dx,$$

where $\mathfrak{M}$ is the same family of functions as above.

In terms of this (so-called *polyharmonic*) capacity, we can derive a growth lemma and a Phragmen-Lindelöf type theorem for elliptic equations of order $2k$, in analogy with Lemma 2.1 and Theorem 1.32 (on this topic, see Landis [1974]).

2.7. The Question of Regularity of Boundary Points. Let us consider the classical Dirichlet problem for *Laplace's equation*. We have a bounded domain G, a continuous function f is given on its boundary, and we have to find a function harmonic in G, continuous in $\bar{G}$ and equal to f on ∂G. As is well known, such a problem is not always soluble. In view of the removable singularity theorem, in the punctured disc, in general, the classical Dirichlet problem cannot be solved.

In 1912 Lebesgue constructed an example of a domain in R^3 which is homeomorphic to a ball and has a boundary that is smooth everywhere except at a single cusp point. For this domain the Dirichlet problem is not soluble for some continuous boundary data.

Let us quote this example. Consider the function

$$v(x) = \sqrt{x_1^2 + x_2^2 + x_3^2} + x_1 \ln(\sqrt{x_1^2 + x_2^2 + x_3^2} - x_1).$$

This function is harmonic everywhere except on the positive part of the x_1 axis. Let us consider the level surface $v(x) = -1$. This is a surface of revolution around the x_1 axis, which touches the x_1 axis at the origin and has a cusp point there.

Let us construct a closed surface which coincides with the surface $v(x) = -1$ in a neighborhood of the origin and does not intersect the positive x_1 semi-axis. Let us denote by G the bounded region this surface encloses. Let us prescribe boundary values on ∂G which coincide with v except at the origin and equal -1 at the origin.

The boundary function is continuous, while $v(x_1, 0, 0) \to 0$ as $x_1 \to 0$, that is, v does not take the prescribed boundary values at the origin. At the same time it is easy to show that any other bounded harmonic function coinciding with v on ∂G outside of the origin, coincides with v everywhere in G. That is, a classical solution of the Dirichlet problem does not exist.

In 1924 Wiener gave necessary and sufficient conditions on the boundary of a domain under which the classical Dirichlet problem is soluble for any continuous boundary function. Moreover, he put every continuous function f in correspondence with a certain harmonic function u_f which nowadays is called a generalized solution in the sense of Wiener, and gave a necessary and sufficient condition in order that for any continuous boundary function f, $u_f(x) \to f(x_0)$ as $x \to x_0$ for any given point x_0 of the boundary. This condition depends locally on the structure of the domain in a neighborhood of the given boundary point x_0. A boundary point x_0 with the property that for any continuous function f defined on ∂G, u_f converges to $f(x_0)$ as $x \to x_0$, is called a *regular* point.

The function u_f is usually constructed by one of the two following methods:

1) Wiener's method – the domain is approximated from the interior by domains G_k with smooth boundary. The function f is continuously extended from the boundary into the interior of the domain. Let us denote the extension by F. Then the Dirichlet problems $u_k|_{\partial G_k} = F|_{\partial G_k}$ are solved. It can be proved that the sequence of such solutions u_k converges to a harmonic function which depends neither on the method of approximation of the domains, nor on the way of extending the function. This limit function is u_f.

2) Perron's method. We call a superharmonic function an upper solution if it is continuous on $\bar{G}$ and larger than f on the boundary. The infimum of all upper solutions is the desired function u_f (it coincides with the function constructed

by the Wiener method, see for example Keldysh [1941]). Perron studied this problem more or less at the same time as Wiener (1923) and independently of him. He constructed a generalized solution and established a sufficient regularity condition.

The necessary and sufficient boundary point regularity conditions given by Wiener (Wiener's criterion) consists of the following (we remind the reader that in this section we consider the case $n > 2$): let us examine a sequence of balls $\{Q_{q^k}^{x_0}\}$ with center at the point x_0 we are interested in, the radii of which decrease in a geometric progression ($0 < q < 1$). If the series

$$\sum_{k=1}^{\infty} \frac{\operatorname{cap}(Q_{q^k}^{x_0} \backslash G)}{q^{(n-2)k}} \tag{61}$$

diverges, the point is regular. Otherwise, it is irregular, that is, there exists a continuous boundary function f such that $u_f(x) \not\to f(x_0)$ as $x \to x_0$. For example, a continuous boundary function that is zero at x_0 and positive otherwise, will have this property.

We shall use the facts that for equations with infinitely differentiable coefficients the generalized solution u_f can be constructed following methods 1) or 2) (see Olejnik [1947]), and that the (boundary point) regularity condition coincides with the regularity condition for Laplace's equation (Olejnik [1949]).

Owing to the fact that for an elliptic operator in divergence form (21) there exists a *fundamental solution*, the ratio of which to the fundamental solution for the Laplacian is finite (at least in a neighborhood of the singularity), boundary points for equations in divergence form are regular (or irregular) if they are so for *Laplace's equation*.

Prior to dealing with this question, we introduce the notion of a *generalized solution in the sense of Wiener* of the Dirichlet problem for such an equation. This can be done using one of the methods indicated above, but it is more conveniently done in a different way.

Thus let $G \subset \mathbb{R}^n$ be a bounded domain, and let a function f be given on ∂G. We extend f to $\bar{G}$ in a continuous fashion, and use the same symbol f for the extension.

Had f belonged to W^1, we would have been able to find a generalized solution u in W^1 such that $u - f \in \mathring{W}^1$. It turns out that this would have been the function u_f we are after. However, not every function continuous on ∂G can be extended in a continuous way to a function in $W^1(G)$.

Let us take a sequence of infinitely differentiable functions on $\mathbb{R}^n$, the restrictions of which to $\bar{G}$, f_k, converge uniformly to f.

Let us solve the generalized Dirichlet problem for each of f_k. Thus we find the solution $u_k(x)$ of equation (26) such that $u_k - f_k \in \mathring{W}^1$. *The sequence* $\{u_k\}$ converges to a function u_f^{gen} which is a generalized solution of the given equation (43) on every subdomain $G' \subset\subset G$.

From the maximum principle for generalized solutions (see Theorem 1.22 of Section 2.1) it is easily seen that u_f^{gen} depends neither on the way we extended f, nor on the approximation of f by smooth functions we used.

Definition. A point $x_0 \in \partial G$ is called *regular* for equation (43) if for each f continuous on ∂G,

$$u_f^{\text{gen}} \to f(x_0)$$

as $x \to x_0$.

Theorem 1.33. *In order that a point $x_0 \in \partial G$ be regular for an equation of form* (43), *it is necessary and sufficient that it is regular for Laplace's equation* (*that is, if Wiener's criterion is satisfied*: *the series* (61) *diverges* (Stampacchia [1965])).

We shall prove sufficiency of this condition. Take any two $q_1, q_2, 0 < q_1 < 1$, $0 < q_2 < 1$. The series (61) either diverges or converges for both of them. Therefore we can choose a specific q. It is convenient to put $q = 1/4$. Thus let

$$\sum_{k=1}^{\infty} \operatorname{cap}(Q_{4^{-k}}^{x_0} \backslash G) 4^{(n-2)k} = \infty. \tag{62}$$

If we prove that for any f that is continuous on ∂G, and every $\varepsilon > 0$, there exists $\delta > 0$ depending only on f and on the constants λ and M of inequality (22), such that for every equation of form (43) with infinitely differentiable coefficients satisfying (22) with the same λ and M, its generalized solution in the sense of Wiener u_f satisfies the inequality

$$|u_f(x) - u_f(x_0)| \leqslant \varepsilon$$

for

$$|x - x_0| < \delta,$$

then we would have proved that for any equation (43) with bounded measurable coefficients that satisfy (22) with given λ and M

$$|u_f^{\text{gen}}(x) - f(x_0)| \leqslant \varepsilon$$

for

$$|x - x_0| < \delta,$$

which follows directly from Theorem 1.30 of Section 2.4.

Thus, let $u_f(x)$ be a generalized solution in the sense of Wiener of equation (43) with infinitely differentiable coefficients. Let us extend f continuously to $\bar{G}$. Let F be the extension.

We assume for simplicity that $c(x) \equiv 0$. The presence of the term $c(x)u$ in the equation, even independently of the sign of $c(x)$ leads to a non-essential complication in the proof (instead of the usual *maximum principle*, a *generalized maximum principle* for domains of small diameter (see Theorem 1.4) has to be used). Thus let $c(x) \equiv 0$. Then

$$v = u_f - f(x_0) - \frac{\varepsilon}{2}$$

is a generalized solution in the sense of Wiener for the function $f_1 = f - f(x_0) - \frac{\varepsilon}{2}$, $v = u_{f_1}$. As the continuous extension of f on $\bar{G}$ we can take $F_1 = F - f(x_0) - \frac{\varepsilon}{2}$. There exists a δ_1 such that $F_1 < 0$ for $x \in Q_{\delta_1}^{x_0} \cap \bar{G}$. Let us consider a sequence G_k of domains that approximates G from the interior and have a smooth boundary. Solutions of (classical) Dirichlet problems $u_k|_{\partial G_k} = F_1|_{\partial G_k}$ converge to $u_{f_1}(x)$. Therefore, if we show that there exists $\delta_2 > 0$, such that

$$u_k(x) < \frac{\varepsilon}{2}$$

for $|x - x_0| < \delta_2$, then

$$u_f \leqslant f(x_0) + \varepsilon \tag{63}$$

for $|x - x_0| < \delta_2$.

Let l_0 be the smallest integer such that $4^{-l_0} \leqslant \delta_1$. Then for $l \geqslant l_0$

$$u_k|_{\partial G_k \cap Q_{4^{-l}}^{x_0}} < 0 \quad (\text{if } G_k \cap Q_{4^{-l}}^{x_0} \neq \varnothing).$$

Let us set $D_l = G_k \cap Q_{4^{-l}}^{x_0}$ and $M_l = \sup_{D_l} u(x)$. Let us select l_1 such that $M_{l_1} \geqslant \varepsilon/2$ (if such $l_1 > l_0 + 1$ does not exist, we can set $\delta_2 = 4^{-(l_0+1)}$). Let $l_0 < l < l_1$. Let us apply Lemma 2.1 (the growth lemma) to $Q_{4^{-l}}^{x_0}$, $Q_{4^{-(l+1)}}^{x_0}$, and D to obtain

$$M_{l+1} > (1 + \xi \operatorname{cap}(Q_{4^{-l}}^{x_0} \backslash G)) 4^{-l(n+2)} M_l.$$

Let us set $\max_{\bar{G}} F = M$. Then

$$2M \geqslant \max_{\bar{G}} F_1 \geqslant \frac{\varepsilon}{2} \prod_{l=l_0}^{l_0} (1 + \xi \operatorname{cap}(Q_{4^{-l}}^{x_0} \backslash G) 4^{-l(n-2)})$$

$$\geqslant \frac{\varepsilon}{2} a \sum_{l=l_0}^{l_1} 4^{-l(n-2)} \operatorname{cap}(Q_{4^{-l}}^{x_0} \backslash G)$$

where $a > 0$ depends on ξ, that is, on λ and M. This gives an upper bound for l_1: $l_1 \leqslant l_2$, and we may set $\delta_2 = 4^{-l_2}$.

Thus we have proved (63). In a similar way we find δ_3 such that

$$u_f \geqslant f(x_0) - \varepsilon$$

for $|x = x_0| < \delta_3$, and put $\delta = \min(\delta_2, \delta_3)$.

We shall not prove here the necessity of condition (61) in order that a point x_0 be regular. The idea of the proof, however, is simple. Let

$$\sum_{k=1}^{\infty} 4^{-k(n-2)} \operatorname{cap}(Q_{4^{-k}}^{x_0} \backslash G) < \infty$$

and let the number k_0 be such that

$$\sum_{k=k_0}^{\infty} 4^{-k(n-2)} \operatorname{cap}(Q_{4^{-k}}^{x_0} \backslash G) < \frac{1}{2}.$$

Let us prescribe a continuous non-negative boundary function f such that $f = 1$ for $x \in \partial G \cap Q^{x_0}_{4^{-(k_0+1)}}$, and $f = 0$ for $x \in \partial G \cap Q^{x_0}_{4^{-k_0}}$. Let μ_0 be the equilibrium measure for the set $Q^{x_0}_{4^{-k_0}} \backslash Q^{x_0}_{4^{-(k_0+1)}} \backslash G$ and let $u_k(x)$ be the potential corresponding to it. Let us set

$$U(x) = \sum_{k=k_0+1}^{\infty} U_k(x).$$

It can be shown that $u_f(x) \leqslant U(x)$ and that in any neighborhood of the point x_0 there are points x in G where $U(x) < 1/2$.

From the proof given above it is seen that at a regular point the *modulus of continuity* of the solution can be determined. It will depend on the *modulus of continuity* of f, on $\max |f|$ and on the rate at which the series (61) diverges. The first to establish this was Maz'ya [1966]. We quote here the relevant theorem.

Theorem 1.34. *Let $x_0 \in \partial G$ be a regular boundary point, let f be a function continuous on ∂G, and such that for some $\delta > 0$*

$$|f(x) - f(x_0)| \leqslant \omega(|x - x_0|) \quad \textit{for} \quad |x - x_0| < \delta,$$

where $\omega(t) \downarrow 0$, $\omega''(t) \leqslant 0$ for $t > 0$.

Then for $|x - x_0| < \delta$

$$|u_f(x) - f(x_0)| \leqslant K\left(e^{-\beta \int_{|x-x_0|}^{\delta} \frac{\operatorname{cap}(Q^{x_0}_\rho \backslash G)\, d\rho}{\rho^{n-1}}} + \omega(\delta)\right) \ldots,$$

where $\beta > 0$ depends on λ, M, δ, and K depends on $\max_{\partial G} |f|$ *as well.*

2.8. Stability of Solutions of the Dirichlet Problem. Let us next consider the question of stability of solutions of the Dirichlet problem for equation (43) under perturbations of the boundary of the domain. In the following study of the question of stability of solutions to the Dirichlet problem we shall consider domains G, the boundary of which contains no interior points of $\bar{G}$, that is domains such that in a neighborhood of any boundary point there are points of the complement of $\bar{G}$. We restrict ourselves to the case of Laplace's equation, that is, when $a_{ij} = \delta_{ij}$; $b_i(x) \equiv 0$, $c(x) \equiv 0$, $f(x) = f_i(x) \equiv 0$. It is easily seen that this discussion extends easily to general equations of form (43). We introduce here definitions given by Keldysh [1941] and describe the main results he obtained in this direction.

Thus, let a bounded domain G have no interior boundary points. Let us consider a sequence of domains G_k, $k = 1, \ldots,$ with infinitely smooth boundary which contain $\bar{G}$ and approximate G from the outside. Let $f(\mathscr{P}) \in C(\partial G)$, $\varphi(\mathscr{P}) \in C(\mathbb{R}^n)$, $f(\mathscr{P}) = \varphi|_{\partial G}$. Let us consider a sequence of harmonic functions $U_{k,\varphi}$ in G_k such that $U_{k,\varphi}|_{\partial G_k} = \varphi$. It is not difficult to show that the sequences $U_{k,\varphi}$ converge in $\bar{G}$, and that moreover, the convergence is uniform in the interior of G, and that the limit function $U_f(\mathscr{P})$ depends neither on the choice of the function $\varphi(\mathscr{P})$, nor on the choice of the sequence G_k.

Let u_f be the generalized solution in the sense of Wiener of the Dirichlet problem in G. We shall say that the Dirichlet problem is stable in G if for every continuous function $f(\mathscr{P})$ in sequence $U_{k,\varphi}(\mathscr{P})$ converges in $\bar{G}$ uniformly, then the Dirichlet problem is stable in $\bar{G}$.

In the study of the stability question of solutions of the Dirichlet problem in a domain, an important role is played by the concept of a *point of stability* of the boundary of a domain. A point $\mathscr{P} \in \partial G$ is called a point of stability if for every continuous function $f(x)$

$$U_f(\mathscr{P}) = \lim_{k\to\infty} U_{k,\varphi}(\mathscr{P}) = f(\mathscr{P}).$$

The following proposition holds: a point of stability is regular in the sense of Wiener. To show that, let $Q \in \partial G$ be a point of stability. To prove regularity of the point $f(\mathscr{P})$ in the sense of Wiener, it is sufficient to consider the case when $f(\mathscr{P})$ coincides on ∂G with the values of some function $\varphi(\mathscr{P})$ defined on the neighborhood of $\bar{G}$ and subharmonic there. From the subharmonicity of $\varphi(\mathscr{P})$ it follows that in G

$$\varphi(\mathscr{P}) \leqslant u_f(\mathscr{P}) \leqslant U_f(\mathscr{P}). \tag{64}$$

From the definition of a point of stability it follows that for k large enough we have the inequality

$$f(Q) \leqslant U_{k,\varphi}(\mathscr{P}) < U_f(Q) + \varepsilon = f(Q) + \varepsilon.$$

Fixing such a k let us select a ρ-neighborhood of the point Q in which the inequality $U_{k,\varphi}(\mathscr{P}) < f(Q) + \varepsilon$ still holds. The functions for $U_{k,\varphi}$ decrease with increasing k, therefore $U_{k^*,\varphi}(\mathscr{P}) < f(Q) + \varepsilon$ for all $k^* > k$. This shows that $\limsup_{\mathscr{P}\to Q} U_f(\mathscr{P}) \leqslant f(Q)$. From (64) we deduce that $\liminf_{\mathscr{P}\to Q} U_f(\mathscr{P}) \geqslant f(Q)$ and therefore $\lim_{\mathscr{P}\to Q} U_f(\mathscr{P}) = f(Q)$, which proves regularity of the point Q.

It is possible to formulate a necessary and sufficient condition of stability of a boundary point in terms of the *Wiener capacity*, which is similar to the necessary and sufficient condition of regularity in the sense of Wiener. Such a condition has been obtained by Keldysh and consists of the following.

Let $Q \in \partial G$, let λ_k be the capacity of the open set 0_k, the points of which belong to the complement of $\bar{G}$ such that their distance from the point Q is between $2^{-(k+1)}$ and 2^{-k}. The point Q will be stable if and only if the series $\sum 2^{-k(n-2)}\lambda_k$ diverges.

Let us move on to the question of stability of solutions of Dirichlet problems in a domain. An important role in this question is played by the concept of *harmonic measure*.

Let us introduce this concept. Let $G \subseteq \mathbb{R}^n$ and let F be a closed set contained in ∂G. Let us consider a family $\{U\}$ of functions that are superharmonic in G and satisfy the following conditions:

1) For any point $Q \in F$ the limiting values as $\mathscr{P} \to Q$ of any function $U(\mathscr{P})$ from this family are less than one.

2) $U(\mathscr{P}) \geqslant 0$, $\mathscr{P} \in G$.

We define the harmonic measure of the set F at a point $\mathscr{P}$ of a domain G to be the infimum of the values taken by functions of the family $\{U\}$ at $\mathscr{P}$:

$$h(F, G, \mathscr{P}) = \inf\{U(\mathscr{P})\}.$$

If $F = \partial G$ and G is a bounded domain, then clearly $h(\mathscr{P}) \equiv 1$. The harmonic measure of an arbitrary set E is defined in a manner analogous to the definition of the Lebesgue measure of a set.

The following distinguishing properties of the harmonic measure are easily proved.

a) the harmonic measure $h(E, G, \mathscr{P})$ of a measurable set is a harmonic function of the point $\mathscr{P}$.

b) If $h(E, G, \mathscr{P})$ is zero at any point, then $h(E, G, \mathscr{P}) \equiv 0$.

The following proposition provides a link between the Wiener capacity and harmonic measure. If a set $E \subset \partial G$ has capacity zero then the harmonic measure of E is zero.

We also have the following form of the maximum principle. If a bounded harmonic function $U(\mathscr{P})$ satisfies the inequality

$$\limsup_{\mathscr{P}\to Q} U(\mathscr{P}) \leqslant M$$

at all points of ∂G except for a set E of harmonic measure zero, then everywhere in G,

$$U(\mathscr{P}) \leqslant M.$$

To see this, let $\mathscr{P}_0$ be a point of the domain G, and let Ω be an open set contained in the boundary such that $E \subset \Omega$, and $h(\Omega, \mathscr{P}_0) < \varepsilon$. Let us set $K = \operatorname{ess\,sup}_G u(\mathscr{P})$. For all regular points of the boundary we have

$$u(\mathscr{P}) \leqslant M + (K - M)h(\Omega, G, \mathscr{P}), \tag{65}$$

In fact, if a boundary point does not belong to Ω, we have (65) by (64), and if a regular point $\mathscr{P}$ belongs to Ω then $h(\Omega, G, \mathscr{P}) = 1$ and (65) holds again. But from (65) it follows that $u(\mathscr{P}_0) \leqslant M + (K - M)\varepsilon$, which proves our assertion.

In particular, if two bounded functions take the same values on the boundary except on a set of harmonic measure zero, then they coincide.

Using the harmonic measure of a set it is possible to give the following integral representation of a generalized in the sense of Wiener solution of the Dirichlet problem:

$$u_f(\mathscr{P}) = \int_{\partial G} f(M)\, dh(E, G, \mathscr{P}).$$

Now we can state the following criterion of stability of solutions of the Dirichlet problem in the interior of the domain G.

Theorem 1.35. *In order that the Dirichlet problem be stable in the interior of the domain G it is necessary and sufficient that the set of points of instability of the boundary of G has harmonic measure zero.*

Theorem 1.36. *If every point of the boundary of G is a point of stability, then the Dirichlet problem is stable in* $\bar{G}$.

The above results can be extended without difficulty to the case of equations of form (43), at least with $b_i \equiv 0, i = 1, \ldots, n, c \equiv 0$, due to the properties of their fundamental solutions. The analog of harmonic measure constructed with the help of the operator L is called the L-harmonic measure.

Cafarelli *et al.* [1987] give an interesting example: G is the unit circle, and the authors construct a set of positive Lebesgue measure, L-harmonic measure of which is zero. They also construct an example of a set of zero Lebesgue measure that has positive L-harmonic measure. In all these examples the coefficients a_{ij} of the operator L are continuous in $\bar{G}$.

2.9. The Neumann Condition. Trichotomy. Let D be a cylindrical domain, $D = G' \times (-\infty < x_n < \infty)$, and let us consider Laplace's equation

$$\Delta u(x) = 0, \quad x \in D, \quad \left.\frac{\partial u}{\partial n}\right|_{\partial G'} = 0,$$

where G' is a bounded region with smooth boundary in (n-1)-dimensional space $\{x'\} = \{(x_1, \ldots, x_{n-1})\}$.

If a harmonic function is defined in the semi-infinite cylinder $D^+ = \{x \in D, x_n > 0\}$ and satisfies the zero *Neumann condition* on its lateral boundary, then as $x_n \to \infty$, it approaches exponentially either a constant or a linear function, or

$$\lim_{t \to \infty} (\ln M(t)/t) > 0,$$

where

$$M(t) = \sup_{x_n = t} |u(x)|.$$

We shall use the word trichotomy to denote such a subdivision of all solutions into three classes according to their behavior as $x_n \to \infty$. It also holds for solutions of uniformly elliptic equations in divergence form.

Here we restrict ourselves to the consideration of the simpler equation

$$\sum_{i,k=1}^{n} \frac{\partial}{\partial x_i}\left(a_{ik}(x)\frac{\partial u}{\partial x_k}\right) = 0.$$

Let $G \subset \mathbb{R}^n$ be a domain, and let Γ be a relatively open set of ∂G. We shall say that a generalized solution $u \in W^1_{\text{loc}}$ of equation (43) satisfies the zero Neumann condition on Γ if $\forall \varphi \in C_0^\infty$, $\operatorname{supp}(\varphi \cdot u) \subset G \cup \Gamma$, we have the relation

$$\int_G \sum_{k=1}^{n} a_{ik} \frac{\partial u}{\partial x_i} \frac{\partial \varphi}{\partial x_k} dx = 0.$$

Let D and D^+ have the same interpretation as above. The boundary of G' is, as above, assumed to be smooth. (Probably, it is sufficient that the Poincaré inequality holds in G', but this case has not yet been considered).

We have the following theorems:

Theorem 1.37 (Lakhturov [1980]). *Let $u \in W^1_{\mathrm{loc}}$ be a generalized solution of equation (26) in D^+ that satisfies the zero Neumann condition on the lateral boundary of the semi-infinite cylinder D^+. Then there exist constants C_1, C_2, $0 < C_1 < C_2$ such that one of the following three cases occurs:*

1) $u(x) \to \mathrm{const}$ *as* $x_n \to \infty$;
2) $C_1 x_n < u(x) < C_2 x_n$ *or* $-C_2 x_n < u(x) < -C_1 x_n$ *for* x_n *sufficiently large;*
3) $\overline{\lim}_{t\to\infty} (\ln M(t)/t) > 0$.

Theorem 1.38 (Nadirashivili [1958]). *For any equation of form (26), all three classes of solutions are non-empty.*

Theorem 1.37 can be sharpened in the following manner.

Theorem 1.39 (Lakhurov [1980]). *There exists a solution $u_0 \in W^1_{\mathrm{loc}}(D^+)$ of equation (26) which satisfies the zero Neumann condition on the lateral boundary of D^+, such that*

$$C_1 x_n < u_0(x) < C_2 x_n, \quad 0 < C_1 < C_2,$$

and for any other solution satisfying the zero Neumann condition on the lateral boundary that doesn't satisfy condition 3), there exist constants a and b such that

$$u(x) = a u_0 + b + o(\exp - (\alpha x_n)),$$

where $\alpha > 0$ is a constant that depends on the ellipticity constant of the equation.

We have the following assertion in the case of a whole cylinder D.

Theorem 1.40 (Lakhturov [1980]). *There exists a solution $v_0(x)$ satisfying the zero Neumann condition on the lateral boundary of D such that $C_1 x_n < v_0(x) < C_2 x_n$ for large positive x_n and $-C_2 x_n < v_0(x) < -C_1 x_n$ for x_n negative with a large absolute value ($0 < C_1 < C_2$); any other solution $u(x)$ of this equation with zero Neumann conditions either satisfies*

$$\overline{\lim_{t\to\infty}} (\ln M(t)/t) > 0,$$

or there exist constants a and b such that $u(x) = a v_0(x) + b$.

The following theorem is similar to Theorem 1.40.

Theorem 1.41. (Lakhturov [1980]). *If in the equation $Lu = f$ the function f is finite then there exists a solution $u_0(x)$ satisfying the zero Neumann condition on the lateral surface of D such that $C_1 x_n \leqslant u_0(x) \leqslant C_2 x_n$ for $x_n > 0$ large enough and $-C_2 x_n \leqslant u_0(x) \leqslant -C_1 x_n$ for $x_n < 0$ of large enough absolute value.*

Furthermore, it is clear that if $\int_{x_n=0} \frac{\partial u}{\partial v}\, dx_1 \dots dx_{n-1}$ is suitably defined for generalized solutions, then $\int_{x_n=-t} \frac{\partial u}{\partial v}\, dx_1 \dots dx_{n-1} - \int_{x_n=t} \frac{\partial u}{\partial v}\, dx_1 \dots dx_{n-1} = \int f\, dx$ for $t > 0$ sufficiently large.

Solutions of class 2) differ from solutions of class 3) also in that for x_n sufficiently large solutions of class 2) do not change sign, while solutions of class 3) necessarily change sign for x_n arbitrarily large.

This allows us to prove trichotomy of solutions also in the case of non-cylindrical domains. Let us state one such theorem for harmonic functions (Ibragimov [1983a]).

In order to avoid introducing cumbersome definitions from (Ibragimov [1983a]) which describe the domain under investigation in terms of isoperimetric inequalities, we consider here a simple particular case that follows from the results of Ibragimov [1983a] (see also Lakhturov [1980]).

Theorem 1.42. *Let* $\varphi(t)$, $0 \leqslant t < 1$ *be a continuously differentiable function* $\psi(0) = 0$, $\psi(t) > 0$, *for* $t > 0$, $|\varphi'(t)| < K$. *Let us consider a function* $u(x)$ *which is harmonic in*

$$\Omega = \left\{ x \in \mathbb{R}^n : 0 < x_n < 1, \left(\sum_{i=1}^{n-1} x_i^2 \right)^{1/2} < \varphi(x_n) \right\}$$

and satisfies the zero Neumann condition on

$$S = \left\{ 0 < x_n < 1, \left(\sum_1^{n-1} x_i^2 \right)^{1/2} = \varphi(x_n) \right\}.$$

Let us use the notation $M(t) = \max_{x_n = t} |u(x)|$. *Then one of the following three possibilities occurs:*

1) *There exists a constant* M_0 *such that*

$$u(x) = M_0 + O\left(\exp\left(-a \int_{x_n}^1 d\tau/(\varphi(\tau))^{n-1} \right) \right),$$

where $a > 0$ *is a constant that depends on* K.

2) $\overline{\lim}_{t \to 0} (\ln M(t) / \int_t^1 d\tau/(\varphi(\tau))^{n-1}) > 0$ *and* u *changes sign in every neighborhood of the point* 0;

3) *There are constants* l_1, l_2, $0 < l_1 < l_2$ *such that for* x_n *sufficiently small*

$$l_1 \int_{x_n}^1 \frac{d\tau}{(\varphi(\tau))^{n-1}} \leqslant u(x) \leqslant l_2 \int_{x_n}^1 \frac{d\tau}{(\varphi(\tau))^{n-1}}$$

or

$$-l_2 \int_{x_n}^1 \frac{d\tau}{(\varphi(\tau))^{n-1}} \leqslant u(x) \leqslant -l_1 \int_{x_n}^1 \frac{d\tau}{(\varphi(\tau))^{n-1}}$$

and $u(x)$ *does not change sign in a neighborhood of* 0.

2.10. Zaremba's Problem. This term describes a mixed boundary value problem for a second order elliptic equation in which a Dirichlet condition is prescribed on a part of the boundary and a Neumann condition is prescribed on the rest of the boundary (see Zaremba [1910]).

Let G be a bounded domain in $\mathbb{R}^n$ and let $E \subset \partial G$ be a closed set. Let us introduce the class of functions $W_E^1(G)$. This is the subspace of $W^1(G)$ obtained by taking the closure in the $W^1(G)$ norm of the set of infinitely differentiable functions in G which vanish in a neighborhood of E. When the domain G is fixed, we simply write W^1, W_E^1 instead of $W^1(G)$, $W_E^1(G)$.

Let a solution of (43) be defined in G and let $\varphi \in W^1$ be some given function. We call a function $u \in W^1$ a generalized solution with Dirichlet data on E defined by the function φ, and with zero Neumann data on $\partial G \setminus E$, if $u - \varphi \in W_E^1$ and $\forall \varphi_0 \in W_E^1$

$$\int_G \sum_{i,k=1}^n a_{ik}(x) \frac{\partial u}{\partial x_i} \frac{\partial \varphi_0}{\partial x_k} \, dx = 0. \tag{66}$$

Let the function φ be continuous in $\bar{G}$. Let us take $x_0 \in E$ and consider the following question: when does $u(x) \to \varphi(x_0)$ for any $\varphi \in W^1 \cap C(\bar{G})$ and when is $u(x)$ continuous on $\partial G \setminus E$.

If the domain has a Lipschitz boundary, that is, if in a neighborhood of each of its points, ∂G is the graph of a function satisfying the Lipschitz condition, then this question is easily soluble. On $\partial G \setminus E$ the function u is always continuous, while the question of continuity of u at a point $x_0 \in E$ is reduced to the question of convergence or divergence of the following series:

$$\sum_{k=1}^{\infty} \frac{\operatorname{cap}(E \cap Q_{q^k})}{q^{(n-2)k}}, \quad 0 < q < 1, \quad n > 2. \tag{67}$$

If this series diverges, then $u(x) \to \varphi(x_0)$ for any φ from the class above, and if it converges, then there exists $\varphi \in W^1 \cap C(\bar{G})$ for which $u(x)$ does not converge to $\varphi(x_0)$.

In order to prove this statement we must apply a "*Lipschitz diffeomorphism*", that is, a bijective Lipschitz mapping $f: G \to G'$ that straightens the boundary in a neighborhood of x_0. Let moreover $f(x) = x'$, $f(x_0) = 0$, let the boundary in a neighborhood of x_0 be mapped into the coordinate hyperplane $x_1' = 0$, and let the image G in a neighborhood of 0 lie in the half-space $x_1' < 0$. Let us examine this neighborhood of the origin (in the new coordinate system).

Let $f(E) = E'$. Set $u'(x') = u(x)$. The function u' is a generalized solution of an equation of the same form as (43) (with some different constants in the inequality (22)). Let this equation be

$$\sum_{i,k} \frac{\partial}{\partial x_i'} \left(a_{ik}'(x') \frac{\partial u'}{\partial x_k'} \right) = 0.$$

Let us extend $u'(x')$ in x_1' in an even fashion, let us extend the coefficients a_{ik} multiplying derivatives of even order in x_1' evenly, and those multiplying first order x_1' derivatives, oddly. This extension of the solution is a generalized solution of the extended equation everywhere (in a neighborhood of the origin) outside of E'. Furthermore, the points $x' \in \{x_1' = 0\} \setminus E'$ are interior points, therefore at these points, due to the a-priori Hölder norms bound, the solution is continuous. The question of convergence of $u'(x')$ to $\varphi'(0) = \varphi(x_0)$ is thus reduced to the question of regularity of the origin, in a neighborhood of which the boundary consists of points of E', that is, we have to consider whether Wiener's series for E' diverges. Since the mapping f is Lipschitz, the corresponding series converges (or diverges), if so does the series (67).

Kerimov *et al.* [1981] consider the case in which the boundary does not satisfy the Lipschitz condition in a neighborhood of x_0, and the point itself is a limit point both for the sets of points on which a Dirichlet condition is prescribed, and for the set of points on which the Neumann condition is prescribed. They obtain sufficient conditions on the boundary in order that the generalized solution of Zaremba's problem satisfy the Hölder condition at x_0.

Kerimov [1982] considers the semi-infinite cylinder $\Omega = G \times [0, \infty)$, with the conditions $\partial u/\partial v = 0, x \in \partial\Omega \backslash F; u = 0, x \in F$ being prescribed on $\partial\Omega$. He obtains a necessary and sufficient condition under which the solution converges to zero as $|x| \to \infty$, which is similar to Wiener's condition of boundary point regularity.

§ 3. Second Order Elliptic Equations with Regular Coefficients

Up to now we considered equations with no restriction on the coefficients, apart from the uniform ellipticity condition. In this section we shall consider equations of the form

$$Lu \equiv \sum_{i,j=1}^{n} a_{ij}(x)\frac{\partial^2 u}{\partial x_i \partial x_j} + \sum_{i=1}^{n} b_i(x)\frac{\partial u}{\partial x_i} + c(x)u = 0 \tag{68}$$

or

$$L_1 u \equiv \sum_{i,j=1}^{n} \frac{\partial}{\partial x_i}\left(a_{ij}(x)\frac{\partial u}{\partial x_j}\right) + \sum_{i=1}^{n} b_i(x)\frac{\partial u}{\partial x_i} + c(x)u = 0, \tag{69}$$

with various restrictions supplementing conditions (2) and (3) imposed on the coefficients.

3.1. Dependence of the Smoothness of the Solution on the Smoothness of the Coefficients. First we mention some classical results.

If the coefficients are analytic functions, then the solutions are also *analytic* (Bernstein, Levi, Gevrey, Petrovskij).

If the coefficients of equation (68) are continuously differentiable $k > 1$ times in a domain G, then the solution is $(k-1)$ times continuously differentiable, and its derivatives up to $(k-1)^{\text{th}}$ order inclusive satisfy in every interior subdomain G' an estimate in terms of the maximum of the modulus of the solution and of the distance between G' and the boundary of G (Bernstein estimates).

A sharper result was obtained by Schauder (see Miranda [1970], Landis [1971]):

If the coefficients of equation (68) are in $C^{k,\alpha}$ $(0 < \alpha < 1)$, $k \geqslant 0$ in a domain $G \subset \mathbb{R}^n$, then in $G_\rho = \{x \in G, \operatorname{dist}(x, \partial G) > \rho\}$ the solution $u \in C^{k+2,\alpha}$ and

$\|u\|_{C^{k+2,\alpha}}$ can be estimated in terms of the $C^{k,\alpha}$ norm of the coefficients, the maximum of the modulus of the solution, and ρ.

Under the same conditions on the coefficients, if the boundary of the domain G is smooth of class $C^{k+2,\alpha}$, and if $u|_{\partial G} = \varphi \in C^{k+2,\alpha}$, then $u \in C^{k+2,\alpha}(\bar{G})$, and the norm of the solution in this function space admits an estimate in terms of the corresponding norms of the coefficients, the boundary function, and of the functions defining the boundary (Schauder estimates up to the boundary). In the case of generalized solutions of equation (69) Morrey [1943] obtained the following result: if the coefficients are in $C^{k,\alpha}(G)$, $k \geqslant 0$, $0 < \alpha < 1$, then the generalized solution belongs to $C^{k+1,\alpha}(G)$.

The results mentioned above have been extended to the case of elliptic equations of arbitrary order (Morrey [1943]).

Instead of imposing the Hölder condition on the kth derivative as in the theorems of Schauder and Morrey, one could require that a repeated Dini condition is satisfied. Then the corresponding order derivative of the solution will also satisfy the *Dini condition.* The most complete result in this direction has been obtained by Ejdel'man and Matijchuk [1970] who proved similar statements in the case of general elliptic and parabolic systems.

When the coefficients of equation (68) do not satisfy the Hölder condition in $\bar{G}$, Gilbarg and Hörmander [1980] obtained the following generalization of the Schauder estimates:

Let $G \subset \mathbb{R}^n$ be a bounded domain. Let us set, as before,

$$G_\rho = \{x \in G\colon \operatorname{dist}(x, \partial G) > \rho\}, \quad \rho > 0$$

Let a and b be real numbers such that $a \geqslant 0$ and $a + b \geqslant 0$. Set

$$\|u\|_a^{(b)} = \sup_\rho \rho^{a+b} \|u\|_{a, G_\rho}, \tag{70}$$

where $\|u\|_{\alpha, G_\rho}$ is the C^α norm of the restriction of u to G_ρ.

It can be verified that the right-hand side of (70) satisfies the norm axioms. Let us denote the completion of $C^\infty(G)$ in this norm by $H_a^{(b)}$. Let us use the notation $H_a^{(-a)} = H_a$, $\|u\|_a^{(-a)} = \|a\|$. Let $H_a^{(b-0)}$ be the set of functions u such that

$$\lim_{\rho \to 0} \rho^{a+b} \|u\|_{a, G_\rho} = 0.$$

Theorem 1.43. *Let us assume that $\partial G \in C^{1+\gamma}$, $\gamma \geqslant 0$ (this means that locally the boundary can be represented as the graph of a function of that class). Let*

$$P = \sum_{|\alpha| \leqslant 2} p_\alpha(x) D^\alpha \tag{71}$$

be a uniformly elliptic operator, let a and b be non-integer numbers such that $a > 2$, $0 < b \leqslant a$, $b \leqslant \gamma$. Let us assume that

$$p_\alpha \in H_{a-2}^{(2-b)} \quad \text{for} \quad |\alpha| \leqslant 2,$$

$$p_\alpha \in H_0 \quad \text{for} \quad |\alpha| = 2,$$

$$p_\alpha \in H_{a-2}^{(2-|\alpha|-0)} \quad \text{for} \quad b < |\alpha|.$$

Then there exists a constant $C > 0$ *such that*

$$C^{-1}(\|u\|_b + \|Pu\|_{a-2}^{(2-b)} + \|u\|_0) \leqslant \|u\|_a^{(-b)}$$
$$\leqslant C(\|u\|_b + \|Pu\|_{a-2}^{(2-b)} + \|u\|_0)$$

for all $u \in H_a^{(-b)}$.

Theorem 1.44. *Let* ∂G *be a Lipschitz surface such that each of its points can be touched from the outside by a cone with the angle of opening* θ. *Let the uniformly elliptic operator* (71) *be such that*

$$p_\alpha \in H_{a-2}^{(2)} \quad \text{for} \quad |\alpha| = 2,$$
$$p_\alpha \in H_{a-2}^{(1-0)} \quad \text{for} \quad |\alpha| = 1,$$
$$p_0 \in H_{a-2}^{(2-b)},$$

where $a > 2$ *is a non-integer number,* $0 < b < \hat{b}$, *where* $\hat{b} \leqslant 1$ *depends on* θ. *Then there exists a constant* $C > 0$ *such that*

$$C^{-1}(\|u\|_0 + \|u\|_b + \|Pu\|_{a-2}^{(2-b)}) \leqslant \|u\|_a^{(-b)}$$
$$\leqslant C\|u\|_0 + \|u\|_b + \|Pu\|_{a-2}^{(2-b)})$$

for all $u \in H_a^{(-b)}$.

If $\partial G \in C^1$ *this statement is true for all* $b < 1$.

If the coefficients are such that the spread of roots of the characteristic equation is small, then the $C^{1,\alpha}(G_\rho)$ norm of the solution can be estimated in terms of maximum of the modulus of the solution, ρ, and constants λ and M of inequalities (2) and (5). This result was obtained by Cordes [1956a].

3.2. Behavior of Solutions in a Neighborhood of a Boundary Point. A necessary and sufficient condition of regularity of boundary points for Laplace's equation was obtained by Wiener in 1924. In 1949 Olejnik [1949] showed that the boundary point regularity condition for equation $Lu = 0$ where L is an operator of form (1) with smooth coefficients, coincides with the regularity condition for Laplace's equation (this was conjectured by Petrovskij [1946]). Krylov [1967] weakened the smoothness condition on coefficients, replacing them by a *Dini condition.*

It appears that the *Dini condition* is the weakest possible under which the boundary point regularity condition coincides with the regularity condition for Laplace's equation. Novruzov [1979] constructed an example demonstrating that for every continuity modulus ω which does not satisfy the Dini condition, it is possible to construct an equation, the coefficients of which have ω as their continuity modulus and for which regularity conditions are different from regularity conditions for Laplace's equation. Novruzov and his students also showed that in the general case of an equation with continuous coefficients, necessary and sufficient conditions stated in terms of the continuity modulus of coefficients, do not coincide; he obtained these conditions. On the other hand, if some additional restrictions are imposed on the structure of the coefficients, then even in the case of discontinuous coefficients it is possible to find conditions, under which necessary and sufficient regularity conditions coincide and are identical to the corresponding conditions for Laplace's equation.

Let us quote some of these results.

We consider in a domain $G \subset \mathbb{R}^n$ the operator

$$L = \sum_{i,k=1}^{n} a_{ik}(x) \frac{\partial^2}{\partial x_i \partial x_k} + \sum_{l=1}^{n} b_i(x) \frac{\partial}{\partial x_i} + c(x) \tag{72}$$

which satisfies uniform ellipticity conditions and the maximum principle

$$\lambda^{-1} |\xi|^2 \leqslant \sum_{i,k=1}^{n} a_{ik}(x) \xi_i \xi_k \leqslant \lambda |\xi|^2, \quad \forall x \in G \quad \forall \xi \in \mathbb{R}^n, \tag{73}$$

$$|b_i(x)| < M, \quad -M < c(x) \leqslant 0 \quad \forall x \in G. \tag{74}$$

Novruzov [1983b] introduces the following function:

$$n(x, y) = \frac{\sum_{i=1}^{n} a_{ii}(x)}{\sum_{i,k=1}^{n} a_{ik}(x_i - y_i)(x_k - y_k)}, \quad x \in G, \quad y \in \partial G, \tag{75}$$

This function is called the *ellipticity function* of operator L.

Let us assume that there exist constants s_1 and s_2 such that

$$n(x, y) - s_1 \leqslant \varphi(|x - y|), \tag{76}$$

$$s_2 - n(x, y) \geqslant -\varphi(|x - y|), \tag{77}$$

where

$$\int_0 dt/\varphi(t) < \infty.$$

Let us consider the functions

$$G_+(r) = \int_r^{\operatorname{diam} G} t^{1-s_1} \exp\left[\int_t^{\operatorname{diam} G} (\varphi(\tau)/\tau)\, d\tau\right] dt,$$

$$G_-(r) = \int_r^{\operatorname{diam} G} t^{1-s_2} \exp\left[\int_t^{\operatorname{diam} G} (\varphi(\tau)/\tau)\, d\tau\right] dt.$$

Using the kernels $G_\pm(v)$, we construct capacities $\gamma_\pm(E)$ of a Borel set E by a standard method (see, for example Landkof [1966]). Let $x_0 \in \partial G$. Let us construct the Wiener series for these capacities. Thus, let $E_k = Q_{2^{-k}}^{x_0} \setminus G$ and $\gamma_k^\pm = \gamma^\pm(E_k)$; we consider the series

$$\sum_{k=1}^{\infty} 2^{k(n-2)} \gamma_k^\pm . \tag{78}$$

Theorem 1.45 (Ibragimov [1983b]). *If condition* (76) *holds and if the series* (78) *with the* $(+)$ *sign diverges, then the point* x_0 *is regular. If condition* (77) *holds and if the series* (78) *with the* $(-)$ *sign converges, then the point* x_0 *is irregular.*

Corollary. *In the particular case* $s_1 = s_2 = n$, *the capacities* $\gamma^{\pm}(E)$ *and the Wiener capacity of the set* E *are finitely related. In this case we obtain a necessary and sufficient boundary point regularity condition for the equation* $Lu = 0$, *which coincides with the analogous condition for Laplace's equation.*

As shown by Ibragimov [1983b], in this particular case the coefficients of the operator L can actually be discontinuous.

In the case of equation $Lu = 0$, where L is an operator of form (72) with continuous higher order terms coefficients and bounded measurable coefficients of lower order terms, Bauman [1985] obtained necessary and sufficient boundary point regularity conditions, which are similar to Wiener's criterion.

Let $n \geqslant 3$, $0 < R' < R$ and let Q_R, $Q_{R'}$ be concentric open balls. Let $x_0 \in Q_R \backslash Q_{R'}$ be some fixed point. Capacity $\operatorname{cap}_L K$ is defined for a compact set $K \subset Q_{R'}$ in the following way:

$$\operatorname{cap}_L K = \inf u(x_0),$$

where the infimum is taken over all supersolutions of the equation $Lu = 0$ in Q_R such that $u|_{\partial G} = 0$, $u|_Q \geqslant 0$, and $u(x) \geqslant 1$ on K. $\operatorname{cap}_L K$ depends, of course, on the choice of the points x_0, but capacities with respect to different points are finitely related.

There exist Green's functions $G(x, y)$ for the operator L in Q_R. Bauman constructs a function $\hat{G}(x, y)$ which she calls the "normalized Green's function" by setting $\hat{G}(x, y) = G(x, y)/G(x_0, y)$. Of course, this function is not a Green's function but it turns out that it has many properties in common with Green's function.

Let now Ω be a domain, $\bar{\Omega} \subset Q_{R'}$ and let $x^* \in \partial\Omega$. Let us consider the family of balls $Q_m = \{x \in \mathbb{R}^n : |x - x^*| < 2^{-k}\}$, and set $K_m = \bar{Q}_{m+1} \backslash Q_{m+2}$.

As is well known, the classical Wiener criterion states that a necessary and sufficient condition for regularity of the point x^* is the divergence of the series

$$\Sigma 2^{(n-2)m} \operatorname{cap} K_m,$$

where $\operatorname{cap} K_m$ is the Wiener capacity of the set K_m. The number $2^{(n-2)m}$ is proportional to the value $G(x^*, x_m)$ of Green's function in Q_R with pole at x^*, at the point $x_m \in \partial Q_m$. Moreover, the proportionality constant is independent of m. Let e be an arbitrary unit vector. Then we can set $x_m = x^* + 2^{-m}e$ so that the Wiener condition can be rewritten in the form

$$\sum_{m=m_0}^{\infty} G(x^*, x^* + 2^{-m}e) \operatorname{cap} K_m = \infty,$$

where m_0 is so large that $K_{m_0} \subset Q_{R'}$.

The necessary and sufficient regularity condition obtained by Bauman is the divergence of the series

$$\sum_{m=m_0}^{\infty} \hat{G}(x^*, x^* + 2^{-m}e) \operatorname{cap}_L K_m.$$

3.3. Questions Related to the Cauchy Problem for an Elliptic Equation. Let us consider the elliptic equation

$$Lu \equiv \sum_{i,k=1}^{n} a_{ik}(x)\frac{\partial u}{\partial x_i \partial x_k} + \sum_{i=1}^{n} b_i(x)\frac{\partial u}{\partial x_i} + c(x)u = f \tag{79}$$

in a domain $\Omega \subset \mathbb{R}^n$. Let us study the Cauchy problem associated with it: let S be a smooth $(n-1)$-dimensional manifold with a boundary; we want to find a solution of equation (79) satisfying

$$u|_s = \varphi, \quad \left.\frac{\partial u}{\partial l}\right|_s = \psi, \tag{80}$$

where l is a direction which is not tangent to S and which changes smoothly along S, and with φ, ψ being given functions.

As is well known, this problem is not well posed in the sense of Hadamard. However, if the coefficients of higher-order terms of L are in C^1, and if the rest are bounded, we have uniqueness of solutions of this problem.

Theorem 1.46 (uniqueness theorem). *If u_1 and u_2 are two solutions of* (79) *in Ω satisfying on S conditions* (80) *with the same functions φ and ψ, then $u_1 \equiv u_2$ in Ω* (see, for example Hörmander [1986–88]).

If a_{ik} do not satisfy the Lipschitz condition, then in general the uniqueness theorem for the solution of the Cauchy problem is no longer valid even if the coefficients satisfy the Hölder condition with any given, arbitrarily close to one, exponent. The first such example was constructed by Pliś [1963]. The equation he employs has the form $u_{tt} + a(t, x, y)u_{xx} + b(t, x, y)u_{yy} + l.o.t = 0$, $0 < \alpha < a(t, x, y), b(t, x, y) < \beta$. The coefficients in this equation satisfy the Hölder condition with exponent arbitrarily close to one, and are C^∞ outside of the coordinate plane $t \equiv 0$. This equation admits a solution $u(t, x, y) \in C^\infty$ which is zero for $t \leqslant 0$ and is non-zero in any neighborhood in the half-plane $t > 0$.

In the example above, the equation is not in divergence form. However, Miller [1974] showed that the uniqueness theorem does not hold for equations in divergence form either. He constructed an example of an equation of the form

$$u_{tt} + \sum_{i,k=1}^{2} (a_{ik}(t, x, y)u_{x_k})_{x_i} = 0,$$

which is uniformly elliptic with coefficients which are periodic in x and y, which satisfy the Hölder condition with any a-priori given constant smaller than one, and are infinitely differentiable for $t \neq 0$. This equation has a solution $u(t, x, y)$ periodic in x and y, which is also zero for $t \leqslant 0$ but not identically zero for $t > 0$.

In these examples it is essential that the number of independent variables is larger than two. In the case of two independent variables, Bers and Nirenberg showed that uniqueness of solutions of the Cauchy problem does take place both in the case of equations not in divergence form and of equations in divergence form. For this to hold, it is enough that coefficients are measurable and bounded.

To have uniqueness in the case when the coefficients of higher order terms are C^1 and those of lower order ones are bounded, it is not necessary to prescribe u and $\partial u/\partial l$ on all of the surface S.

Let $f \equiv 0$ in equation (79). Then the following theorem, due to Nadirashvili [1986], holds:

Let $E \subset S$ be such that its Hausdorff dimension is larger than $n - 2$, and $u|_E = \operatorname{grad} u|_E = 0$. Then $u \equiv 0$ in Ω.

It is essential that the surface S belong to the interior of the domain Ω in which the solution is defined. If, on the other hand, S belongs to a smooth piece of the boundary of Ω and has positive $(n - 1)$-dimensional measure, then the question of uniqueness of solution to the Cauchy problem in the case of dimensional $n \geqslant 3$, is unresolved even in the simplest case when Ω is a ball and S is its boundary. Let $u \in C^\infty(\bar{\Omega})$, $E \subset S$ is a set of positive $(n - 1)$-dimensional measure and $u|_E = \operatorname{grad} u|_E = 0$. Does it follow that $u \equiv 0$?[1]

Uniqueness of the solution for the Cauchy problem for the equation $Lu = 0$ is equivalent to the problem of extending the solution from a subdomain to the entire domain: if $\Omega' \subset \Omega$ and $u|_{\Omega'} = 0$, then $u = 0$ in Ω. A strengthened form of this result states that if $x_0 \in \Omega$ and $u(x)$ is a solution of the equation $Lu = 0$ which decays as $x \to x_0$ faster than any power, that is, $\forall k \lim\limits_{x \to x_0} \dfrac{u(x)}{|x - x_0|^k} = 0$, then $u \equiv 0$. Under assumption of sufficient smoothness of coefficients, this was first proved by Cordes [1956b] (this is also a corollary of the three ball theorem we formulate below).

If the point x_0 belongs to the boundary, then as $x \to x_0$, the solution can already decay exponentially without being identically zero. Let Ω be a disc in the plane x_1, x_2

$$\Omega = \{(x_1, x_2): (x_1 - 1)^2 + x_2^2 < 1\}, \quad x_1 + ix_2 = z.$$

Then as z approaches (0, 0) the function $u' - = \operatorname{Re} e^{-1/z}$ decays exponentially.

Borai [1968] obtained the following result for the equation $Lu = 0$, with assumptions on coefficients being as follows: $a_{i,k} \in C^2$, $b_i \in C^1$, $c(x)$ is a bounded function.

If $u(x)$ is a solution of the equation in a domain Ω with a C^1 boundary, $x_0 \in \partial\Omega$, and there exists an $\varepsilon > 0$ such that $\lim_{x \to x_0}(u(x)/e^{-1/(|x - x_0|^{1+\varepsilon})}) = 0$, then $u \equiv 0$ in Ω.

The question of uniqueness of solutions to the Cauchy problem is closely related to the question of *continuous dependence* of the solution on the Cauchy initial data in a class of a-priori bounded functions. We have the following theorem:

[1] Recently, T.H. Wolf announced that he had constructed a counterexample for Laplace's equation in a ball in $\mathbb{R}^3$.

Theorem 1.47 (on continuous dependence of solutions to the Cauchy problem). *Let $\Omega \subset \mathbb{R}^n$ be a domain, let $S \subset \partial\Omega$ be a smooth $(n-1)$-dimensional surface; let Ω' be a subdomain of Ω adjoining S, that is $S = \partial\Omega' \cap \partial\Omega$.*

Let $u(x)$ be a solution of the equation $Lu = 0$ in $\Omega \cup S$, and assume that $a_{ik} \in C^1$, and that b_i and c are Hölder continuous. Let M be a positive number. Then there exist numbers $C > 0$ and $\sigma > 0$ such that if $|u|_{|\Omega} < M$, and for arbitrary $\varepsilon > 0$ $|u|_{|S} < \varepsilon$ and $\left|\dfrac{\partial u}{\partial n}\right|_{|S} < \varepsilon$ then $|u| < C\varepsilon^\sigma$ in Ω'.

This theorem was proved, under somewhat stronger assumptions on the coefficients than those given above, by one of the authors (Landis [1956b]). Subsequently these requirements were weakened as indicated by Borai [1968]. He derived this theorem from the *Hörmander inequality* (Hörmander [1986–88]). As this is an important result, we quote it here.

Let $P = \sum_{|\alpha| \leqslant m} a_\alpha(x) D^\alpha$ be a differential operator of order m, and let $P_m = \sum_{|\alpha|=m} a_\alpha(x) D^\alpha$ be elliptic. Assume that the coefficients of P_m are in C^1, and that the rest of coefficients of P are bounded. Let a function $\varphi \in C^2$ be defined in a neighborhood of $x_0 \in \mathbb{R}^n$, and grad $\varphi(x_0) \neq 0$. The *level surface* $\varphi(x) = \varphi(x_0)$ is called *strongly quasiconvex* (in a sufficiently small neighborhood of x_0) if the equation $\mathscr{P}_m(\xi + i\tau \operatorname{grad} \varphi(x_0)) = 0$ has no multiple roots τ for any real ξ. If this condition is satisfied and if $u \in \mathring{H}_m$ such that the support of u is contained in a small enough neighborhood of x_0, then for sufficiently large positive τ

$$\sum_{|\alpha|<m} \tau^2(m - |\alpha|) - 1 \int |D^\alpha u|^2 e^{2\tau\varphi}\, dx \leqslant K \int |Pu|^2 e^{2\tau\varphi}\, dx,$$

$$K = \text{const} > 0. \tag{81}$$

Any surface is strongly quasiconvex for a second order elliptic operator. Therefore inequality (81) holds. The derivatives in (81) are generalized derivatives in the sense of Sobolev. The Hölder condition on the lower order terms in the statement of the theorem is only needed in order that a classical solution may be obtained via Schauder's theorem.

The following theorem on continuous dependence in a class of a-priori bounded functions in the problem of continuation of a solution from a subdomain to the entire domain parallels the theorem on continuous dependence of solutions of the Cauchy problem.

Theorem 1.48. *Let Ω, Ω', and Ω'' be domains in $\mathbb{R}^n$ such that $\Omega'' \subset\subset \Omega' \subset\subset \Omega$, and let there be defined in Ω a solution $u(x)$ of the equation $Lu = 0$ with the same assumptions on the coefficients as in the previous theorem. Let M be a positive number. Then there exist constants $C > 0$ and $\sigma > 0$ depending on Ω, Ω', Ω'', and M, and L such that for any $\varepsilon > 0$, it follows from $|u|_{|\Omega} < M$ and $|u|_{|\Omega''} < \varepsilon$, that $|u|_{|\Omega'} < Ce^\sigma$.*

This theorem is contained in the paper of Borai [1968] mentioned above.

If stronger smoothness conditions $a_{ik} \in C^2$, $b_i \in C^1$, $c(x)$ is a bounded function, are imposed on the coefficients of the operator L, then a sharper theorem can be obtained.

Theorem 1.49 (three balls theorem). *Let Q_r^0 be the ball of radius r with center at the origin. Let there be defined in Q_1^0 a solution $u(x)$ of the equation $Lu = 0$, coefficients of which satisfy the conditions detailed above. Let $|u|_{|Q_1^0} < 1$. Then there exists a constant $C > 0$ depending on L such that if $r_1, r_2, 0 < r_1 < r_2 < 1$, and $\alpha > 0$ are arbitrary numbers and $|u|_{|Q_{r_1}^0} < r_i^\alpha$, then*

$$|u|_{|Q_{r_2}^0} < (Cr_2)^\alpha. \tag{82}$$

This theorem is proved in Landis [1963] under slightly stronger smoothness requirements on coefficients, and instead of (82) the estimate $|u|_{|Q_{r_2}^0} < (Cr_2)^\alpha |\ln r_2|$ is given. Gusarov [1975] weakened the requirements on coefficients as indicated in the theorem.

Since in general the constant C is large, the theorem is meaningful for small r_2 (and large α). Gusarov, Landis [1982] also considered a different version of this theorem, which provides an estimate in the case when r_2 is close to 1.

The three ball theorem can be considered as a generalization to solutions of elliptic equations of the Hadamard three circle theorem from the theory of analytic functions of a complex variable.

Nadirashvili [1976] proved the following version of the *three balls theorem* in the case when the coefficients of the operator are analytic functions.

Theorem 1.50 (modified three balls theorem). *Let equation $Lu = 0$ with analytic coefficients be defined in Q_1^0. There exist constants $\sigma > 0$ and $\alpha_0 > 0$ such that if $0 < r_1 < r_2 < 1/2$, $E \subset Q_{1/2}^0$ is a set such that meas $E =$ meas $Q_{r_1}^0$ and $u(x)$ is a solution in Q^0 satisfying conditions $|u|_{|Q_1^0} < 1$ and $|u|_{|E} < r_1^\alpha$, where $\alpha > \alpha_0$, then $|u|_{|Q_{r_2}^0} < r_2^{\gamma\alpha}$.*

Whether a similar theorem holds in the case when L has non-analytic coefficients, is an open question.

3.4. A Phragmen-Lindelöf Type Theorem Connected with the Cauchy Data. Apart from the generalization of the classical Phragmen-Lindelöf theorem from the theory of analytic functions, which is related to the transition from an analytic function $f(z)$ to the harmonic function $u(x, y) = \ln |f(z)|$, that was given above, there exists a different generalization, related to the transition from an analytic function $f(z)$ to the harmonic function Re $f(z)$.

We consider the equation $Lu = 0$ with coefficients satisfying the following conditions: $a_{ik}(x) \in C^1$, $b_i(x)$ and $c(x)$ are Hölder continuous.

Theorem 1.51. *Let Ω be a cylinder defined by the inequalities $\sum_{i=2}^n x_i^2 \leqslant h^2$, $x_1 \geqslant x_1^0$ and let S be its lateral surface. Suppose that $u(x)$ is a solution of equation $Lu = 0$ satisfying $|u|_{|S} < 1$, $\left|\frac{\partial u}{\partial n}\right|_S < 1$. Then either $u(x)$ is uniformly bounded in Ω for large x_i, or*

$$\varlimsup_{x_1 \to \infty} (|u(x)|/\exp(Cx_1)) > 0, \tag{83}$$

where $C > 0$ is a constant depending on the equation and on h.

Proof. For arbitrary $a \geqslant x_1^0 + 1$, let us denote by Z_a and Z'_a cylinders defined, respectively, by inequalities

$$\sum_{i=2}^{n} x_i^2 \leqslant h^2, \quad a-1 \leqslant x_1 \leqslant a+2$$

and

$$\sum_{i=2}^{n} x_i^2 \leqslant h^2, \quad a \leqslant x_1 \leqslant a+1.$$

Let S_a be the lateral surface of the cylinder Z_a.

From the theorem of Borai [1968] on continuous dependence of solutions of the Cauchy problem quoted above, it follows that there exist constants $\varepsilon_0 > 0$ and μ, $0 < \mu < 1$, depending on the equation and on h, such that if ε, $0 < \varepsilon < \varepsilon_0$, is an arbitrary number, and if $v(x)$ is a solution of the equation $Lu = 0$ in Z_a, which satisfies the conditions

$$v| < 1 \text{ in } Z_a, \quad |v|_{|S_a} < a, \quad \left|\frac{\partial u}{\partial u}\right|_{|S_a} < \varepsilon, \tag{84}$$

then

$$|v| < \varepsilon^{\mu} \text{ in } Z_a. \tag{85}$$

Let us set $M_a = \max_{x \in Z_a} |u|$ and $M'_a = \max_{x \in Z'_a} |u|$. $M = \max(1/\varepsilon_0, 1)$, and assume that $u(x)$ is unbounded in the cylinder Ω as $x_1 \to \infty$. Then we can find $a > x_1^0 + 2$ such that

$$M'_a > M, \quad M'_a > M'_{a-1}. \tag{86}$$

Consider in Z_a the function $v(x) = u(x)/M_a$. This function solves the equation $Lv = 0$ and satisfies the inequalities $|v(x)| \leqslant 1$ in Z_a, $|v|_{|S} < 1/M_a$, $\left|\frac{\partial u}{\partial u}\right|_{|S} < 1/M_a$. Therefore, by (84) and (85)

$$|v| < (1/M_a)^{\mu} \text{ in } Z_a,$$

that is,

$$M_a > (M'_a)^{1/(1-\mu')}. \tag{87}$$

As $M_a = \max(M'_{a-1}, M'_a, M'_{a+1})$, it follows from (86) and (87) that $M_a = M'_{a+1}$, and thus

$$M'_{a+1} > (M'_a)^{1/(1-\mu)}.$$

Now we can repeat the argument, substituting first $a+1$ and then $a+2$, etc. for a, to obtain

$$M'_{a+k} > (M'_a)^{(1/(1-\mu))^k} = \exp\left(\ln M' \exp\left(k \ln \frac{1}{1-\mu}\right)\right),$$

hence inequality (83) follows; we can take for C any number smaller than ln $(1/(1-\mu))$. □

Let us give another example of a theorem in the same vein. Its proof is similar to the above, and uses the theorem on continuous dependence of solutions of the problem of continuation from a subdomain to the entire domain.

Theorem 1.52. *Let Ω be a ball punctured at the center: $\Omega = \{x: 0 < |x| \leqslant R\}$, and let K be the cone defined by inequalities*

$$\sum_{i=2}^{n} x_i^2 < a^2 x_1^2, \quad a \neq 0, \quad x_i > 0.$$

Let $u(x)$ be a solution in Ω of the equation $Lu = 0$ with the same assumptions on the coefficients of L as above, and suppose that $u(x)$ is bounded in $\Omega \cap K$. Then either $u(x)$ is bounded everywhere in Ω, or

$$\overline{\lim_{|x|\to 0}} \, (|u(x)|/\exp(1/|x|^C)) > 0,$$

where $C > 0$ is a constant depending on the equation and on a.

Let us quote, finally, another theorem of this class. It was obtained by Nadirashvili [1972] and uses the modified three balls theorem.

Theorem 1.53. *Let L be a uniformly elliptic operator defined on $\mathbb{R}^n$, the coefficients of which are analytic, can be analytically continued into the region $\{z \in \mathbb{C}^n$: $\operatorname{Re} z \in \mathbb{R}^n$, $|\operatorname{Im} z| < \delta\}$, and are bounded in modulus in this region.*

Let R and μ be positive numbers, and let E be a set with the following property: for any point $x_0 \in \mathbb{R}^n$, $\operatorname{meas}(E \cap Q_R^{x_0}) > \mu$, where $Q_R^{x_0}$ is the ball of radius R with center at x_0.

Let us consider $u(x)$, a solution of the equation $Lu = 0$ in $\mathbb{R}^n$, which is bounded on E. Then either u is bounded everywhere in $\mathbb{R}^n$, or

$$\overline{\lim_{R\to\infty}} \left(\max_{|x|=R} |u(x)|/e^{e^{CR}} \right) > 0,$$

where $C > 0$ depends on R, μ, and the operator L.

It seems quite plausible that the analyticity requirement in this theorem can be relaxed, as it is related to the method of proof (the modified three balls theorem has, up till now, been proved only under analyticity assumption on the coefficients).

3.5. On the Possible Rate of Decay of Solutions of Elliptic Equations. Let G_k be the semi-infinite cylinder

$$G_h = \left\{ x \in \mathbb{R}^n : x_1 > 0, \sum_{i=2}^{n} x_i^2 < h^2 \right\}$$

and let a solution of equation (79) be defined in G_k. Let us set $M(t) = \max_{x_1=t} |u(x)|$. Then there exists a constant $a > 0$, depending on the equation, such that if for large t

$$M(t) < \exp(-\exp(at/h)),$$

then necessarily $u \equiv 0$.

This theorem follows immediately from the three balls theorem.

If $u(x)$ is a solution defined in the cone

$$K_h = \left\{ x \in \mathbb{R}^n : x_1 > 0, \left(\sum_{i=2}^{n} x_i^2 \right)^{1/2} < hx_1 \right\},$$

and in equation (79), $b_i \equiv c \equiv 0$, then from the same three balls theorem it follows that there exists a constant b depending on the equation such, that if the solution decays at infinity faster than $\exp(-b|x|/h)$, then it is identically zero.

Up to the constants a and b, these estimates are the best possible, as shown by the following examples: $u(x, y) = \operatorname{Re} e^{-e^{z/h}}$ in $G_{(\pi-\varepsilon)/h}$ and $u(x, y) = \operatorname{Re} e^{z/h}$ in $K_{(\pi-\varepsilon)/h}$.

In the remainder of this section we shall consider solutions of (79) in domains exterior to some ball.

The following theorem was obtained by Kato [1959].

Theorem 1.54. *Let $u(x)$ ($x \in \mathbb{R}^n$, $|x| > R_0$) be a solution of the equation*

$$\Delta u + (k^2 - p(x))u = 0,$$

where k^2 is a positive constant and

$$\frac{1}{2k} \overline{\lim_{r\to\infty}}\, r \sup_{|x|=r} |p(x)| = \mu < 1.$$

Then if $u(x) = o(|x|^{-(n+1)/2})$ $u \equiv 0$.

This theorem was extended to the case of elliptic equations with variable principal part coefficients by Shifrin [1972]. We shall quote statements of theorems from that work.

Change of Variable Theorem. *Let L_0 be the operator*

$$L_0 = \sum_{i,j=1}^{n} \frac{\partial}{\partial x_i}\left(a_{ij} \frac{\partial}{\partial x_j} \right),$$

the coefficients of which are defined in a neighborhood of infinity and satisfy the following conditions:

1) $a_{ij} = a_{ji}$

2) $a_{ij} \in C^2$; $a_{ij} = a_{ij}^c + \varphi_{ij}(x)$ *with* $\varphi_{ij} = O(|x|^{-2})$, $\dfrac{\partial \varphi_{ij}}{\partial x_k} = O(|x|^{-3})$, $\dfrac{\partial \varphi_{ij}}{\partial x_k \partial x_l} = O(|x|^{-4})$, $k, l, = 1, \ldots, n$ *and the constant matrix* $\|a_{ij}^c\|$ *is positive definite.*

Then in a neighborhood of infinity we can choose new coordinates, denoted again by $x_1, \ldots, x_n$, such that in polar coordinates introduced in the usual way, L_0 has the form

$$L_0 = \Phi(\theta, r)\left[\frac{\partial^2}{\partial r^2} + \frac{n-1}{r}\frac{\partial}{\partial r} - \frac{N_r}{r^2} \right] + Q(\theta, r)\frac{\partial}{\partial r},$$

and moreover $|\Phi(\theta, r) - 1| \leqslant C_1/r$ *and* C_1 *can be made arbitrarily small if* $|x|^2 \varphi_{ij}(x)$ *and* $|x|^3 \partial\varphi_{ij}/\partial x_k$ *are small enough; for every function* $v \in C^2(S^{n-1})$ *we have that* $\dfrac{d}{dv}(N_r v, v) \leqslant 0$; *the operator* $-N_r$ *is uniformly elliptic (in* θ; *with ellipticity constants independent of* r); $rQ(\theta, r) \to 0$ *as* $r \to \infty$.

Theorem 1.55. *Let the operator*

$$L = \sum_{i,j=1}^{n} \frac{\partial}{\partial x_i}\left(a_{ij}(x)\frac{\partial}{\partial x_j}\right) + \sum_{i=1}^{n} b_i(x)\frac{\partial}{\partial x_i} + [k^2 - p(x)],$$

be defined in a neighborhood of infinity and suppose it satisfies the following assumptions:

1) *the principal part satisfies the conditions of the change of variables theorem*;
2) k^2 *is a positive constant*;
3) $b_i(x)$ *and* $p(x)$ *are Hölder continuous.*
4) $\overline{\lim}_{|x|\to\infty} |x|\,p(x) = 2k\mu,\ \mu < 1$;
5) $|x|(\sum_{i=1}^{n} |b_i(x)|)^{1/2} \to 0$ *as* $|x| \to \infty$;
6) *the constant* C_1 *in the change of variables theorem is so small that* $C_1 + \mu < 1$.

Let $u(x)$ *be a twice differentiable solution of the equation* $Lu = 0$. *Let us set*

$$M(r) = \left(\int_{S_r} |u(x)|^2\, d\omega_r\right)^{1/2}.$$

Then, if $M(r) = O(r^{-1})$ *and* $u(x)$ *grows not faster than an exponential,* $u(x) \equiv 0$.

Corollary. *If* $Lu = 0$ *and* $u(x) = O(|x|^{-(n+1)/2})$, *then* $u \equiv 0$.

The assumptions on the smoothness of a_{ij} and on the decay rates of first and second partial derivatives of the coefficients a_{ij} are used only for the change of variables. It could be that these restrictions are not essential.

However, another restriction, viz. $\overline{\lim}_{|x|\to\infty} |x|\,p(x) = 2k\mu$, $\mu < 1$, is essential. More precisely, it is essential that this limit superior is finite. It is possible to construct examples showing that this requirement cannot be relaxed. If it is not fulfilled, the solution may decay faster than any power.

An example also exists which demonstrates that the requirement of decay of the coefficients of the first derivatives is essential as well.

If $u(x)$ is a solution of the equation $-\Delta u + k^2 u = 0$, defined in $\mathbb{R}^n$ outside of some compact set K, then, as is well known, if $u(x)$ decays as $|x| \to \infty$ at the rate $e^{-(k+\varepsilon)|x|}$, $\varepsilon > 0$, then $u \equiv 0$.

Now let $u(x)$ be a solution of the equation $-\Delta u + q(x) = 0$ also defined on $\mathbb{R}^n \setminus K$. Is it true that if $|q(x)| \leqslant k^2$ and $|u(x)| < e^{-(k+\varepsilon)|x|}$ for some $\varepsilon > 0$ and for $|x|$ sufficiently large, then $u \equiv 0$? So far, this is an open question. The answer to the weaker question, whether it is true that if $u(x)$ decays as $|x| \to \infty$ faster than $e^{-\mu|x|}$ for all μ, then $u \equiv 0$, is also not known.

A closely related problem is the following one.

It is well known that if a harmonic function $u(x)$ is defined outside of a compact set K and $u(x)$ decays as $|x| \to \infty$ faster than any power, then $u \equiv 0$.

Let us again consider the equation $-\Delta u + q(x) = 0$. How fast must $|q(x)|$ decay as $|x| \to \infty$ in order that this property of solutions is conserved? This problem was solved by Meshkov [1986]. He shows that the relation $q(x) = O(|x|^{-2})$ must hold; this estimate is the best possible one.

3.6. The Behavior of Solutions in a Neighborhood of an Irregular Boundary. The survey paper of Kondratiev and Olejnik [1982] is devoted specifically to this question. Here we only discuss briefly some particular cases. Namely, in this section we will study the question of behavior of solutions in a neighborhood of a boundary point $O \in \partial G$. Moreover, we shall assume that in no neighborhood U of O does there exist a smooth non-degenerate mapping $U \to \mathbb{R}^n$ that maps $\partial G \cap U$ into the $(n-1)$-dimensional ball. Such a point will be called irregular, otherwise it is regular. The degree of smoothness of the mapping needed in the definition of a regular point is defined separately in every concrete case.

Let $u(x)$ be a generalized solution of the problem $Lu = fu|_{\partial G}, u|_{\partial G} = 0$, and at the point O, which we take to be at the origin, the coefficients $a_{ij}(x)$ are continuous. Without loss of generality we may assume that $a_{ij}(O) = \delta_{ij}$. The following theorem gives an estimate of the modulus of continuity of the solution in a neighborhood of O.

Theorem 1.56. *Let S_ρ be the intersection of the sphere $|x| = \rho$ with G. Let us denote by S'_ρ the set of points x on the unit sphere, such that $\rho x \in S_\rho$. Let us denote by λ_ρ the smallest in absolute value eigenvalue of the Beltrami operator in S'_ρ with zero Dirichlet data on $\partial S'_\rho$. Let us assume that $|\lambda_\rho| \geqslant \Lambda$, $\Lambda = \text{const.} > 0$, $f \equiv 0$ in a neighborhood of O. Then*

$$|u(x)|^2 \leqslant C_\varepsilon |x|^{\gamma-\varepsilon} \int_G |\text{grad } u|^2 \, dx, \tag{88}$$

if $|x|$ is small enough, $\varepsilon > 0$ is arbitrary,

$$\gamma = 2 - n + ((n-2) - 4\Lambda^2)^{1/2}.$$

It suffices to prove this theorem in the case when ∂G is an infinitely smooth surface, and $a_{ij}(x)$ are infinitely smooth functions, since only an approximation argument is needed in order to pass to the general case. For the sake of simplicity, let us assume that $b_i(x) \equiv 0$, $i = 1, \ldots, n$; $c(x) \equiv 0$, and that the support of $u(x)$ is contained in the ball $|x| < \delta$, where $\delta = \text{const} > 0$ is small enough.

Let us rewrite the equation in the form

$$\Delta u = f + \sum_{i,j=1}^{n} ((a_{ij}(0) - a_{ij}(x))u_{x_i})_{x_j}. \tag{89}$$

Let us multiply both parts of equations (89) by $r^\alpha u$; $\alpha = \text{const.} > 0$ which will be chosen later. After integration by parts we obtain

$$-\int_G r^{\alpha+n-1} u_r^2 \, dr \, d\omega + \frac{\alpha+n-1}{2} \int_G r^{\alpha+n-3} u^2 \, dr \, d\omega$$

$$-\int_G r^{\alpha+n-3} (\nabla_\omega u)^2 \, dr \, d\omega$$

$$= \int_G r^\alpha u f \alpha x - \sum_{i,j=1}^n \int_G r^\alpha u_{x_j}(a_{ij}(0) - a_{ij}(x))u_{x_i}\, dx$$

$$- \sum_{i,j=1}^n \int_G \alpha r^{\alpha-1} \frac{x_j}{r} u(a_{ij}(0) - a_{ij}(x))u_{x_i}\, dx, \tag{90}$$

where (r, ω) is the spherical system of coordinates, $(\Delta\omega^n)^2$ is the spherical part of $\text{grad}^2 u$. From the classical Hardy inequality $\int_0^\infty x^s y'^2\, dx \geqslant \frac{(s-1)^2}{4} \int_0^\infty x^{s-2} y^2\, dx$ for $y(0) = y(\infty) = 0$, and using (90) we obtain

$$\int_G \left[\frac{(\alpha + n - 2)\cdot 2 \cdot (1 + \kappa)}{4} - \frac{\alpha(\alpha + n - 2)}{2} \right] r^{\alpha+n-3} u^2\, dr\, d\omega$$

$$- \kappa \int_G r^{\alpha+n-1} u_r^2\, dr\, d\omega + (\lambda_0 + \kappa) \int_G r^{\alpha+n-3} u^2\, dr\, d\omega$$

$$+ \kappa \int_G r^{\alpha+n-3} (D_\omega u)^2\, dr\, d\omega \leqslant \int_G r^\alpha |u|\,|f|\, dx$$

$$+ \eta \int_G r^\alpha\, \text{grad}^2 u\, dx + \eta \int_G r^{\alpha-2}\, \text{grad}^2 u\, dx$$

$$\forall \kappa > 0. \tag{91}$$

To get the estimate (91) we used the variational principle

$$\int_{S_\rho^r} u^2\, d\omega \leqslant \frac{1}{\lambda_\rho} \int_{S_\rho^r} (\Delta_\omega u)^2\, d\omega.$$

If $|\alpha| < \sqrt{(n-2)^2 + 4\lambda}$, then from inequality (91) for κ, η small enough it follows that

$$\int_G r^{\alpha-2} u^2\, dx + \int_G r^2\, \text{grad}^2 u\, dx \leqslant C \left[\int_G r^{\alpha+2} f^2\, dx + \int_G \text{grad}^2 u\, dx \right]. \tag{92}$$

From Theorem 1.56 it follows that

$$|u(x)|^2 \leqslant C\lambda^{-n} \int u^2\, dx, \tag{93}$$

if $\lambda/2 < |x| < 2\lambda$, and λ is so small that $f \equiv 0$ for $|x| < 2\lambda$. Inequality (93) means that

$$|u(x)|^2 \leqslant |x|^{2-n+\alpha} \int r^{2-\alpha} u^2\, dx,$$

which, together with (92), concludes the proof of the theorem.

Theorem 1.56 is sharp in the sense that the exponent $\varepsilon > 0$ in estimate (88) cannot be taken to be zero without additional restrictions. Verzhbinskij and

Maz'ya [1971] showed that estimate (88) holds with $\varepsilon = 0$ if the coefficients $\alpha_{ij}(x)$ satisfy the Hölder condition. There are many works dealing with the case in which there is only one (or a finite number of) singular point(s) on the boundary of the domain. The case of $n = 2$, and of the boundary containing a finite number of corner points was considered by Nikol'skij [1956] and by Fufaev [1982]. They considered Laplace's equation and *Poisson's equation*.

$$\Delta u = f(x_1, x_2), \tag{94}$$

The following properties of the solution u of (94) with zero Dirichlet conditions on the boundary of the domain G were obtained. Let ∂G be an infinitely smooth curve everywhere apart from the point O which is placed at the origin. In a neighborhood of O, let the boundary ∂G be composed of two straight intersecting segments, with intersection angle being ω. The smoothness of the solution depends on the magnitude of ω.

Theorem 1.57. *If $f \in C^{k,\lambda}(G)$, $\omega < \dfrac{\pi}{k+2+\lambda}$, then $u \in C^{k+2,\lambda}$.*

Proof. Let us assume initially that all derivatives of f up to order k inclusively, are zero at O. Then $|f| \leqslant C|x|^{k+1}$. Let us choose coordinate axes in such a way that in a neighborhood of O, G is contained in the infinite angle $0 < r < \infty$, 0, $0 < \varphi < \varphi_0 = \pi/(k+2+\lambda)$. For $\varepsilon > 0$ sufficiently small and for $C > 0$ sufficiently large, the function $u_1 = Cr_0^{(\pi/\varphi)-\varepsilon} \cdot \sin(\pi\varphi/\varphi_0)$ satisfies the condition $\Delta u_1 - f < 0$ for $r < r_0$. Moreover for C sufficiently large, $u_1 > u$ on the boundary of the sector $0 < r < r_0$, $0 < \varphi < \omega$. From the maximum principle we obtain that in the interior of this sector $u \leqslant u_1$. Similarly, we have that $u \geqslant -u_1$. Thus $|u| \leqslant u_1$ and therefore everywhere in G

$$|u| \leqslant Cr^{(\pi/\varphi_0)-\varepsilon} \leqslant C_1 r^{k+2+\lambda}. \tag{95}$$

Next we use the classical Schauder theorem (see Miranda [1970]) concerning the regularity of solutions close to a smooth section of the boundary. Namely, let $\Omega \subset \mathbb{R}^n$, $\Omega' \subset \Omega$, let $\Gamma \subset \partial\Omega$ be a $C^{k+2,\lambda}(\bar{\Omega})$ surface with $k > 0$, $\lambda > 0$, let $\partial\Omega' \cap \partial\Omega \subset\subset \Gamma$. Assume that the coefficients of equation (89) are in $C^{k,\lambda}(\bar{\Omega})$, $f \in C^{k,1}$, $u|_\Gamma = \psi(x) \in C^{k+2,\lambda}(\Gamma)$. Then $C^{k+2,\lambda}$ and moreover

$$\|u\|_{C^{k+2,\lambda}(\Omega)} \leqslant C\left[\|f\|_{C^{k,\lambda}(\bar{\Omega})} + \|\psi\|_{C^{k+2,\lambda}(\Gamma)} + \max_{\bar{G}} |u|\right], \tag{96}$$

where the constant C depends on the ellipticity constant of equation (1), on the $C^{k,\lambda}$ norms of its coefficients on Ω, and on Ω'. □

Let us apply this theorem to the solution of equation (79). Let

$$\Omega = \{x \in G, h/2 \leqslant |x| \leqslant 2h\}, \quad \Omega' = \{x \in G, \tfrac{3}{4}h \leqslant |x| \leqslant h\}.$$

Inequality (96) leads to the estimate

$$\sup_{\substack{|\alpha|=k+2\\ x\in\Omega'\\ y\in\Omega'}} \frac{|D^\alpha u(x) - D^\alpha u(y)|}{|x-y|^\lambda} \leqslant C \sup_{\substack{|\alpha|=k\\ x\in\Omega\\ y\in\Omega}} \frac{|D^\alpha f(x) - D^\alpha f(y)|}{|x-y|^\lambda}$$

$$+ C_1 \sup_{x\in\Omega} \frac{|f(x)|}{|x|^{k+\lambda}} + \sup_{x\in\Omega} \frac{|u(x)|}{|x|^{k+2+\lambda}}.$$

From the assumption $|f(x)| \leqslant C|x|^{k+\lambda}$ and inequality (95) it follows that the number

$$\sup_{\substack{|\alpha|=k+2\\ x\in\Omega'\\ y\in\Omega'}} \frac{|D^\alpha u(x) - D^\alpha u(y)|}{|x-y|^\lambda}$$

is bounded uniformly in h. In a similar manner we obtain from inequality (95) that $|D^\alpha u(x)|/(|x|)^\lambda$ is bounded uniformly in h in Ω'.

From these statements it follows easily that $D^\alpha u(x) \in C^\lambda(\bar{G})$. If the condition $|f(x)| = C|x|^{k+\lambda}$ is not satisfied, then one proceeds as follows.

Let us use Taylor's formula to represent $f(x)$ in the form $f(x) = \mathscr{P}(x) + f_1(x)$ where $\mathscr{P}(x)$ is a polynomial of order k, and $|f_1(x)| \leqslant C|x|^{\lambda+k}$. Let us find a particular solution of the problem

$$\Delta u_2 = \mathscr{P}(x), \quad u_2|_{\varphi=0} = 0, \quad u_2|_{\varphi=\varphi_0} = 0, \quad 0 < r < \infty,$$

in the form of a polynomial of order $k+2$. Since π/φ_0 is a non-integer number by assumption, such a solution exists. The function $v(x) = u - u_2$ satisfies the equation $\Delta v = f_1(x)$, where $f_1 \in C^{k,\lambda}$, $|f_1| \leqslant C|x|^{\lambda+k}$.

The situation is slightly different if π/ω is an integer. In this case $u \in C^k(\bar{G})$ if $f \in C^\infty(\bar{G})$ and satisfies a finite number (which depends on k) of compatibility conditions at the point O. These compatibility conditions are framed in terms of linear combinations of values of f and of some of its derivatives at O.

Let us give an example of compatibility conditions. Let the domain Ω coincide in a neighborhood of O with the straight angle $\{x_1 > 0, x_2 > 0\}$. Let us consider solutions of *Poisson's equation* (94) with the boundary condition $u|_{\partial\Omega} = 0$. Let us assume that the boundary $\partial\Omega$ is an infinitely smooth curve everywhere except at O. Then the following proposition holds: if $f \in C^\lambda(\bar{\Omega})$, $0 < \lambda < 1$, $f(0) = 0$ and $f_{x_1x_1}(0) - f_{x_2x_2}(0) = 0$, then $u \in C^{2,\lambda}(\bar{\Omega})$. The condition $f_{x_1x_1}(0) - f_{x_2x_2}(0) = 0$ is the compatibility condition necessary in order that $u \in C^{4,\lambda}(\bar{\Omega})$.

From the argument above it follows that if π/ω is an integer and $f \equiv 0$ in a neighborhood of 0, then $u \in C^\infty(\bar{\Omega})$. Proof of this statement was first obtained by Fufaev [1963] who used a conformal mapping of the plane $\omega = z^{\pi/\omega}$ to transform Ω into a domain with a smooth boundary.

As we can see, the solution of Dirichlet's problem for Poisson's equation in Ω has, in general, a finite number of derivatives that are continuous in $\bar{\Omega}$. Moreover, this number depends on the intersection angle ω of arcs of the boundary curve. The smaller ω is, the smoother is the solution (under sufficient smoothness of the data of the problem). There exist exceptional angles (such that π/ω is an integer) which present no obstacle to the differentiability of solutions. Solutions of both the second and of the mixed boundary value problems for Poisson's equation have the same properties. *Schauder* type *estimates* for the aforementioned problems were obtained under certain conditions on the boundary by Grisvard [1981].

Let us consider whether Bernstein's estimate

$$\|u\|_{W^2(G)} \leqslant C[\|f\|_{L^2(G)} + \|u\|_{L^2}] \tag{97}$$

holds for solutions of equation $Lu = f$, satisfying the boundary condition

$$u|_{\partial G} = 0.$$

If ∂G is composed of a finite number of smooth arcs that intersect at angles smaller than π, then the generalized solution of the problem $Lu = f$, $u|_{\partial G} = 0$ is in $W^2(G)$ and inequality (97) holds. Moreover, coefficients $a_{ij}(x)$ have to be in $C^2(\bar{G})$, $f \in L^2(G)$. If the boundary contains angles larger than π, the generalized solution does not necessarily belong to $W^2(G)$. There is an example in which the boundary of domain G consists, in a neighborhood of the point O, of straight segments contained in the rays $\varphi = 0$ and $\varphi = \omega$ ($\omega > \pi$), such that the solution in a neighborhood of O coincides with the function $r^{\pi/\omega} \sin \dfrac{\pi}{\omega} \varphi$.

The survey paper of Kondratiev and Olejnik [1983] deals in detail with the results of papers devoted to questions of asymptotic behavior of solutions of elliptic equations in a neighborhood of a corner or of a conical point, as well as in a neighborhood of a rib.

Iosefian and Olejnik [1971] consider the mixed problem for the equation $Lu = 0$ in a domain Ω, with boundary conditions of the type

$$u|_{\gamma_1} = \varphi_1, \quad \left.\frac{\partial u}{\partial v}\right|_{\gamma_2} = \varphi_2, \quad \left.\left(\frac{\partial u}{\partial v} + bu\right)\right|_{\gamma_3} = \varphi_3, \tag{98}$$

where $\gamma_1 \cup \gamma_2 \cup \gamma_3 = \partial\Omega$, $\dfrac{\partial}{\partial v} = \sum\limits_{i,j=1}^{n} a_{ij} v_j \dfrac{\partial}{\partial x_i}$, $v = (\varphi_1, \ldots, \varphi_n)$ is the unit normal vector, φ_1, φ_2, φ_3, and ω are given functions. It is assumed that the Gauss-Ostrogradskij formula holds in Ω. Let $\Omega_\tau = \Omega \backslash \sigma_\tau$ where σ_τ is a neighborhood of some set σ of points in $\partial\Omega$. Under some restrictions on the coefficients of the equation $Lu = f$, the authors obtain estimates for solutions of the problem $Lu = f$ with (98), $\varphi_1, \varphi_2, \varphi_3 = 0$ and $f = 0$ in a neighborhood of σ. These are estimates of the Dirichlet integral over Ω_{τ_1} in terms of the Dirichlet integral over Ω_{τ_2} under the condition $\Omega_{\tau_1} \subset \Omega_{\tau_2}$ (τ is a parameter), with a coefficient depending on the geometric structure of $\partial\Omega \cap \partial(\Omega_{\tau_2} \backslash \Omega_{\tau_1})$. From these estimates the authors deduce uniqueness theorems for solutions of the problem $Lu = f$ with (98) on the boundary of the domain in the class of functions having unbounded Dirichlet integral over Ω, as well as theorems dealing with removable singularities of solutions of the problem $Lu = f$ with condition (98) on the boundary of the domain. These theorems indicate that either the Dirichlet integral over Ω_τ grows fast enough as meas $\sigma_\tau \to 0$ or that $u(x)$ has a bounded Dirichlet integral over Ω. Examples are given to show that the obtained estimate and uniqueness classes are the best ones possible. Similar assertions can be made for certain classes of second order equations with non-negative characteristic form.

Chapter 2
Parabolic Equations

§ 1. Parabolic Equations in Non-divergence Form

Let

$$L = \sum_{i,k=1}^{n} a_{ik}(x, t)\frac{\partial^2}{\partial x_i \partial x_k} + \sum_{i=1}^{n} b_i(x, t)\frac{\partial}{\partial x_i} + c(x, t) \tag{1}$$

be a uniformly elliptic operator defined on some domain $G \subset \mathbb{R}^{n+1} = \mathbb{R}^n_x \times \mathbb{R}^1_t$.

In this section we shall consider *parabolic operators*

$$L - \frac{\partial}{\partial t}.$$

A solution of the equation

$$Lu - \frac{\partial u}{\partial t} = 0 \tag{2}$$

is a function $u \in C^{2,1}$ for which (2) is identically satisfied. We call a function $u \in C^{2,1}$ such that $Lu - \dfrac{\partial u}{\partial t} \geqslant 0$ ($\leqslant 0$) a subparabolic (superparabolic) function.

In what follows, we shall use the symbol $Z^{t_0, t_1}_{x_0, r}$ ($x \in \mathbb{R}^n$, $r > 0$, $t_1 > 0$) to denote the cylinder

$$Z^{t_0, t_1}_{x_0, r} = \{(x, t) \in \mathbb{R}^{n+1}: |x - x_0| < r, t_0 < t < t_1\}$$

Let G be a domain in $\mathbb{R}^{n+1}$. We call a set $\gamma(G) \subset \partial G$ the upper top of the set G if for each point $(x_0, t_0) \in \gamma(G)$ there exists $\varepsilon > 0$ such that $Z^{t_0-\varepsilon, t_0}_{x_0, \varepsilon} \subset G$ and $Z^{t_0+\varepsilon, t_0}_{x_0, \varepsilon} \cap G = 0$. The set $\Gamma(G) = \partial G \backslash \gamma(G)$ is called the parabolic boundary of the set G.

1.1. The Maximum Principle. The following statement is elementary (see for example Landis [1971]). Let G be a bounded domain and let an operator of form (1) be defined in G with $c(x, t) \leqslant 0$; let $u(x, t)$ be a subparabolic (superparabolic) function. Then if $\sup_G u > 0$ ($\inf_G u < 0$),

$$\sup_G u = \varlimsup_{\substack{(x,t)\to\Gamma(G)\\(x,t)\in G}} u(x, t)$$

$$\left(\inf_G u = \varliminf_{\substack{(x,t)\to\Gamma(G)\\(x,t)\in G}} u(x, t)\right).$$

If the condition $c(x, t) \leqslant 0$ is not satisifed, there exists a constant K depending on λ and M of inequalities (2) of Chapter 1, such that for every solution of equation (2) in G we have the inequality

$$|u(x,t)| \leqslant K \varlimsup_{\substack{(x,t)\to\Gamma(G)\\(x,t)\in G}} |u(x,t)|.$$

Nirenberg (see for example Landis [1971]) proved the following *strong maximum principle.*

Let $G \subset \mathbb{R}^{n+1}$ and $(x_0, t_0) \in G \cup \gamma(G)$. We shall say that a subdomain $G' \subset G$ is subordinate to the point (x_0, t_0) if every point $(x, t) \in G'$ can be connected to the point (x_0, t_0) by a curve belonging to $G \cup (x_0, t_0)$ which can be projected in a one to one fashion onto the interval $[t, t_0]$, $t < t_0$.

Theorem 2.1 (the strong maximum principle). *Let operator* (1) *with* $c(x,t) \leqslant 0$ *be defined in* $G \subset \mathbb{R}^{n+1}$. *If* $u(x,t)$, *a subparabolic (superparabolic) function in* $G \cup \gamma(G)$, *reaches a positive maximum (negative minimum) at some point* $(x_0, t_0) \in G \cup \gamma(G)$, *then* $u \equiv$ const *in a subdomain* G' *subordinate to the point* (x_0, t_0).

1.2. Sub- and Superparabolic Functions of Potential Type. Let the operator (1) be defined in a domain $G \subset \mathbb{R}^{n+1}$. Let us consider in $\mathbb{R}^{n+1}\backslash 0$ the function

$$\mathscr{E}_{s,\beta}(x,t) = \begin{cases} \dfrac{1}{t^s} e^{-(|x|^2)/4\beta t} & \text{for } t > 0 \\ 0 \quad \text{for } t \leqslant 0 & \text{away from } x = 0,\ t = 0, \end{cases}$$

$$s = \text{const} > 0, \quad \beta = \text{const.} > 0.$$

Let

$$M_1 = \sup_{(x,t)\in G} \sum_{i=1}^n a_{ii}(x,t), \quad M_2 = \inf_{(x,t)\in G} \sum_{i=1}^n a_{ii}(x,t),$$

$$\alpha_1 = \inf_{(x,t)\in G} \min_{|\xi|=1} \sum_{i,k=1}^n a_{ik}(x,t)\xi_i\xi_k, \quad \alpha_2 = \sup_{(x,t)\in G} \max_{|\xi|=1} \sum_{i,k=1}^n a_{ik}(x,t)\xi_i\xi_k. \tag{3}$$

The following lemmas are easily proved by direct computation.

Lemma 1. *If the coefficients of the operator L satisfy the condition*

$$\begin{aligned} |b_i(x,t)| &\leqslant K_1(|x| + 1), \\ |c(x,t)| &\leqslant K_2(|x|^2 + 1), \end{aligned} \tag{4}$$

$K_1, K_2 = \text{const} > 0$, *and we have that*

$$\begin{aligned} &\beta > \alpha_2 \ (\beta < \alpha_1), \\ &s < \frac{M_2}{2\beta}\left(s > \frac{M_1}{2\beta}\right), \end{aligned} \tag{5}$$

then there exists $\eta > 0$ *such that* $\mathscr{E}_{s,\beta}(x,t)$ *is a superparabolic (subparabolic) function for* $x \in \mathbb{R}^n$, $0 \leqslant t \leqslant \eta$, $(x,t) \neq (0,0)$.

Lemma 2. *If the coefficients of operator L are such that*

$$\sum_{i=1}^{n} b_i(x,t)x_i \geqslant 0 \quad \left(\sum_{i=1}^{n} b_i(x,t)x_i \leqslant 0\right), \qquad c(x,t) \leqslant 0 \quad (c(x,t) \geqslant 0) \tag{6}$$

and

$$\beta \geqslant \alpha_2 \quad (\beta \leqslant \alpha_1), \qquad s \leqslant \frac{M_2}{2\beta} \quad \left(s \geqslant \frac{M_1}{2\beta}\right), \tag{7}$$

then the function $\mathscr{E}_{s,\beta}$ is superparabolic (subparabolic) everywhere except at the origin.

The lemmas are proved by direct substitution. Using $\mathscr{E}_{s,\beta}$ we can construct potential-type functions that are super- or subparabolic for suitable chosen parameters.

Let $E \subset \mathbb{R}^{n+1}$ be a Borel set on which a measure μ, $\mu(E) < \infty$, is defined. In the complement of E let us consider the function

$$U_{s,\beta}(x,t) = \int_E \mathscr{E}_{s,\beta}(x-\xi, t-\tau)\, d\mu(\xi,\tau).$$

From lemma 1 it follows that if relations (3) and (4) hold, and if E is contained in the layer $t_0 < t < t_0 + \eta$, then $U_{s,\beta}$ is a superparabolic (subparabolic) function in this layer outside of $\bar{E}$. If, on the other hand, conditions (6) and (7) are satisfied, then it follows from Lemma 2 that $U_{s,\beta}$ is superparabolic (subparabolic) in $\mathbb{R}^{n+1} \setminus E$.

1.3. Uniqueness for the Cauchy Problem. The Cauchy problem for the equation

$$Lu - u_t = 0 \tag{8}$$

consists of finding a solution of equation (8) in the layer $t_0 < t \leqslant T < \infty$, which is continuous up to the hyperplane $t = t_0$ and satisfies the condition

$$u(x, t_0) = f(x)$$

where $f(x)$ is a given continuous function.

It is not hard to prove uniqueness of the solution to the Cauchy problem if the operator L is uniformly elliptic and it is a priori known that the solution is bounded for $t_0 \leqslant t \leqslant T$. Precise conditions that have to be imposed on the solution under which uniqueness holds, were pointed out by Tikhonov [1935].

We say that a function $v(x,t)$ defined in the layer $t \leqslant t \leqslant T$ belongs to the Tikhonov class if there exists $C_1 > 0$, $C_2 > 0$ such that $|v(x,t)| \leqslant C_1 e^{C_2|x|^2}$.

Theorem 2.2. *If coefficients of the operator L satisfy conditions* (3) *and* (4), *the Cauchy problem has unique solutions in the Tikhonov class.*

This theorem derives from the following maximum principle for a layer.

Theorem 2.3. *Let $u(x)$ be a subparabolic (superparabolic) function for equation (8), the coefficients of which satisfies conditions (3) and (4) in the layer $t_0 < t < T$. Let $u(x, t)$ be continuous up to the hyperplane $t = t_0$ and non-positive (non-negative) for $t = t_0$. If $u(x, t)$ belongs to the Tikhonov class, then $u(x, t) \leqslant 0$ ($u(x, t) \geqslant 0$) for $t_0 \leqslant t \leqslant T$.*

Proof of Theorem 2.3. For definiteness, let u be a subparabolic function. Let us set $\beta = 2\alpha_2$ and $s = M_2/8\alpha_2$ where the constants α_2, M_2 are defined in inequality (3). Let η be the constant from Lemma 1 corresponding to these β, s. Let us set

$$\varepsilon = \min\left(T - t_0, \frac{1}{64C_2\beta}, \eta/2\right)$$

and for arbitrary $R > 1$ let us consider the function

$$v_R(x, t) = Me^{C_2R^2} \int_{|\xi|=R} \mathscr{E}_{s,\beta}(t - t_0 + \varepsilon, x - \xi)\, ds_\xi,$$

where ds_ξ is a surface element of the $(n-1)$-dimensional sphere $|\xi| = R$. If $t > t_0$, then

$$v_R(x, t) = \frac{Me^{C_2R^2}}{(t - t_0 + \varepsilon)^s} \int_{|\xi|=R} e^{-(|x-\xi|^2)/(4\beta(t-t_0+\varepsilon))}\, ds_\xi.$$

Below we shall choose the constant $M > 0$ in such away that it is independent of R.

From Lemma 1 it follows that v_R is superparabolic for $t_0 < t < t_0 + \varepsilon$.

Let us consider the cylinder $Z_{0,R}^{t_0,t_0+\varepsilon}$. On its lower base $v_R > 0$. Let us choose M such that on the lateral surface $v_R > C_1e^{C_2R^2}$. To do this, note that for $0 < t - t_0 \leqslant \varepsilon$, we have the inequality

$$v_R \geqslant \frac{Me^{C_2R^2}}{(2\varepsilon)^s} \int_{|\xi|=R} e^{-(|x-\xi|^2)/4\beta\varepsilon}\, ds_\xi.$$

If $|x| = R$, $R > 1$, $\int_{|\xi|=R} e^{-(|x-\xi|^2)/4\beta\varepsilon}\, ds_\xi > a$, where $a = \text{const} > 0$ independent of R. Therefore, setting $M = C_2(2\varepsilon)^s/a$ we see that on the lateral surface of $Z_{0,R}^{t_0,t_0+\varepsilon} v_r \geqslant C_1e^{C_2R^2}$. From the maximum principle we deduce that $u \leqslant v_R$ inside this cylinder.

Let (x', t') be an arbitrary point in the interior of the layer $t_0 < t \leqslant t_0 + \varepsilon$, and let $R > 1$ be an arbitrary number such that $R > 2|x'|$. Then

$$u(x', t') \leqslant v(x', t') \leqslant \frac{Me^{C_2R^2}}{\varepsilon^s} \int_{|\xi|=R} e^{-(R/2)^2/4\beta\cdot 2\varepsilon}\, ds_\xi$$

$$\leqslant \frac{M}{\varepsilon^s} e^{C_2R^2} \omega_n R^{n-2} e^{-2C_2R^2} = \frac{\omega_n M}{\varepsilon^s} e^{-C_2R^2} R^{n-1}.$$

Here ω_n is the surface area of the n-dimensional sphere. Letting R go to infinity, we get that $u(x', t') \leqslant 0$. Thus $u(x, t) \leqslant 0$ for $t_0 < t \leqslant t_0 + \varepsilon$. Taking t_0 to be $t_0 + \varepsilon$, we see that $u(t, x)\, t \leqslant 0$ for $t_0 < t \leqslant t_0 + 2\varepsilon$. After a finite number of such steps we get that $u(x, t) \leqslant 0$ for $t_0 \leqslant t \leqslant T$ which is exactly what had to be proved. □

Täcklind (see for example Kamynin and Khimchenko [1981]) found a wider class of functions in which uniqueness of solutions of Cauchy problems holds. He showed that uniqueness holds in the class of functions such that $|u(x, t)| \leqslant e^{C|x|h(|x|)}$, where $h(t) > 0$ is a monotone non-decreasing function such that $\int^{\infty} \frac{dt}{h(t)} = \infty$. We shall call this class of functions the Täcklind class. Täcklind also showed that this class of functions is the widest one possible: if $h(t) > 0$ is monotone non-decreasing and such that

$$\int^{\infty} dt/h(t) < \infty,$$

then there exists a non-zero solution $u(x, t)$ of the Cauchy problem for the heat equation such that $u|_{t=0}$ and $|u(x, t)| \leqslant e^{C|x|h(|x|)}$. Kamynin and Khimchenko [1981] showed that uniqueness of solutions of the Cauchy problem holds in the Täcklind class of functions for parabolic equations with bounded coefficients.

Petrovskij [1946] posed the question of finding uniqueness classes for solutions of the Cauchy problem for parabolic equations of high order and for parabolic systems. The first sharp results of this kind were obtained by Gel'fand and Shilov [1953]. This result turned out to be one of the most interesting applications of the theory of distributions. Gel'fand, Shilov, and their students considered quite general, not necessarily parabolic, equations and systems with constant coefficients. The case of the Cauchy problem for parabolic systems with variable coefficients was studied by Ejdel'man [1964].

The proof of uniqueness given above employed the maximum principle. Kamymin and Khimchenko [1981] also use this tool. This method is comparitively easy, but only applicable for equations of second order. In Gel'fand and Shilov's [1953] paper the question of uniqueness is reduced to the question of existence of solutions of the adjoint problem, which is tackled by Fourier transform methods. Ejdel'man [1964] obtained estimates for the fundamental solution that are used to prove solubility of the adjoint problem. Olejnik and Radkevich [1978] study parabolic problems by the method of introducing a parameter.

1.4. A Weak Stabilization Theorem for the Solution of the Cauchy Problem as $t \to \infty$. We quote here a stabilization theorem for the solution of the Cauchy problem for equation (8), the coefficients of which satisfy only conditions (3) and (4). Later on, during the discussion of equations in divergence form, we shall return to the question of stabilization.

Theorem 2.4. *Let the coefficients of operator L in* (8) *satisfy conditions* (3) *and* (4). *Let $u(x, t)$, $t_0 < t < \infty$ be the Tikhonov class solution of the Cauchy problem*

$u|_{t=t_0} = f(x)$, *and let* $f(x) \to 0$ *as* $|x| \to \infty$. *Then* $u(x, t) \to 0$ *as* $|x| \to \infty$ *uniformly in* x.

Proof. Let $\varepsilon > 0$ be arbitrary. We fix R such that $f(x) < \frac{\varepsilon}{2}$ for $|x| > R$. Let us consider an arbitrary function

$$v = M\mathscr{E}_{s,\beta}(x, t - t_0 + 1) + \frac{\varepsilon}{2} = \frac{M}{(t - t_0 + 1)^s} e^{-|x|^2/(4\beta(t-t_0+1))} + \frac{\varepsilon}{2},$$

where $s = M_2/2\alpha_2$, $\beta = \alpha_2$ (M_2 and α_2 are the constants defined in equality (3)). By Lemma 2 of Section 3.5 the function v is superparabolic. Let us choose the constant M so large that for $|x| \leqslant R$ the inequality $v(x, t_0) > f(x)$ holds. Then $v(x, t_0) > f(x)$ everywhere in $\mathbb{R}^n$ and therefore $u < v$ for $t > 0$. If we put $t_1 = (2M/\varepsilon)^{1/s} + t_0 - 1$, then $u(x, t) < \varepsilon$ for $t > t_1$, $x \in \mathbb{R}^n$. In a similar manner we obtain that for $t > t_2$, $x \in \mathbb{R}^n$, $u(x, t) > -\varepsilon$. The theorem is proved. □

1.5. The Growth Lemma. An a-priori Bound of the Hölder Norm and the Harnack Inequality. The results of this section were first obtained by Krylov and Safonov [1979]. As before, we consider equation (8), $Lu - u_t = 0$, the coefficients of which satisfy conditions (3) and also the following condition

$$|b_i(x, t)| \leqslant K_1, \quad -K_2 < c(x, t) < 0. \tag{9}$$

Lemma 3 (the growth lemma). *Let there be given a number* r, $0 < r < 1$, *and a point* $(x_0, t_0) \in \mathbb{R}^{n+1}$. *Let us consider the following three cylinders*

$$Z_1 = Z_{x_0,r}^{t_0,t_0+r^2},$$
$$Z_2 = Z_{x_0,(3/4)r}^{t_0+(3/4)r^2,t_0+r^2}$$
$$Z_3 = Z_{x_0,(1/2)r}^{t_0+(1/4)r^2,t_0+(1/2)r^2}$$

Let D *be an open set in* $\mathbb{R}^{n+1}$ *contained in* Z_1 *and having non-empty intersection with* Z_2. *Let us consider* $E \subset Z_2 \setminus D$ *such that* meas $E > 0$. *We denote* $\Gamma(D) \cap Z_1 = \Gamma_1$. *Let, finally, there be defined in* D *a solution* $u(x, t)$ *of equation* (8) *and*

$$\overline{\lim_{\substack{(x,t)\to\Gamma \\ (x,t)\in D}}}\, u(x, t) \leqslant 0, \quad \sup_{D\cap Z_2} u > 0$$

Then

$$\sup_D u > (1 + \xi) \sup_{D\cap Z_2} u$$

where $\xi > 0$, $\xi = \text{const.}$ and depends on $K_1, K_2, \alpha_2, M_1, M_2$ (these are constants from inequality (3)) and on meas Z_3/meas E.

The proof of this theorem is similar to the proof of the corresponding lemma in the elliptic case (Section 1.4, Chapter 1). From the growth lemma we derive the theorem about an a-priori estimate of the Hölder norm.

Theorem 2.5. *Let* $G \subset \mathbb{R}^{n+1}$, *let* $u(x, t)$ *be a solution of equation* (8) *in* G. *For each* $\rho > 0$ *the following inequality holds*

$$\|u\|_{C^\alpha(G_\rho)} \leqslant M \sup_G |u|,$$

where $G_\rho = \{(x, t) \in G\colon \rho((x, t), \partial G) > \rho\}$, $\alpha > 0$ *and depends on* $\alpha_1 M_1, K_1, K_2, n$, *and* M, *moreover depends on* ρ *as well.*

The *Harnack inequality* in the formulation below, also follows from this lemma.

Theorem 2.6. (the Harnack inequality). *Let* $0 < r < 1$ *and let* u *be a positive solution of equation* (8) *in* Z_1. *Then*

$$\sup_{Z_3} u / \inf_{Z_2} u < C$$

where C *is a positive constant that depends on* $\alpha_1, M_1, K_1, K_2, n$; Z_1, Z_2, *and* Z_3 *have the same meaning as in the statement of Lemma* 3 *of Section* 1.5.

The method of proof of the Harnack inequality is similar to the method we used in deriving the Harnack inequality from the growth lemma in the elliptic case (see Section 1.4, Chapter 1). It resembles the method used by Landis [1971] to prove the Harnack inequality for parabolic equations with a small spread of the roots of the characteristic equation of the matrix $\|a_{ik}\|$.

Remark 1. Let in the operator L, $b_i \equiv c \equiv 0$. Then the number r does not enter Lemma 3. Therefore the Harnack inequality holds in Z_1, Z_2, Z_3 for any value of r.

In view of the a-priori Hölder norm bound both for parabolic, and for elliptic equations, the following question presents itself. Let us consider a cylindrical region, for example the cylinder

$$\Omega = Z_{0,1}^{0,1}$$

Let Γ be its parabolic boundary. Let L be a uniformly elliptic operator (1) with measurable bounded coefficients a_{ik}, b_i, c. Let f be the restriction to Γ of a smooth, for example, infinitely differentiable, function defined on $\mathbb{R}^{n+1}$. Let moreover $\{a_{ik}^{(n)}\}$, $\{b_i^{(n)}\}$ and $(c^{(n)}\}$ be sequences of infinitely differentiable functions converging, respectively, to a_{ik}, b_i and c almost everywhere, and comprising the coefficients of uniformly elliptic operators L_n with the same constant of ellipticity and with the same bounded for the coefficients. Let u_n be the solution of the boundary-value problem with the value f on Γ for the equation $L_n u_n - \dfrac{\partial u_n}{\partial t} = 0$. By the a-priori bound for the Hölder norm, we can extract a subsequence of $\{u_n\}$ which converges in Ω, and uniformly on every compact set. Moreover, we can extract s subsequence converging uniformly in Ω. The question is whether all such subsequences converge to the same limit.

1.6. Phragmen-Lindelöf Type Theorems. In this section we shall assume that in equation (8) $b_i \equiv 0$, $i = 1, \ldots, n$, $c(x, t) \leqslant 0$. The two theorems below are obtained from the growth lemma (Lemma 3, Section 1.5).

Theorem 2.7. *Let* $G \subset \mathbb{R}^{n+1}$ *be a bounded domain in the layer* $0 < t < T$, *having limit points on both boundaries of the layer. Let* $u(x, t)$ *be a solution of equation*

(8) which is continuous in $\bar{G}$ and zero the part of the boundary of G contained in the interior of the layer. Let meas $G = \sigma$. *Let us use the notation* $M = \max|u(x, 0)|$, $m = \max|u(x, T)|$. *Then* $M > m \exp(CT^{(n-2)/2}/\sigma^{2/n})$, *where* $C > 0$ *is a constant dependent on* M_1 *and* α_1.

Theorem 2.8. *Let $\varphi(t)$, $0 < t < \infty$ be a continuously differentiable function such that $|\varphi'| < k =$ const. Let $G \subset \mathbb{R}^{n+1}$ be a domain lying in the half-space $t > 0$, and such that for every $\tau > 0$ section of G by the hyperplane $t = \tau$, G_τ, is bounded. Moreover,*

$$\text{meas } G_\tau < \varphi(\tau), \quad 0 < \tau < \infty.$$

Let a function $u(x, t)$ be a solution of equation (8) in G, continuous in $\bar{G}$ and zero on the part of the boundary lying in the half-space $t > 0$. Let us use the notation

$$M(t) = \max_x |u(x, t)|, \quad 0 \leqslant t < \infty.$$

Then

$$M(t) \leqslant M(0)e^{-C\int_0^t \varphi(\tau)^{-2/n}\, d\tau},$$

where $C > 0$ is a constant which depends on α, M, n, and k.

This theorem was first proved by Cheremnykh [1959] in the case of smooth coefficients (and of one space variable).

We assumed here that $b_i(x, t) \equiv 0$, $i = 1, \ldots, n$. If $\varphi(t)$ is non-increasing in t, then a statement identical to Theorem 2.8 is true for bounded coefficients b_i, and if $\varphi(t) \to 0$, then the coefficients b_i can grow at a certain rate. Of great interest is the case when the coefficients b_i are different from zero, and do not tend to zero as $t \to \infty$, while $\varphi(t) \to \infty$. In this regard, the first result was obtained by Cheremnykh [1959] who showed that logarithmic expansion of the domain G is permissible here. In a certain sense, Il'in [1985] obtained the final result in this direction: $\varphi(t)$ must grow not faster than logarithmically.

The Phragmen-Lindelöf type theorems, in particular for domains that can become unbounded in x for finite values of t, were obtained in a series of papers of Olejnik, Radkevich and Iosifian (Iosifian and Olejnik [1977], Olejnik and Iosifian [1977], Olejnik and Radkevich [1978]). They used estimates of the energy integral of the solution using Green's formula (the principal part of the equation must, therefore be in divergence form). The same authors also obtained similar results for higher order equations and for systems. We shall return to this question when we talk about equations in divergence form.

1.7. Liouville Type Theorems. In this section we shall asume that in the operator L $b_i(x, t) \equiv c(x, t) \equiv 0$, $i = 1, \ldots, n$. Then Lemma 1 gives the following oscillation theorem.

Theorem 2.9. *Let $(x_0, t_0) \in \mathbb{R}^{n+1}$ be an arbitrary point, let Z_1 and Z_2 have the same meaning as in the statement of Lemma 1. Let a solution $u(x, t)$ of equation (8) be defined in Z_i. Then*

$$\operatorname*{osc}_{Z_1} u > p \operatorname*{osc}_{Z_2} u$$

where $p > 1$ depends on M_1, α_1, and n.

The following version of the two-sided Liouville theorem is immediately obtained from this theorem.

Theorem 2.10. *Let $u(x, t)$ be a solution of the equation $Lu - u_t = 0$ in the half-space $t \leqslant 0$ and $|u| \leqslant M$. Then $u \equiv$ const.*

The one-sided Liouville theorem does not hold for parabolic equations, as the following example shows: e^{x+t} is a solution of the equation $u_{xx} - u_t = 0$. However, the following theorems, that are obtained from the Harnack inequality, can be considered as analogs in a certain sense of the one-sided Liouville theorem.

Theorem 2.11. *Let $u(x, t)$ be a solution of equation* (8) $Lu - u_t = 0$, *which is defined and non-negative for $t \leqslant 0$. Let $u(x, 0) < M$. Then $u \equiv$ const.*

Theorem 2.12. *Let $u(x, t)$ be a solution of equation* (8) $Lu - u_t = 0$, *in the half-plane $t \leqslant 0$. Let.*

$$\inf_{t\leqslant 0} u = m > -\infty \left(\sup_{t\leqslant 0} u = M < +\infty \right).$$

Then $u(x, t) \to m$ $(u(x, t) \to M)$ as $t \to -\infty$ for each x.

These last two theorems were proved by Glagoleva (see Landis [1971]).

§2. Parabolic Equations in Divergence Form

Let Ω be a domain in $\mathbb{R}^{n+1} = \mathbb{R}^n_x \times \mathbb{R}^1_t$. Let us consider the simplest parabolic equation in divergence form

$$Lu - \frac{\partial u}{\partial t} \equiv \sum_{i,k=1}^{n} \frac{\partial}{\partial x_i}\left(a_{ik}(x, t) \frac{\partial u}{\partial x_k} \right) - \frac{\partial u}{\partial t} = 0, \tag{10}$$

where $a_{ik}(x, t)$ are bounded measurable functions that satisfy the following conditions:

$$\begin{gathered} a_{ik}(x, t) = a_{ki}(x, t), \\ \forall (x, t) \in \Omega \forall \xi \in \mathbb{R}^n \lambda^{-1}|\xi| \leqslant \sum_{i,k=1}^{n} a_{ik}(x, t)\xi_i\xi_k \leqslant \lambda|\xi|^2. \end{gathered} \tag{11}$$

A generalized solution of equation (10) in Ω in a function $u \in V_2^{1,0}$ such that

$$\forall \varphi \in C_0^\infty(\Omega) \int_\Omega \left(\sum_{i,k=1}^{n} \frac{\partial u}{\partial x_i} \frac{\partial \varphi}{\partial x_k} - u \frac{\partial \varphi}{\partial t} \right) dx\, dt = 0 \tag{12}$$

(for a definition of the space $V_2^{1,0}$ see Ladyzhenskaya, Solonnikov, Uraltseva [1967]).

2.1. The Harnack Inequality. The Oscillation Theorem. An a-priori Bound for the Hölder Norm. We shall begin our study with the case when $a_{ik} \in C^\infty$ and $u(x, t)$ is the classical solution of equation (10).

The following theorem was proved by Moser [1964].

Theorem 2.13. *Let* $(x, t) \in \mathbb{R}^{n+1}$, $r > 0$. *Let us use the notation*

$$Z_1 = Z_{x,r}^{t,t+r^2}$$
$$Z_2 = Z_{x,r/2}^{t+(3/4)r^2,t+r^2}$$
$$Z_3 = Z_{x,r/2}^{t+(1/4)r^2,t+(1/2)r^2}$$

Let a positive solution equation (10) *be defined in* Z_1. *Then*

$$\sup_{Z_3} u/\inf_{Z_2} u < c, \tag{13}$$

where $C > 0$ *is a constant depending on* λ.

The following two facts are obtained from inequality (13).

1. There exists a constant ξ depending on λ such that for a solution u of equation (10) defined in Z_1, we have that

$$\operatorname{osc}_{Z_1} u > (1 + \xi) \operatorname{osc}_{Z_2} u \tag{14}$$

(for an equation containing first and zeroth order terms, this theorem was first proved by Ivanov [1983]).

2. (A-priori bound for the Hölder norm). Let G be a domain in $\mathbb{R}^{n+1}$ and let ρ be a positive number. We set

$$G_\rho = \{(x, t) \in G \cup \gamma(G): Z_{x,\rho}^{t-\rho^2,t} \subset G\}$$

($\gamma(G)$ is the "top" boundary of G). Let a solution u of equation (10) be defined in $G \cup \gamma(G)$. Then

$$\|u\|_{C^\alpha(G_\rho)} \leqslant C\|u\|_{C^0(G)},$$

where α depends on λ and C, depends on ρ as well. This theorem was first obtained by Nash [1958].

Let us consider a cylinder $Z_{x_0,r}^{t_1,t_2}$. We have.

Theorem 2.14. *Let a generalized solution u of equation* (10) *be defined in* $Z_{x_0,r}^{t_1,t_2}$. *Then there exists a sequence of matrices* $\{\|a_{ik}^{(m)}(x, t)\|\}$ *with infinitely differentiable coefficients, which satisfy conditions* (11) *with the same constant* λ *and a sequence of classical solutions* $u^{(m)}(x, t)$ *of equations* $\sum_{i,k=1}^n \frac{\partial}{\partial x_i}\left(a_{ik}^{(m)} \frac{\partial u^{(m)}}{\partial x_k}\right) - \frac{\partial u^{(m)}}{\partial t} = 0$ *such that* $u^{(m)} \to u$, *and furthermore, the covergence is uniform on every compact set in* $Z_{x_0,r}^{t_1,t_2} \cup \gamma(Z_{x_0,r}^{t_1,t_2})$.

This theorem can be proved, for example, by employing Galërkin's method and using the uniform a-priori estimate of the Hölder norm (see Ladyzhenskaya, Solonnikov and Uraltseva [1967]).

Hence it follows that Theorems 2.11 and 2.12 of this section hold for generalized solutions as well; in particular, each generalized solution satisfies the Hölder condition in every interior subdomain (the top boundary of the original domain can be taken as belonging to the subdomain).

2.2. Green's Function and its Estimate. Let equation (10) be defined in $\mathbb{R}^{n+1}$. A function $G(x, t)$ that is a generalized solution of equation (10) everywhere apart from the origin, is called Green's function if it has the following properties:

a) $G(x, t) \to 0$, as $t \to 0$, $x \neq 0$.

b) $G(x, t) = 0$ for $t < 0$.

c) $G(x, \tau)$ forms a family of functions converging to the δ-function in the variables $x \in \mathbb{R}^n$ as $\tau \to +0$.

Aronson [1967] proved the following theorem:

Theorem 2.15. *Let equation* (10) *be defined everywhere in* $\mathbb{R}^{n+1}_+$. *Then Green's function exists and admits the estimates*

$$\frac{C_1}{t^{n/2}} e^{-k_1(|x|^2/t)} \leqslant G(x, t) \leqslant \frac{C_2}{t^{n/2}} e^{-k_2(|x|^2/t)}, \tag{15}$$

where k_1, k_2, C_1, C_2 are positive constants dependent on λ.

Remark to Sections 2.1, 2.2. Porper and Ejdel'man [1984] noticed that results of these sections are still valid if instead of equation (10), the equation $Lu - p(x)u_t = O$ is considered, where $p(x)$ is a measurable function bounded both from below and above by positive constants. Definition of a generalized solution will change in an obvious manner, since p is independent of t.

2.3. The Maximum Principle. *G*-Capacity. The Growth Lemma. The Strong Maximum Principle

Theorem 2.16 (the weak maximum principle). *Let $G \subset \mathbb{R}^{n+1}$ be a bounded domain and let Γ be its parabolic boundary. If $u(x, t)$ is the generalized solution of equation* (10) *in G, then*

$$\sup_G u \leqslant \overline{\lim_{\substack{(x,t)\to\Gamma \\ (x,t)\in G}}}\, u(x, t)$$

and

$$\inf_G u \geqslant \overline{\lim_{\substack{(x,t)\to\Gamma \\ (x,t)\in G}}}\, u(x, t).$$

The proof follows directly from the fact that it is possible to approximate $u(x, t)$ by solutions $u_m(x, t)$ of equations with smooth coefficients uniformly in every interior subdomain.

Let $E \subset \mathbb{R}^{n+1}$ be a bounded Borel set. Let us consider in E all possible measures μ such that $\mu(E) < \infty$ and

$$\int_E G(x - \xi, t - \tau)\, d\mu(\xi, \tau) \leqslant 1 \quad \text{outside of } \bar{E}. \tag{16}$$

We call the number sup $\mu(E)$, where the supremum is taken over all measures satisfying condition (16), G-capacity of the set E and denote it by $C_G(E)$. It can be shown that there exists a measure μ_0 that satisfies inequality (16) and realizes this extremum, so that $C_G(E) = \mu_0(E)$. This measure is called the equilibrium measure (see for example Watson [1978]). We shall not use this fact.

Note that from Theorem 2.15 of Section 2 it follows that

$$C_G(E) \geqslant \kappa \text{ meas } E, \tag{17}$$

where $\kappa > 0$ is a constant depending on λ.

Furthermore, under the transformation

$$x_i' = kx_i, \quad i = 1, \ldots, n, \quad t' = k^2 t$$

$G'(x', t') = \dfrac{1}{k^n} G(x, t)$ is Green's function in the new coordinates. Therefore if E' is the image of E under the transformation,

$$C_{G'}(E') = k^n C_G(E). \tag{18}$$

Let $(x_0, t_0) \in \mathbb{R}^{n+1}$. Let us consider the three cylinders

$$Z_1 = Z_{x_0, r}^{t_0, t_0 + br^2},$$

$$Z_2 = Z_{x_0, ar}^{t_0 + (b-a)r^2, t_0 + br^2},$$

$$Z_3 = Z_{x_0, ar}^{t_0, t_0 + ar^2},$$

where $0 < a < b$, $0 < r \leqslant 1$. Let S_1 be the lateral surface of the cylinder Z_1, let

$$m = \sup_{\substack{(x,t) \in S_1 \\ (\xi, \tau) \in Z_3 \\ (x_0, t_0) \in \mathbb{R}^{n+1}}} G(x - \xi, t - \tau)$$

and

$$M = \inf_{\substack{(x,t) \in S_1 \\ (\xi, \tau) \in Z_3 \\ (x_0, t_0) \in \mathbb{R}^{n+1}}} G(x - \xi, t - \tau)$$

In view of the inequality (13), the numbers a and b, that depend on k_1, k_2, c_1, c_2, and therefore on λ, can be found in such a way that

$$M > 2m. \tag{19}$$

These numbers will depend neither on r, nor on the point (x_0, t_0).

Lemma 1 (the growth lemma). (Compare with Lemma 3 of Section 1.4, Chapter 1). *Let $(x_0, t_0) \in \mathbb{R}^{n+1}$, $r > 0$, and let a, b be the numbers we fixed above, and let Z_1, Z_2, and Z_3 be the above defined cylinders. Let us assume that $D \subset Z_1$. Let us consider the generalized solution of equation* (10) *in D, which is continuous in $\bar{D}$, positive in D and zero on Γ, the parabolic boundary of D. Let $E = Z_3 \backslash D$. Then*

$$\max_{\bar{D}} u \geqslant \left(1 + \xi \frac{C_G(E)}{r^n}\right) \max_{\overline{D \cap Z_2}} u, \tag{20}$$

where $\xi > 0$ depends on λ.

To prove this theorem let us choose an arbitrary $\varepsilon > 0$ and find on E the measure μ that satisfies (16) and such that $\mu(E) > C_G(E) - \varepsilon$. Let us set

$$U(x, t) = \int_E G(x - \xi, t - \tau)\, d\mu(\xi, \tau).$$

Consider the function

$$V(x, t) = \max_{\bar{D}} u(1 - U(x, t)/r^n + (C_G(E) + \varepsilon)/r^n). \tag{21}$$

By the maximum principle, $u > V$ and D and we obtain inequality (20) from (19) and (21). □

A corollary of this lemma is

Theorem 2.17 (the strong maximum principle). *If $u(x, t)$ is the generalized solution of equation* (10) *in a domain Ω and u takes either its maximum or minimum value either at an interior point of the domain or on its top boundary, then this solution is a constant in any domain subordinate to this point* (see the definition of subordination in Section 1.1 of Chapter 2).

2.4. Stabilization of Solutions as $t \to \infty$. A sizable number of recent works is devoted to the question of stabilization of solutions of parabolic equations. We take the stabilization of solutions at a point x to mean the existence of a limit $\lim_{t\to\infty} u(x, t)$. If the limit is uniform on a compact set K (or in $\mathbb{R}^n$) we say that the solution stabilizes uniformly on K (respectively, in $\mathbb{R}^n$). The first study of the stabilization of solutions of the heat equation was performed by Tikhhonov [1950].

Let us state the result obtained in that work. Tikhonov considered the equation

$$\frac{\partial u}{\partial t} = \frac{\partial^2 u}{\partial x^2}$$

for $0 < x < \infty$, $t > 0$, with the boundary condition

$$\sum_{k=0}^{m} \alpha_k \frac{\partial^k}{\partial x^k} u(0, t) = f(t).$$

He proved the following proposition:

In order that the solution of the boundary value problem above has a limit as $t \to \infty$ for every function $f(t)$, it is necessary and sufficient that all the roots of the equation $\sum_{k=0}^{m} \alpha_k q^k = 0$ lie in the region $-\frac{3\pi}{4} < \arg q_j < \frac{3\pi}{4}$. Furthermore, $\lim_{t\to\infty} u(x, t) = \lim_{t\to\infty} f(t)$.

In the early fifties, Krzyzanski initiated interesting studies of stabilization of the solution of the Cauchy problem. In particular, he (Krzyzanski [1957]) constructed an example of a bounded initial function $u_0(x)$ for which the solution of the Cauchy problem for the heat equation has no limit as $t \to \infty$.

Necessary and sufficient conditions of uniform stabilization on compact sets were obtained in the work of Repnikov and Ejdel'man [1967] in the case of a bounded initial function $u_0(x)$. Their condition is given by the existence of the finite limit $\lim_{\rho\to\infty} \dfrac{1}{\text{meas } V_\rho} \displaystyle\int_{V_\rho} u_0(x)\,dx$, where V_ρ is the ball $\sum_{i=1}^n x_i^2 \leqslant \rho^2$.

Necessity of the condition given above was first established by Repnikov [1963] using one of Wiener's *Tauberian theorems* (Wiener [1932]). Let us consider in detail the question of uniform, in $\mathbb{R}^n$, stabilization of solutions of the Cauchy problem for the heat equation. Let us introduce the concept of the uniform limit of the average over cubes.

A measurable function $u_0(x)$ has a *uniform limiting average* l if for every $\varepsilon > 0$ there is $R_0(\varepsilon)$ such that for any cube V_R^η with side $2R$ and center at η we have the inequality

$$\left|\frac{1}{(2R)^n}\int_{V_R^\eta} u_0(y)\,dy - l\right| < \varepsilon$$

for all η and $R \geqslant R_0$. Repnikov [1963] showed that if the initial function $u_0(x)$ is bounded, then in order that the solution of the Cauchy problem stabilize uniformly in $\mathbb{R}^n$, it is necessary and sufficient that $u_0(x)$ has a uniform limit of the average over cubes.

The question of stabilization in he case of an unbounded initial function, is an interesting one. The survey paper of Denisov and Repnikov [1984] is devoted to this question. Let us examine the question of stabilization in the case of variable coefficients.

Consider the equation

$$p(x)\frac{\partial u}{\partial t} = \sum_{i,j=1}^n \frac{\partial}{\partial x_j}\left(a_{ij}(x,t)\frac{\partial u}{\partial x_i}\right), \tag{22}$$

where $0 < \gamma_1 < p(x) < \gamma_2$, $\gamma_1, \gamma_2 = \text{const.}$, while the matrix $A = \|a_{ij}\|$ is symmetric, bounded, and positive definite,

$$\lambda^{-1}(\xi,\xi) \leqslant (A\xi,\xi) \leqslant \lambda(\xi,\xi), \quad \lambda = \text{const} \geqslant 1.$$

Drozhzhinov [1962], and Valitskij and Ejdel'man [1976] obtained sufficient conditions of pointwise stabilization of the solution as $t \to \infty$.

Following Porper and Ejdel'man [1978], we shall say that $f(x)$ has zero corner limit average at a point x_0 with speed not less than $\delta(N, x_0)$, if

$$\left|\left(\sum_{i=1}^n a_i\right)^{-1}\int_{R_{a,x_0}} f(x)\,dx\right| < \delta(N, x_0) \quad \forall a \in \mathbb{R}^n,$$

$$a = (a_1, \ldots, a_n), \quad |a_i| > N,$$

where we denote by R_{a,x_0} the parallelipiped $\{x = (x_1, \ldots, x_n) \in \mathbb{R}^n\colon |x_{0i}| \leqslant |x_i| \leqslant |x_{0i}| + a, i = 1, \ldots, n, x_0 = (x_{0i}, \ldots, x_{0n}), (x_i - x_{0i})a_i \geqslant 0\}$.

Porper and Ejdel'man [1975] prove that if the function $p(x)u(0, x)$ has zero limit average at x_0 with speed $\delta(N, x_0)$ then there exists $\varepsilon(t)$, $\varepsilon(t) \to 0$ as $t \to \infty$, defined by the constants $\gamma_1, \gamma_2, \lambda$, and $\delta(N, x_0)$, such that $|u(x_0, t)| \leqslant \varepsilon(t)$.

Various forms of necessary and sufficient stabilization conditions were obtained by Porper and Ejdel'man [1975] and by Mikhailov [1970].

Zhikov [1977] and Zhikov, Kozlov and Olejnik [1981] use *averaging methods* to obtain stabilization results. In this method an important role is played by Nash-type estimates.

Gushchin [1982] studies the question of stabilization of the Neumann problem. He also considers the question of dependence of the speed of stabilization on the geometry of the boundary of the domain. His results are, in particular, applicable to the Cauchy problem.

2.5. On Uniqueness Classes of the Solution of the Second Initial-boundary Value Problem for Parabolic Equations in Unbounded Domains. Let $G \subset \mathbb{R}^n$ be an unbounded domain, $\Omega = G \times [0, T]$, and let ∂G be piecewise continuously differentiable, $\partial G \times [0, T] = \Gamma$. Consider in Ω the equation

$$\sum_{i,k=1}^{n} \frac{\partial}{\partial x_i}\left(a_{ik}(x)\frac{\partial u}{\partial x_k}\right) + \sum_{i=1}^{n} b_i(x)\frac{\partial u}{\partial x_i} + c(x) - \frac{\partial u}{\partial t} = 0, \tag{23}$$

where $a_{ij}(x) = a_{ki}(x)$ are continuously differentiable and satisfy the following conditions: there exist $\lambda \geqslant 1$ and $M \geqslant 0$ such that

$$\lambda^{-1}|\xi|^2 \leqslant \sum_{i,k=1}^{n} a_{ik}(x)\xi_i\xi_k \leqslant \lambda|\xi|^2 \quad \forall \xi \in \mathbb{R}^n \quad \forall x \in G, \tag{24}$$

$$|b_i(x)| \leqslant M, \quad |c(x)| \leqslant M.$$

Let us consider the second homogeneous initial-boundary value problem

$$u|_{t=0} = 0, \quad \left.\frac{\partial u}{\partial \nu}\right|_{\Gamma} = 0 \tag{25}$$

$\left(\frac{\partial}{\partial \nu} \text{ -- is the co-normal derivative to } \partial G\right)$.

Let a function $h(r)$, $0 < r < \infty$ be monotone increasing and such that

$$\int^{\infty} \frac{dr}{h(r)} = \infty.$$

Let us set $F_0(r) = e^{rh(r)}$. The so-called Täcklind condition demands that for $|x|$ sufficiently large,

$$|u(x)| \leqslant F_0(|x|). \tag{26}$$

In Section 1.3, Chapter 2, we saw that this is precisely the condition satisfied in the uniqueness class for the Cauchy problem. Olejnik and Radkevich [1978] and Gushchin [1982] showed that if a solution of the problem (23), (25) in Ω for an arbitrary domain G, satisfies condition (26), then it is necessarily identically zero. In this section we investigate the dependence of uniqueness classes of the problem (23), (25) on the geometry of the domain. We shall consider classical solutions $u \in C^{2,1}$ of equation (23).

Let an arbitrary unbounded domain $G \subset \mathbb{R}^n$ be given. Let us consider a sequence of pairs of numbers $\{r_k', r_k''\}$ such that $0 < r_k' < r_k'' \leqslant r_{k+1}'$ $k = 1, 2, \ldots$ and $r_k' \to \infty$ as $k \to \infty$.

Let us denote by G_k the part of G contained between the spheres $|x| = r_k'$ and $|x| = r_k''$.

Theorem 2.18. *Let $F(r)$, $r > 0$ be an arbitrary positive monotone increasing function, and let the domain G satisfy the condition*

$$\text{meas } G_k \cdot F^2(r_k'')/(r_k'' - r_k')^2 \to 0 \text{ as } k \to \infty, \tag{27}$$

and let $u(x, t)$ be a solution of the problem (23), (25) *n* $\Omega = G \times [0, T]$. *Then if for large $|x|$ the inequality*

$$|u(x, t)| \leqslant F(|x|) \tag{28}$$

holds, then $u \equiv 0$.

Proof. It suffices to consider the case when T is sufficiently small. Let t_0, $0 < t_0 < T$ be arbitrary. Let us set $V(x) = \int^{t_0} u^2(x, t)\, dt$. Let us consider the restriction of V to G_n. By the integral mean-value theorem (cf. Theorem 1.25) there exists a piecewise smooth surface Σ_k separating the spheres $|x| = r_k'$ and $|x| = r_k''$ in G_k, such that

$$\int_{\Sigma_k} |\partial V/\partial v|\, ds \leqslant C \text{ meas } G_n \cdot \text{osc } V/(r_k'' - r_k')^2. \tag{29}$$

Let us denote by D_k the portion of G which is contained in the ball $|x| < r_k''$; and which is not separated from the sphere $|x| = r_k''$ by the surface Σ_k. Multiplying equation (23) by u and integrating over $D_k \times [0, t_0]$, we obtain, having used Green's formula and having taken into account (24) and (29), that

$$\int_{D_k} u^2(x, t_0)\, dx + \lambda^{-1} \int_t^{t_0} \int_{D_k} |\text{grad}_x u|^2\, dx\, dt - M \int_0^{t_0} |\text{grad}_x u| \cdot u\, dx\, dt$$

$$- M \int_0^{t_0} \int_{D_k} u^2\, dx\, dt \leqslant \int_{\Sigma_k} |\partial V/\partial v|\, ds \leqslant C \text{ meas } G_k \cdot F^2(r_k')(r_k'' - r_k')^2.$$

Integrating with respect to t_0 from 0 to T and taking T sufficiently small, we find that

$$\frac{1}{2} \int_0^T \int_{D_k} u^2(x, t)\, dx\, dt \leqslant T \cdot C \text{ meas } G_k \cdot F^2(r_k'')/(r_k'' - r_k')^2 \to 0$$

as $k \to \infty$. That is, $u \equiv 0$. □

Let us now consider a particular form of domain G. Let $f(r) \in C^1(0, \infty)$, $f(r) > 0$ and $f(r) \downarrow 0$, $f'(r) \downarrow 0$ as $r \to \infty$. Let us use the notation

$$G = \{r \in \mathbb{R}^n \colon x_1 > 0, |x_i| < f(x_1), i = 2, \ldots, n\}. \tag{30}$$

Let $h(r)$, $0 < r < \infty$ be a positive continuous monotone decreasing function, and let $F(r)$, $r > 0$ be monotone increasing and such that $F(r) = o(h(r)^{2/n-1}(f(r-h(r)))^{2/n-1})$ as $r \to \infty$.

Theorem 2.19. *For G and F as above the following claim is true: If $u(x, t)$ is a solution of the problem* (23), (25) *in Ω such that*

$$u(x, t) < F(|x|),$$

then $u \equiv 0$.

Obviously, this theorem is a particular case of the preceeding one.

In particular, for a domain G defined by $f(r) = \exp(-r^k)$, $k > 2$, according to (30), the uniqueness class is defined by the function $F(r) = o(r^{2/n}) \exp\left(\frac{n-1}{2} r^k\right)$; for a domain G defined by $f(r) = \exp(-\exp)$, $F(r) = o(\exp(-r)) \exp(-\exp r))$.

In the more general case when the coefficients of the equation depend on x and t, a similar theorem was proved by Gagnidze [1984], who used the method of introducing a parameter suggested by Olejnik and Radkevich. We also note that earlier Olejnik [1983] constructed an example of an equation of the form $\Delta u + \sum b_i \frac{\partial u}{\partial x_i} - \frac{\partial u}{\partial t} = 0$ for which, for $f(r) \geqslant \exp(-Cr)$, the Täcklind class cannot be made larger.

§ 3. Parabolic Equations with Smooth Coefficients

3.1. On the Regularity of Boundary Points for Parabolic Equations. Let us first consider the heat equation

$$\Delta u - u_t = 0. \tag{31}$$

Let G be a bounded domain in $(n + 1)$-dimensional space (x, t), $x = (x_1, \ldots, x_n)$. Let us denote by γ the set of boundary points (x_0, t_0) of this domain for each of which there exists a positive number ε such that the cylinder $\{(x, t): t_0 - \varepsilon < t < t_0, |x - x_0| < \varepsilon\}$ is contained in G, while the cylinder $\{(x, t): t_0 < t < t_0 + \varepsilon, |x - x_0| < \varepsilon\}$ lies outside G. The set γ is called the top boundary of the domain G, and the set $\Gamma = \partial G \backslash \gamma$ is its parabolic boundary (see § 1, p. 159).

The first boundary value problem for equation (31) is posed as follows. Let f be a function continuous on the parabolic boundary. It is required to find a solution of the heat equation in $G \cup \gamma$, which is continuous on $\bar{G}$ and equal to f on Γ. Naturally, this problem is not soluble for all domains G. But as in the case of Laplace's equation, it is possible to find a solution of equation (31) corresponding to any function f continuous on Γ. This solution $u_f(x, t)$ is called the generalized solution of the boundary value problem at hand. It can be obtained either by the method of supersolutions of Perron (see, for example, Petrovskij [1935]) or by Wiener's method of approximating the domain from the interior by step-shaped domains (see, for example, Landis [1971]). Both methods lead to the same function $u_f(x, t)$.

A point (x_0, t_0) belonging to the parabolic boundary is called *regular* if for every function f which is continuous on Γ, $u_f(x, t) \to f(x_0, t_0)$ as $(x, t) \to (x_0, t_0)$, $(x, t) \in G$.

In 1934, Petrovskij (see Petrovskij [1935]) undertook a thorough study of regularity conditions in the case of G being a planar domain. His criterion is as follows.

Let $(x_0, t_0) \in \partial G$, and suppose there is a continuous curve with an endpoint in this point, which is contained in the intersection of the half-plane $t < t_0$ with G. Let $p(\eta)$ be a positive function defined for $0 < \eta < \varepsilon < 1$, which goes to zero as $\eta \to 0$. Let us use the notation

$$H = \{(x, t): |x - x_0| < 2\sqrt{(t_0 - t)|\ln \rho(t_0 - t)|}, 0 < t_0 - t < \varepsilon < 1\},$$

$$H^- = \{(x, t): x_0 - \varepsilon < x < x_0, 0 < t_0 - t < \varepsilon\}\backslash H,$$

$$H^+ = \{(x, t): x_0 < x < x_0 + \varepsilon, 0 < t_0 - t < \varepsilon\}\backslash H.$$

If

$$\int_0 \frac{d\eta}{\eta\rho(\eta)} < \infty \tag{32}$$

and either H^+ or H^- belong to the complement of G, then the point (x_0, t_0) is regular. If

$$\int_0 \frac{d\eta}{\rho\eta(\eta)} = \infty$$

and H belongs to G, then the point (x_0, t_0) is irregular.

Using the methods of Petrovskij's work quoted above, it is possible to obtain a quite general regularity condition in the multidimensional case. Let us consider, in the space $x_1, \ldots, x_n, t$ the surface

$$\{(x, t): r = 2\sqrt{|t||\ln \rho(|t|)|}, |x| = r < r_0, -\varepsilon < t < 0\},$$

where ρ satisfies condition (32).

Let us cut out of this surface a piece S. Using the hyperplane $t = -H < 0$ and the cylinder

$$\left\{(x, t): x_n = \varepsilon\left(\sum_{i=1}^{n-1} x_i^2\right)^{1/2}, \varepsilon > 0\right\}.$$

Let us call the point (0, 0) the vertex of S. Let us consider rigid body translations of S composed of translations in the x space and along the t axis. If there exist H and ε such that S can be translated so that its vertex coincides with $(x_0, t_0) \in \partial G$, while all other points of S belong to the complement of G, then the point (x_0, t_0) is regular (for the part of G contained in the half-space $t < t_0$, which is assumed to be non-empty).

Tikhonov [1938] showed that if G is the cylindrical domain $G = \Omega \times (0, T]$ where Ω is a domain in the n-dimensional x-space, a boundary point $(x_0, t_0) \in$

$\partial\Omega \times [0, T]$ is regular if and only if x_0 is a regular point for Laplace's equation defined in Ω. (The points of the bottom base of the cylinder are, naturally, always regular.)

Continuing Petrovskij's studies. Pini [1954] considered the equation

$$u_{xx} - u_t = 0 \tag{33}$$

in a planar domain, assuming that in a neighborhood of the boundary point under consideration, the boundary is a graph of a continuous function $x = \varphi(t)$. He found a necessary regularity condition and also gave a certain sufficient regularity criterion. Significantly, Pini stated his conditions in terms similar to the ones used in stating Wiener's criterion of regularity of boundary points for Laplace's equation.

As is well known, Wiener's condition consists of the following. Let $\Omega \subset \mathbb{R}^n$ be a bounded domain and let $x_0 \in \partial\Omega$. For simplicity, let us assume that $n \geqslant 3$. Let q, $0 < q < 1$ be an arbitrary number. Let us denote the ball of radius q^k with center at the point x_0, by Q_k, and the part of the complement of Ω which is contained in the annular $Q_k \backslash Q_{k+1}$, by E_k. Let us consider the series

$$\sum_{k=1}^{\infty} \frac{\operatorname{cap} E_k}{q^{(n-2)k}}, \tag{34}$$

where cap E_k denotes the capacity of the set E_k. Then divergence of the series (34) is a necessary and sufficient condition for regularity of the point x_0.

The terms of series (34) can be given different interpretations. One of these goes as follows. The annulus $Q_k \backslash Q_{k+1}$ is a domain in which the fundamental solution of the Laplacian (with a pole at x_0) is bounded from below and from above by the two constants λ_k and λ_{k+1} such that

$$\lambda_{k+1} = \lambda_k \cdot \lambda, \quad \lambda > 1$$

($\lambda = 1/q^{(n-2)}$). Every term of the series (34) is the product of the capacity of the part of the complement of the domain that is contained in the relevant annulus and of the value of the fundamental solution on the outer boundary of the annulus (or at any other point of the annulus: it doesn't make any difference, since values of the fundamental solution at any two points of the annulus are finitely related).

Thermal capacity can be defined in a natural way for the heat equation. Thus, let $E \subset \mathbb{R}^{n+1} = \{(x, t)\}$ be a Borel set; a measure μ on E is called admissible if

$$U(x, t) = \int_E G(x, t; \xi, \tau)\, d\mu(\xi, \tau) \leqslant 1,$$

where $G(x, t; \theta, \tau)$ is the fundamental solution:

$$G(x, t; \xi, \tau) = \begin{cases} \dfrac{1}{2^n(\pi(t-\tau))^{n/2}} e^{-(|x-\xi|^2/(4(t-\tau)))}, & t > \tau, \\ 0, \quad t \leqslant \tau, \quad (x, t) \neq (\xi, \tau), \end{cases}$$

The number $C(E) = \sup \mu(E)$, where the supremum is taken over all admissible measures is called the thermal capacity of the set E. (We note that the measure for which the supremum is obtained, always exists; it is called the equilibrium measure).

Thermal capacity was introduced by Pini [1954] in order to study questions of boundary point regularity. (At present, much more is known about the thermal capacity, see, for example, Watson [1978]).

Pini's condition is stated in the following way: let us consider the annulus lying between level curves of the fundamental solutions of the equation $u_{xx} + u_t = 0$ (which is adjoint to (33)), corresponding to the values λ^k and λ^{k+1}, where $\lambda > 1$. Let e_k be the part of the curve $x = \varphi(t)$ that is contained in this annulus. Then Pini's necessary regularity condition is the divergence of the series

$$\sum_{k=1}^{G} C(e_k)\cdot\lambda^k. \tag{35}$$

As in the case of Laplace's equation, it would have been more natural to consider the part of the complement to the domain contained in the annulus, and its capacity, instead of considering the part of the boundary contained in the annulus. When the boundary is the graph of a continuous function $x = \varphi(t)$, the two approaches are equivalent: the resulting series converge or diverge simultaneously. However, in the multi-dimensional case (and even in the planar case of a more complicated boundary), this is no longer so. It turns out that it is exactly the capacity of the part of the complement of the domain that is contained in the annulus, that has to be considered. That was exactly what Lanconelli [1973] did. He proved the necessity of such a regularity condition for a general $(n + 1)$-dimensional domain. Lanconelli himself could not prove sufficiency of his condition (in the work quoted above he gave a more restrictive sufficient condition that does not coincide with the necessary one). However, Evans and Gariepi [1982] showed that Lanconelli's condition is both a necessary and a sufficient regularity condition.

Let us return to Wiener's criterion for Laplace's equation. A different interpretation of the terms of series (34) is possible.

Let μ_k be the equilibrium measure corresponding to the set E_k, and let

$$u_k(x) = \int_{E_k} \frac{d\mu_k(y)}{|x - y|^{n-2}}$$

be the potential corresponding to it. Then each term of series (34) is finitely related to the value of this potential u_k at x_0: there exist positive constants C_1 and C_2 such that

$$C_1 \operatorname{cap} E_k/q^{(n+2)k} \leqslant u_k(x_0) \leqslant C_2 \operatorname{cap} E_k/q^{(n-2)k}.$$

Therefore, series (34) converges and diverges if and only if the series

$$\sum_{k=1}^{\infty} u_k(x_0). \tag{36}$$

does.

Instead of the balls Q_k, radii of which decrease in geometric progression, we could consider balls $\hat{Q}_k$ with centers at x_0, the radii of which decrease faster than that. Let us denote by $\hat{E}_k$ the intersection of the complement of the domain G with the annulus $\hat{Q}_k \setminus \hat{Q}_{k+1}$, let us find the equilibrium measure $\hat{\mu}_k$ for $\hat{E}_k$, the corresponding potential $\hat{u}_k(x)$ and let us construct the series

$$\sum_{k=1}^{\infty} \hat{u}_k(x_0).$$

Divergence of this series is also a necessary and sufficient if regularity of the point x_0. However, the faster is the rate of decrease of the radii of the balls $\hat{Q}_k$, the closer this necessary and sufficient condition comes to becoming a tautology. In order that the criterion contain as much information as possible, the rate of decrease of radii of the balls must be made as low as possible. On the other hand, it can be shown that radii have to decrease at such a rate that $\hat{u}_k$ on $\hat{E}_{k+p}$, $p = \text{const.}$ is commensurate with $\hat{u}(0)$. Hence it follows that radii cannot decrease slower than in geometric progression, and we conclude that Wiener's criterion is the best possible criterion of this type.

Interpreted as the divergence of a series of the form (36), this criterion was extended to the heat equation by one of the authors (Landis [1969]) in the following way.

Let $\{t_k\}$, $k = 1, 2, \ldots$ be an increasing sequence converging to t_0. Let us denote the intersection of the layer $\{(x, t): t_k < t < t_{k+1}, x \in \mathbb{R}^n\}$ with the complement of domain G by E_k. Let us consider μ_k, the equilibrium measure of the set E_k for a parabolic potential, and let $u_k(x, t)$ be the corresponding potential:

$$u_k(x, t) = \frac{1}{(2\sqrt{\pi})^n} \int_{E_k} \frac{1}{(t-\tau)^{n/2}} e^{-(|x-\xi|^2/(4(t-\tau)))} \, d\mu_k(\xi, \tau), \quad t > t_k.$$

Let us construct the series

$$\sum_{k=1}^{n} u_k(x_0, t_0).$$

If t_k converges to t_0 fast enough, then the divergence of this series is a necessary and sufficient condition of regularity of the point (x_0, t_0).

Here it is important to find the slowest admissible rate of decrease of the thickness of the layer $t_k < t < t_{k+1}$. The appropriate sequence $\{t_k\}$ was found in the work (Landis [1969]) quoted above.

The sequence $\{t_k\}$ is defined as follows. By recursion, an auxiliary sequence $\{\rho_k\}$ is constructed: $\rho_0 = 1/2$, $\rho_{k+1} = \rho_k \sqrt{|\ln \rho_k|}$; t_k is found from the condition $4t_k \cdot \ln |t_k| = \rho_{k-2}^2$.

The following generalizations deal with the extension of the results for the heat equation (31) to linear second order parabolic equations with variable coefficients.

Let us consider the equation

$$\frac{\partial u}{\partial t} = \sum_{i,k=1}^{n} a_{ik}(x, t) \frac{\partial^2 u}{\partial x_i \partial x_k} + \sum_{i=1}^{n} b_i(x, t) \frac{\partial u}{\partial x_i} + c(x, t) u.$$

If

$$a_{ik}(0, 0) = \delta_{ik}, \quad i, k = 1, \ldots, n, \tag{37}$$

where δ_{ik} is the Kronecker delta, and the coefficients satisfy the Dini condition (with respect to the parabolic distance) in a neighborhood of a point (x_0, t_0), then, as Novruzov [1973] showed, regularity conditions for the point (x_0, t_0) coincide with regularity conditions for this point for equation (31). He also showed (Novruzov [1973]), that if the Dini condition is violated, a point that is regular for the heat equation is not necessarily regular for the equation with variable coefficients.

Let us note that here condition (37) is essential. Unlike the elliptic case, when under sufficient smoothness conditions on the coefficients, boundary point regularity conditions coincide for all elliptic second order equations, in the parabolic case the situation is different. For example, equations

$$\frac{\partial u}{\partial t} = \frac{\partial^2 u}{\partial x^2} \tag{38}$$

and

$$\frac{\partial u}{\partial t} = a\frac{\partial^2 u}{\partial x^2}, \quad a = \text{const} > 0, \tag{39}$$

for $a \neq 1$ have different regularity conditions. To see that, let $a = 1/(1 + \varepsilon)$, $\varepsilon > 0$. Let us consider the domain

$$G = \left\{(x, t) \colon x^2 < 4 \ln \frac{1}{\ln|t|}, \ -1 < H < t < 0\right\}.$$

The point (0, 0) is regular for equation (38).

Let us make a change of variables in equation (38): $x = x_1$, $t = \frac{1}{a}t_1$, so that if we set $u(x, t) = u_1(x_1, t_1)$ we get that $\frac{\partial u_1}{\partial t_1} = \frac{\partial^2 u_1}{\partial x_1^2}$. Under this change of variables, G is mapped into the domain

$$\left\{(x_1, t_1) \colon x_1^2 < 4(1 + \varepsilon)t_1 \ln \frac{1}{\ln|t_1(1 + \varepsilon)|}, \ aH < t_1 < 0\right\}.$$

As for $|H|$ and ε sufficiently small, $4(1 + \varepsilon)t_1 \ln \frac{1}{\ln|t_1(1+\varepsilon)|} > 4(1 + \varepsilon)t_1 \ln \frac{1}{\ln|t_1|}$, the point (x_0, t_0) will be irregular for the domain G and for equation (39).

Therefore if one wants to find necessary regularity conditions and sufficient regularity conditions that should hold for the whole family of uniformly parabolic second order equations one cannot expect these conditions to coincide.

Such sufficient conditions were found for equations with principal part in divergence form,

$$\frac{\partial u}{\partial t} = \sum_{i,k=1}^{n} \frac{\partial}{\partial x_i}\left(a_{ik}(x, t)\frac{\partial u}{\partial x_k}\right) + \sum_{i=1}^{n} b_i(x, t)\frac{\partial u}{\partial x_i} + c(x, t)u \tag{40}$$

by Lanconelli [1975]. He only required measurability and boundedness of coefficients, as well as that the inequality

$$\lambda^{-1}|\xi|^2 \leqslant \sum_{i,k=1}^{n} a_{ik}(x, t)\xi_i\xi_k \leqslant \lambda|\xi|^2$$

holds for some $\lambda \geqslant 1$.

Using Aronson's [1967] estimate of the *fundamental solution* of such an equation in terms of the *fundamental solution* of the heat equation (see Section 2.2), Lanconelli obtained the following regularity condition for a boundary point (x_0, t_0) of a domain $\Omega \subset \mathbb{R}^n$, which holds for any equation of the form (40).

Let Ω' be the complement of Ω, and let λ, $0 < \lambda < 1$ be an arbitrary number. Let us set

$$\Omega'_{h,k} = \{(x, t) \in \Omega' \colon \lambda^{k+1} \leqslant t_0 - t < \lambda^k, \lambda^{h+1} \leqslant e^{-(|x-x_0|^2/(4(t-\tau)))} \leqslant \lambda^h\}.$$

A sufficient condition of regularity is that for any $a > 0$

$$\sum_{h,k=0}^{\infty} \frac{C(\Omega'_{h,k})}{\lambda^{k\frac{h}{2}-ha}} = \infty,$$

where $C(\Omega'_{h,k})$ is the thermal capacity of the set $\Omega'_{h,k}$.

Landis [1971] obtained sufficient boundary point regularity conditions for uniformly parabolic equations not in divergence form with, in general, discontinuous coefficients. These conditions are stated in terms of the divergence of a series similar to Wiener's series. Using these conditions, sufficient geometric criteria are found. For example, a point (x_0, t_0) will always be regular if it can be reached from outside the domain by the vertex of a "squashed cone". (K is a "squashed cone" if in a neighborhood of x_0 in $\mathbb{R}^n$ an orthogonal system of coordinates $x'_1, \ldots, x'_n$ can be introduced such that for some $a > 0$, $b > 0$

$$K = \left\{(x', t) \colon t \leqslant t_0, 0 < x'_1 < b, \sum_{i=2}^{n} (x_i'^2 + t_0 - t)^{1/2} < ax_1\right\}.$$

3.2. Behavior of Solutions on Parabolic Equations on the Characteristic. We saw that solutions of elliptic equations share properties with analytic functions even in the case where the coefficients of these equations are differentiable only a finite number of times. It turns out that bounded solutions of parabolic equations on the characteristic also have similar properties.

Let us consider a parabolic equation of the form

$$Lu - \frac{\partial u}{\partial t} \equiv \sum_{i,k=1}^{n} \frac{\partial}{\partial x_i}\left(a_{ik}(x)\frac{\partial u}{\partial x_k}\right) + c(x)u - \frac{\partial u}{\partial t} = 0 \tag{41}$$

with coefficients depending on x and having some degree of smoothness ($a_{ik} \in C^2, \in L^\infty$).

In order to study this equation on the characteristic, Landis and Olejnik [1974] employed the following method, which they call the elliptic continuation of solutions of parabolic equations.

Theorem 2.20. *Let a bounded solution $u(x, t)$ of equation* (41) *be defined in the cylinder $\Omega = \{(x, t): 0 < t \leqslant t_0, |x| < R\}$. Then for every $p > 0$ there exist constants $C > 0$ and $\delta > 0$ depending on L, R, t_0, and ρ such that there exists a solution $U(x, y)$ of the elliptic equation*

$$\frac{\partial^2 U}{\partial y^2} + LU = 0, \tag{42}$$

in the cylinder $Q = \{(x, y), x \in \mathbb{R}^n, y \in \mathbb{R}^1, |x| < R - \rho, |y| < \delta\}$. $U(x, y)$ satisfies

$$U|_{y=0} = u(x, t_0), \left.\frac{\partial U}{\partial y}\right|_{y=0} = 0$$

and

$$\sup_Q |U| \leqslant C \sup_\Omega |u|.$$

In the work quoted above, this theorem is proved for parabolic equations of high order.

A number of corollaries follow from this theorem.

Theorem 2.21 (uniqueness theorem). *Let Ω be the same cylinder as above, and let $u(x, t)$ be a solution of equation* (41) *in Ω. Suppose $|x_0| < R$, and that as $x \to x_0$ the function $u(x, t_0)$ decays faster than any polynomial, that is, for each k there exists C_k such that $|u(x, t_0)| < C_k|x - x_0|^k$. Then $u(x, t_0) \equiv 0$.*

As a particular case of this theorem, we have a uniqueness theorem for the extension of a solution on the characteristic from a subdomain to the whole of the domain. We also have a stronger claim.

Theorem 2.22 (the three balls theorem). *Let r_1, r and r_2 be three arbitrary numbers satisfying $0 < r_1 < r < r_2 < 1$. Let a solution $u(x, t)$ of equation* (41) *be defined in the cylinder $\Omega = \{(x, t): 0 < t < r_2^2, |x| < r_2\}$. Let us set $M_1 = \sup_\Omega |u(x, t)|$ and $M(r) = \sup_{|x|<r} |u(x, r_2^2)|$. Then there exist positive constants C_1, C_2 and C_3 that depend only on the operator L, such that for any $\varkappa > C_3$, it follows from the inequality*

$$M(r_1)/M_1 < r_1^\varkappa$$

that

$$M(r)/M_1 < (C_1 r)^{\varkappa/C_2}.$$

Theorem 2.23 (continuous dependence in the problem of extension from a subdomain to the whole of the domain on the characteristic). *Let Ω be a domain of the space (x, t), and let $(\hat{x}, t_0) \in \Omega$. We say that the parabolic distance from $(\hat{x}, t_0)$ to the boundary of Ω is larger than r, if the cylinder $\Omega_r = \{(x, t): t_0 - r^2 < t \leqslant t_0, |x - \hat{x}| \leqslant r\}$ lies in Ω.*

Let a solution $u(x, t)$ of equation (41) *be defined in Ω, let $(\hat{x}, t_0) \in \Omega$, and let the parabolic distance from $(\hat{x}, t_0)$ to the boundary of Ω be larger than r. Then there* *exist constants* $C' > 0$ *and* $\sigma > 0, 0 < \sigma < 1$, *that depend only the operator* L, *such that if* $\sup_\Omega |u| < M$ *and* $\sup_{|x-\hat{x}|<r/4} |u(t_0, x)| = \varepsilon$, *then*

$$\sup_{|x-\hat{x}|<r/2} |u(t_0, x)| < (C'M)^{1-\sigma} \cdot \varepsilon^\sigma.$$

From this theorem we can derive a Liouville type theorem.

Theorem 2.24. *Let equation* (41) *be uniformly elliptic in the semi-infinite cylinder*

$$\Omega = \left\{(x, t): 0 \leqslant t \leqslant r^2, 0 \leqslant x_1 < \infty, \sum_{i=2}^n x_i^2 \leqslant r^2\right\}, \quad r < 1,$$

Assume that derivatives of its coefficients to some order are uniformly bounded, and let $u(x, t)$ *be its solution which is uniformly bounded in* Ω. *Then there is a constant* $H > 0$ *depending only on* L, *such that if for some constant* $C > 0$

$$|u(x, r^2)| \leqslant Ce^{-e^{Hx_1/r}},$$

then $u(x, r^2) \equiv 0$.

In the case when Ω is the layer $0 < t < t_0$, $x \in \mathbb{R}^n$, Gusarov [1975] proved the following rather sharp Liouville type theorem. He considers equation (41) with the following restrictions imposed on its coefficients: $D^\rho a_{ik}(x) = o(|x|^{-|\rho|})$, $\rho = 1, 2$, $c(x) = o(|x|^{-2})$ as $|x| \to \infty$.

Theorem 2.25 (sharp Liouville type theorem). *Let* $u(x, t) = O(e^{K|x|^2})$ *uniformly in* $t \in (0, t_0]$, *where* K *is some positive constant.*

Then there exists a constant $C > 0$, *such that if in a cone contained in the characteristic* $t = t_0$ *we have the estimate*

$$u(x, t_0) = O\left(\exp\left(-C\left(K + \frac{1}{t_0}|x|^2\right)\right)\right),$$

then $u \equiv 0$.

Here C depends on the constant of ellipticity of the operator L, on the dimension of the space and on the angle of opening of the cone.

Remark. If the coefficients are assumed to be somewhat smoother, we can consider the full operator $\sum_{i,k=1}^n a_{ik}(x)\frac{\partial^2}{\partial x_i \partial x_k} + \sum_{i=1}^n b_i(x)\frac{\partial}{\partial x_i} + c(x)$ in Theorems 2.20–2.25 instead of the operator L.

In order for the methods of Gusarov [1975] to work, it is essential that the coefficients be independent of the variable t. It would be interesting to know whether similar results hold in the case, where the coefficients depend both on x and t.

Glagoleva proved the following three cylinder theorem:

Let equation

$$\sum_{i,k=1}^n a_{ik}(x, t)\frac{\partial^2 u}{\partial x_i \partial x_k} + \sum_{i=1}^n b_i(x, t)\frac{\partial u}{\partial x_i} + c(x, t)u - \frac{\partial u}{\partial t} = 0 \tag{43}$$

be defined in $Z_{0,R}^{0,T}$. Assume that its coefficients satisfy

a) $a_{ik}(x, t) \in C^{3,1}$;

b) $\sum_{i,k=1}^{n} a_{ik}(x, t)\xi_i\xi_k \geqslant \lambda|\xi|^2, \lambda > 0$

c) $a_{ik}(x, t)$, their derivatives to indicated order, $b_i(x, t)$ and $c(x, t)$ are bounded.

For a function defined on $Z_{0,\rho}^{t_1,t_2}$ let us set

$$M_f(\rho, t_1, t_2) = \sup_{Z_{0,\rho}^{t_1,t_2}} |f|.$$

Theorem 2.26 (the three cylinder theorem). *Let* $0 < r_0 < R$. *Let* $u(x, t)$ *be a solution of equation* (43) *in* $Z_{0,R}^{0,T}$ *satisfying the condition*

$$M_u(r_0, 0, T) = \Delta, \quad M_u(R, 0, T) = M.$$

Then for any r, $r_0 < r < R/2C$ *and any positive number* a *for sufficiently small* Δ, *the following inequality holds*:

$$\ln M_u(r, a, T-a) \leqslant \ln \Delta \frac{\ln\left(\dfrac{Cr}{R}\right)}{\ln\left(\dfrac{Br}{R}\right)} + \ln M \frac{\ln\left(\dfrac{Br_0}{Cr}\right)}{\ln\left(\dfrac{Br_0}{R}\right)} - \ln\left(\frac{Cr}{R}\right), \tag{44}$$

where $B > 0$, $C > 0$ *are constants depending on the equation and on* a.

It seems plausible that smoothness of the coefficients a_{ik} can be weakened to $C^{2,1}$, and that the term $\ln\left(\dfrac{Cr}{R}\right)$ of inequality (44) can be neglected (compare this with the three balls theorem for elliptic equations (theorem 1.49)).

Kudryavtseva [1971] proved the following theorem concerning the *rate of decay of a solution* in a layer over a cone.

Let K be the cone

$$K = \left\{x \in \mathbb{R}^n \colon x_1 > 0, \sum_{i=2}^{n} x_i^2 < Cx_1^2, C > 0\right\}$$

and

$$\Omega = \{(x, t) \in \mathbb{R}^{n+1} \colon 0 < t < T, x \in K\}.$$

Let equation (43) be defined in Ω. Assume that its coefficients satisfy condition b), that all the coefficients are bounded, that the coefficients a_{ik} are differentiable with $Da_{ik}(x, t) = O(|x|^{-1})$.

Theorem 2.27. *Let a solution of equation* (43) *with coefficients satisfying the conditions indicated above, be defined in* Ω. *Then there exists a constant* N_0 *that depends on the equation, such that if any solution* $u(x, t)$ *in* Ω *satisfies for* x *large enough the inequality*

$$|D^\alpha u(x, t)| \leqslant e^{-(x_1)^{N_0}}, \quad |\alpha| \leqslant 2, \tag{45}$$

then $u \equiv 0$.

Remark. It follows from the Schauder estimates that if the coefficients satisfy the Dini condition, then we can have the inequality $|u(x, t)| \leqslant e^{-(x_1)^{N_0}}$ instead of (45). The method of elliptic continuation of parabolic equations can be used to obtain interesting theorems concerning the stabilization of solutions of the Cauchy problem. Groza [1976] proved the following theorem.

Theorem 2.28. *Let a solution $u(x, t)$ of equation* (43) *be defined in the half-space $\{(x, t)|x \in \mathbb{R}^n, t > 0\}$, and assume that $c(x) \leqslant 0$. If $E \subset \mathbb{R}^n$ is a Borel set of positive measure, and if $u(x, t)$ is bounded in modulus and converges to zero as $t \to \infty$ at every point $x \in E$, then $u(x, t)$ converges to zero everywhere, moreover, the convergence is uniform on compact sets.*

(A similar theorem in the case of parabolic equations of high order was proved by Nutsubidze [1983]).

Comments on the References

We comment here on the monographs and survey articles referred to in the text.

Monographs

Ladyzhenskaya, Solonnikov and Uraltseva [1967] is one of the most complete monographs dealing with second order parabolic linear and quasilinear equations. At the end of the book some results concerning parabolic systems are to be found as well.

Ladyzhenskaya and Uraltseva [1973] is a basic monograph in the theory of linear and quasilinear elliptic equations.

Landkof [1966] is an exposition of potential theory mainly in connection with its applications to elliptic equations.

Landis [1971]. A substantial part of this monograph is devoted to questions of the qualitative theory of second-order elliptic and parabolic equations.

Miranda [1970]. This monograph deals with various questions in the theory of elliptic equations. In its breadth of scope it has no equals. It is invaluable as a reference book.

Sobolev [1962]. This is a book dealing with embedding theorems and their applications to the theory of partial differential equations, written by the originator of the method.

Hörmander [1983–85] is a four volume monograph that faithfully reflects the contemporary state of the art in questions dealing with linear differential operators. § 17, Volume 3 of this monograph comes closest to the topics covered in our paper. There the author proves what we called "Hörmander's inequality."

Ejdel'man [1964] is a book mainly devoted to systems that are parabolic in the sense of Petrovskii. Results concerning a single second order parabolic equation can be found in the book, appearing as particular cases of results for systems.

Gilbarg and Trudinger [1977] is devoted to second order elliptic equations. The book is in two parts, dealing respectively, with linear and quasilinear equations. The book has the character of a textbook and is very clearly written.

Grisvard [1981] deals with solutions of boundary value problems in non-smooth domains.

Morrey [1943], as seen from the title, does not deal directly with qualitative theory. We used some embedding theorems contained in it.

Survey Articles

Gel'fand and Shilov [1953] contains results concerning uniqueness classes of the Cauchy problem with constant coefficients.

Denisov and Repnikov [1984] is a survey of questions of stabilization of solutions of parabolic equations as $t \to \infty$.

Zhikov, Kozlov, and Olejnik [1981] deal with questions of G-convergence, that is, with the question of the possibility of the description of heterogeneous media by equations describing homogeneous media.

Iosefian and Olejnik [1977] consider the character of growth and decay of solutions of parabolic equations in unbounded domains of different types.

Keldysh [1941] has not, in spite of the fact that it is an old paper, lost its importance. This is especially true with regard to the second part of the article, which deals with stability of solutions of the Dirichlet problem under perturbation of the boundary.

Kondrat'ev and Olejnik [1983] survey results of the theory of boundary value problems for partial differential equations in domains with irregular points on the boundary.

Landis [1963] is partially contained in (Landis [1971]). At the same time, many of the results contained in this paper (such as the three balls theorem, the connection between the number of domains in which a solution of an elliptic equation does not change sign and its growth) are not to be found in that monograph.

Olejnik and Radkevich [1978] survey results obtained with the help of the method of introducing a parameter proposed by the authors. It deals with widely varying aspects of the theory of elliptic and parabolic equations.

Porper and Ejdel'man [1984] is devoted to estimates of fundamental solutions and their applications. In particular, it contains a lucid exposition of Aronson's estimates.

Stampacchia [1985] can be considered as a survey by the author of his results on generalised solutions of non-selfadjoint second-order elliptic equations with principal part in divergence form.

Watson [1978] is a survey of works dealing with thermal capacity.

References*

Adel'son-Velskij, G.M. [1945] A generalization of a geometric theorem of S.N. Bernstein. Dokl. Akad. Nauk SSSR *19*, 391–392. Zbl.61,373

Aleksandrov, A.D. [1966] Upper bounds for solutions of second order linear equations. Vestn. Leningr. Univ. Ser. Mat. Mech. Astrom. *21*, No. 1, 5–25. English transl.: Am. Math. Soc., Transl., II. Ser. 68, 120–143 (1968). Zbl.146,347

Aronson, D.G. [1967] Bounds for the fundamental solution of a parabolic equation. Bull. Am. Math. Soc. *73*, 890–896. Zbl.153,420

Bauman, P.A. [1985] A Wiener test for nondivergence structure second order elliptic equations. Indiana Univ. Math. J. *34*, 825–843. Zbl.583.35034

Borai, M.M. [1968] On the correctness of the Cauchy problem. Vestn. Mosk. Univ., Ser. I 23, No. 4, 15–21. Zbl. 157,167

Caffarelli, L.A., Fabes, E.B., Kenig, C.E. [1981] Completely singular elliptic-harmonic measures. Indiana Univ. Math. J. *30*, 917–924. Zbl.482.35020

* For the convenience of the reader, references to reviews in Zentralblatt für Mathematik (Zbl.), compiled using the MATH database, and Jahrbuch über die Fortschritte der Mathematik (Jrb.) have been included as far as possible.

Cheremnykh, Yu. N. [1959] On the asymptotics of solutions of parabolic equations. Izv. Akad. Nauk SSSR, Ser. Mat. *23*, 913–924. Zbl.87,300

Chicco, M. [1972] An a priori inequality concerning elliptic second order partial differential equations of variational type. Matematiche *26*, 173–182. Zbl.234.35023

Cordes, H.O. [1956a] Die erste Randwertaufgabe bei Differentialgleichungen zweiter Ordnung in mehr als zwei Variabeln. Math. Ann. *131* 278–313. Zbl.70,96

Cordes, H.O. [1956b] Über die eindeutige Bestimmtheit der Lösungen elliptischer Differentialgleichungen durch Anfangsvorgaben. Nachr. Akad. Wiss. Göttingen *11*, 239–258. Zbl.74,80

De Giorgi, E. [1957] Sulla differenziabilità e l'analiticità delli estremali degli integrali multipli regolari. Mem. Acad. Sci. Torino, Ser. III 3, Pt. 1, 25–43. Zbl.84,319

Denisov, V.N., Repnikov, V.D. [1984] On the stabilization of solutions of the Cauchy problem for parabolic equations. Differ. Uravn. *20*, No. 1, 20–41. English transl.: Differ. Equations 20, 16–33 (1984). Zbl.589.35054

Drozhzhinov, Yu.N. [1962] On the stabilization of solutions of the Cauchy problem for parabolic equations. Dokl. Akad. Nauk SSSR *142*, 17–20. English transl.: Sov. Math., Dokl. 3, 8–12 (1962). Zbl.123,69

Ejdel'man, S.D. [1964] Parabolic Systems. Nauka: Moscow. English transl.: North-Holland Co.: Groningen (1969). Zbl.121,319

Ejdel'man, S.D., Matijchuk, M.I. [1970] The Cauchy problem for parabolic systems, coefficients of which have a low order of smoothness. Ukr. Mat. Zh. *22*, 22–36. English transl.: Ukr. Math. J. 22, 18–30 (1970). Zbl.208,366

Evans, L.C., Gariepi, R.F. [1982] Wiener's criterion for the heat equation. Arch. Ration. Mech. Anal. *78*, 293–314. Zbl.508.35038

Fufaev, V.V. [1963] On conformal mappings of domains with corners. Dokl. Akad. Nauk SSSR *152*, 838–840. English transl.: Sov. Math., Dokl. 4, 1457–1459 (1964). Zbl.173,85

Gagnidze, A.G. [1984] On uniqueness classes of solutions of boundary value problems for second-order parabolic equations in unbounded domains. Usp. Mat. Nauk *39*, No. 6, 193–194. English transl.: Russ. Math. Surv. 39, No. 6, 209–210 (1984). Zbl. 596.35057

Gel'fand I.M., Shilov, G.E. [1953] Fourier transforms for functions of rapid growth and questions of uniqueness of the solution for the Cauchy problem. Usp. Mat. Nauk *8*(6), 3–54. Zbl.52,116

Gilbarg, D., Trudinger, N. [1977] Elliptic Partial Differential Equations of Second Order. Springer-Verlag: Heidelberg-Tokyo-New York-London. Zbl.361.35003

Gilbarg, D., Hörmander, L. [1980] Intermediate Schauder estimates. Arch. Ration. Mech. Anal. *74*, 197–218. Zbl.454.35022

Gilbarg, D., Serrin, J. [1954/56] On isolated singularities of second order elliptic differential equations. J. Anal. Math. *4*, 309–340. Zbl.71,97

Grisvard, P. [1981] Boundary Value Problems in Non-smooth Domains. Univ. de Nice: Nice

Groza, G.G. [1976] Stabilization as $t \to \infty$ of the solution of a parabolic equations that decays on a set of positive measure. Usp. Mat. Nauk *31*, No. 3, 209–210. Zbl.329.35007

Gusarov, A.L. [1975] On a sharp Liouville theorem for solutions of a parabolic equation on the characteristic. Mat. Sb., Nov. Ser. *97* (139), 379–394. English transl.: Math. USSR, Sb. 26, 349–364 (1976). Zbl.308.35058

Gusarov, A.L., Landis, E.M. [1982] On a version of the three balls theorem for the solutions of an elliptic equation and on the Phragmen-Lindelöf type theorem connected with it. Tr. Semin. Im. I.G. Petrovskogo *8*, 168–186. English transl.: J. Sov. Math. 32, 349–364 (1986). Zbl.584.35037

Gushchin, A.K. [1982] On uniform stabilization of solutions of the Cauchy problem for parabolic equations. Mat. Sb., Nov. Ser. *119*, No. 4, 451–508. English transl.: Math. USSR, Sb. 47, 439–498 (1984). Zbl.544.35055

Heinz, E. [1955] Über die Eindeutigkeit beim Cauchyschen Anfangwertproblem einer elliptischen Differentialgleichung zweiter Ordnung. Nachr. Akad. Wiss. Göttingen *1*, 1–12. Zbl.67,75

Hörmander, L. [1983–85] The Analysis of Linear Partial Differential Operators. (I–IV) Springer-Verlag. Zbl.521.35001, Zbl.521.35002, Zbl.601.35001, Zbl.612.35001

Hopf, E. [1929] Bemerkungen zu einem Satze von S. Bernstein aus der Theorie der elliptischen Differentialgleichungen. Math. Z. *29*(5), 744–745. Jrb.55,289

Ibragimov, A.I. [1983a] Some qualitative properties of solutions of the mixed problem for equations of elliptic type. Mat. Sb., Nov. Ser. *122*, 166–181. English transl.: Math. USSR, Sb. 50, 163–176 (1985). Zbl.545.35030

Ibragimov, A.I. [1983b] On some qualitative properties of solutions of elliptic equations with continuous coefficients. Mat. Sb., Nov. Ser. *12*, No. 4, 454–468. English transl.: Math. USSR, Sb. 49, 447–460 (1984). Zbl.553.35021

Il'in, A.M. [1985] On a sufficient stabilization condition for solutions of parabolic equations. Mat. Zametki *37*, No. 6, 851–856. English transl.: Math. Notes 37, 466–469 (1985). Zbl.601.35046

Iosifian, G.A., Olejnik, O.A. [1971] An analog of Saint-Venant's principle and uniqueness of solutions of boundary value problems for parabolic equations in unbounded domains. Usp. Mat. Nauk *31*, No. 6, 142–166. English transl.: Russ. Math. Surv. 31, No. 6, 153–178 (1976), Zbl.342.35026

Ivanov, A.V. [1967] Harnack's inequality for parabolic equations. Tr. Mat. Inst. Steklova *102*, 54–84. English transl.: Proc. Steklov Inst. Math. 102, 55–94 (1970). Zbl.199,165

Kamynin, L.I., Khimchenko, B.N. [1972] On the maximum principle for elliptic-parabolic second order equations. Sib. Mat. Zh. *13*, 773–789. English transl.: Sib. Math. J. 13, 533–545 (1973). Zbl.245.35060

Kamynin, L.I., Khimchenko, B.N. [1981] On the Tikhonov-Petrovskii problem for second order parabolic equations. Sib. Mat. Zh. *22*, No. 5, 78–109. English transl.: Sib. Math. J. 22, 709–734 (1982). Zbl.501.35040

Kato, T. [1959] Growth properties of solutions of the reduced wave equation with variable coefficients. Commun. Pure Appl. Math. *12*, 403–425. Zbl.91,95

Keldysh, M.V. [1941] On the solubility and stability of the Dirichlet problem. Usp. Mat. Nauk *VIII*, 171–231

Kerimov, T.M. [1982] On the regularity of a boundary point for Zaremba's problem. Dokl. Akad. Nauk Azerb. SSR *264*, 815–818. English transl.: Sov. Math., Dokl. 25, 742–745 (1982). Zbl.508.35029

Kerimov, T.M., Maz'ya, V.G., Novruzov, A.A. [1981] An analog of Wiener's criterion for Zaremba's problem in a cylindrical domain. Funkts. Anal. Prilozh. *16*, No. 4, 70–71. English transl.: Funct. Anal. Appl. 16, 301–303 (1983). Zbl.512.35017

Kondrat'ev, V.A. [1967] On solubility of the first boundary value problem for strongly elliptic equations. Tr. Mosk. Mat. O-va *16*, 293–318. Zbl.163,348

Kondrat'ev, V.A. [1970] On the smoothness of the solution of the Dirichlet problem for a second order elliptic equation in a neighborhood of a rib. Differ. Uravn. *6*, 1831–1843. Zbl.209,411

Kondrat'ev, V.A., Olejnik, O.A. [1983] Boundary value problems for partial differential equations in non-smooth domains. Usp. Mat. Nauk *38*, No. 2, 3–76. English transl.: Russ Math. Surv. 38, No. 2, 1–86. (1983). Zbl.523.35010

Krylov, N.V. [1967] On the first boundary value problem for elliptic equations. Differ. Uravn. *3*, 315–326. English transl.: Differ. Equations 3, 158–164 (1971). Zbl.147,95

Krylov, N.V., Safonov, M.V. [1979] An estimate for a diffusion process hitting a set of positive measure. Dokl. Akad. Nauk SSSR *245*, 18–20. English transl.: Sov. Math., Dokl. 20, 253–256 (1979). Zbl.459.60067

Krylov, N.V., Safonov, M.V. [1980] A certain property of solutions of parabolic equations with measurable coefficients. Izv. Akad. Nauk SSSR, Ser. Mat. *44*, 161–175. English transl.: Math. USSR, Izv. 16, 151–164 (1981). Zbl.439.35023

Krzyzanski, M.K. [1957] Sur l'allure asymptotique des potentiels de chaleur et de l'integrale de Fourier-Poisson. Ann. Pol. Math. *6*(6), 288–299. Zbl.84,301

Kudryavtseva, I.A. [1971] On solutions of parabolic equations, that decrease in space variables. Mat. Sb., Nov. Ser. *84*, 3–13. English transl.: Math. 3–13. USSR, Sb. 13, 1–11 (1971). Zbl.211,132

Ladyzhenskaya, O.A., Solonnikov, V.A., Uraltseva, N.N. [1967] Linear and Quasilinear Equations of Parabolic Type. Nauka: Moscow. English transl.: Acad. Press., New York/London (1968). Zbl.143,336

Ladyzhenskaya, O.A., Uraltseva, N.N. [1973] Linear and Quasilinear Equations of Elliptic Type. Nauka: Moscow. Zbl.269.35029

Lakhturov, S.S. [1980] On the asymptotic behavior of solutions of the second boundary value problem in unbounded domains. Usp. Mat. Nauk *35*, No. 4, 195–196. English transl.: Russ. Math. Surv. 35, No. 4, 175–176 (1980). Zbl.439.35007

Lanconelli, E. [1973] Sul problema di Dirichlet per l'equazione de calore. Ann. Mat. Pura Appl., IV. Ser. *97*, 83–117. Zbl.277.35058

Lanconelli, E. [1975] Sul problema di Dirichlet per equaziono parabolico del secondo ordine a coeffizienti discontini. Ann. Mat. Pura Appl., IV. Ser. *106*, 11–34. Zbl.321.35043

Landis, E.M. [1956a] On the Phragmen-Lindelöf principle for solutions of an elliptic equation. Dokl. Akad. Nauk SSSR *107*, 508–511. Zbl.70,324

Landis, E.M. [1956b] On some properties of solutions of elliptic equations. Dokl. Akad. Nauk SSSR *107*, 640–643. Zbl.75,282

Landis, E.M. [1963] Some questions of qualitative theory of elliptic equations. Usp. Mat. Nauk *18*, No. 1, 3–62. English transl.: Russ. Math. Surv. 18, No. 1, 1–62 (1963). Zbl.125,58

Landis, E.M. [1967] A new proof of De Giorgi's theorem. Tr. Mosk. Mat. O-va *16*, 319–328. English transl.: Trans. Mosc. Math. Soc. 16, 343–353 (1968). Zbl.182.434

Landis, E.M. [1969] Necessary and sufficient boundary point regularity conditions for the Dirichlet problem for the heat equation. Dokl. Akad. Nauk SSSR *185*, 517–520. English transl.: Sov. Math. Dokl. 10, 380–384 (1969). Zbl.187,38

Landis, E.M. [1971] Second Order Equations of Elliptic and Parabolic Types. Nauka: Moscow. Zbl.226.35001

Landis, E.M. [1974] On the behavior of solutions of elliptic equations in unbounded domains. Tr. Mosk. Mat. O-va *31*, 35–58. English transl.: Trans. Mosc. Math. Soc. 31, 30–54 (1976). Zbl.311.35008

Landis, E.M., Bagotskaya, N.V. [1983] On the structure of inessential sets relative to the Dirichlet problem for elliptic equations with discontinuous coefficients. Tr. Mosk. Mat. O-va *46*, 125–135. English transl.: Trans. Mosc. Math. Soc. 1984, No. 2, 127–137 (1984). Zbl.566.35039

Landis, E.M., Nadirashvili, N.S. [1985] Positive solutions of second order elliptic equations in unbounded domains. Mat. Sb., Nov. Ser. *126*, No. 1, 133–139. English transl.: Math. USSR, Sb. 54, 129–134 (1986). Zbl.583.35035

Landis, E.M., Olejnik, O.A. [1974] Generalized analyticity and some properties of solutions of elliptic and parabolic equations connected with it. Usp. Mat. Nauk *29*, No. 2, 190–206. English transl.: Russ. Math. Surv. 29, No. 2, 195–212 (1974). Zbl.293.35011

Landkof, N.S. [1966] Foundations of Modern Potential Theory. Nauka: Moscow. English transl.: Springer, Grundlehren der math. Wissenschaften 180 (1972). Zbl.148,103

Lax, P.D. [1957] A Phragmen-Lindelöf theorem in harmonic analysis and its application to some questions in the theory of elliptic equations. Commun. Pure Appl. Math. *10*(3), 361–389. Zbl.77,315

Littman, W., Stampacchia, G., Weinberger, H.F. [1963] Regular points for elliptic equations with discontinuous coefficients. Ann. Sc. Norm. Super. Pisa, Cl. Sci., III. Ser. *17*, 43–77. Zbl.116,303

Maz'ya, V.G. [1963] On the regularity on the boundary of solutions of elliptic equations, and of a conformal mapping. Dokl. Akad. Nauk SSSR *152*, 1297–1300. English transl.: Sov. Math. Dokl. 4, 1547–1551 (1964). Zbl.166,379

Maz'ya, V.G. [1965] Polyharmonic capacity in the theory of the first boundary value problem. Sib. Mat. Zh. *6*, 127–148. Zbl.149,325

Maz'ya V.G. [1966] On the continuity modulus of solutions of the Dirichlet problem close to an irregular boundary. In Probl. Mat. Analyza, 45–58. English transl.: Probl. Math. Analysis 1, 41–54 (1968). Zbl.166,380

Maz'ya, V.G. [1969] On weak solutions of the Dirichlet and the Neumann problems. Tr. Mosk. Mat. O-va *20*, 137–172. Zbl.179.433

Maz'ya, V.G. Plamenevskii, B.A. [1978] W_p^l and Hölder class estimates, and the Miranda-Agmon maximum principle for solutions of elliptic boundary value problems. Math. Nachr. *81*, 25–82. English transl.: Transl., II. Ser., Am. Math. Soc. 123, 1–56 (1984). Zbl.371.35018

Meshkov, V.Z. [1986] On the decay rate of solutions of one type of equations. Differ. Uravn. *22*, No. 11, 2005–2007. Zbl.662.35002

Mikhailov, V.P. [1970] On the stabilization of solutions of the Cauchy problem for the heat equation. Dokl. Akad. Nauk SSSR *190*, 38–41. English transl.: Sov. Math. Dokl. 11, 34–37 (1970). Zbl.197,75

Miller, K. [1966] Nonexistence of an a priori bound at the center in terms of an L_1 bound on the boundary for solutions of uniformly elliptic equations on a sphere. Ann. Mat. Pura Appl., IV. Ser. *73*, 11–16. Zbl.144,145

Miller, K. [1974] Nonunique continuation for elliptic equations in self-adjoint divergence form with Hölder continuous coefficients. Arch. Rat. Mech. Anal. *54*, 105–117. Zbl.289.35046

Miranda, C. [1970] Partial Differential Equations of Elliptic Type (2nd ed.) Springer, New York. Zbl.198,141

Morrey, C.B. [1943] Multiple integral problems in the calculus of variations and related topics. Univ. California Publ. Math. *1*, 1–130

Moser, J. [1960] A new proof of de Giorgi's theorem concerning the regularity problem for elliptic equations. Commun. Pure Appl. Math. *13*, 457–468. Zbl.111,93

Moser, J. [1961] On Harnack theorem for elliptic differential equations. Commun. Pure Appl. Math. *14*, 577–591. Zbl.111,93

Moser, J. [1964] A Harnack inequality for parabolic differential equations. Commun. Pure Appl. Math. *17*, 101–134. Zbl.149,69

Nadirashvili, N.S. [1976] On a generalization of the Hadamard three circles theorem. Vestn. Mosk. Univ., Ser. I 31, No. *3*, 39–42. English transl.: Mosc. Univ. Math. Bull. 31, No. 314, 30–32 (1976). Zbl.328.31002

Nadirashvili, N.S. [1979] On an estimate for a solution, which is bounded on a set, of an elliptic equation with analytic coefficients. Vestn. Mosk. Univ., Ser. I, 1979 No. *2*, 42–46. English transl.: Mosc. Univ. Math. Bull. 34, No. 2, 44–48 (1979). Zbl.399.35005

Nadirashvili, N.S. [1983] On the question of uniqueness of the solution of the second boundary value for second order elliptic equations. Mat. Sb., Nov. Ser. *122*, No. 3, 341–358. English transl.: Math. USSR, Sb. 50, 325–341 (1985). Zbl.563.35026

Nadirashvili, N.S. [1985] On a general Liouville type theorem on a Riemannian mainfold. Usp. Mat. Nauk *40*, No. 5, 259–260. English transl.: Russ. Math. Surv. 40, No. 5, 235–236 (1985). Zbl.602.31007

Nadirashvili, N.S. [1986] On uniqueness and stability of the continuation from a set to the entire domain of solutions of second order elliptic equations. Mat. Zametki *40*, No. 2, 218–225. English transl.: Math. Notes 40, 623–627 (1986). Zbl.657.35045

Nash, J. [1958] Continuity of solutions of parabolic and elliptic equations. Amer. J. Math. *80*, 931–954. Zbl.96,69

Nikol'skij, S.M. [1956] The Dirichlet problem in domains with corners. Dokl. Akad. Nauk SSSR *109*, 33–35. Zbl.74,92

Nirenberg L. [1953] A strong maximum principle for parabolic equations. Commun. Pure Appl. Math. *6*(2), 167–177. Zbl.50,96

Novruzov, A.A. [1973] On some regularity criteria for boundary points of linear and quasilinear parabolic equations. Dokl. Akad. Nauk SSSR *209*, 785–787. English transl.: Sov. Math. Dokl. 14, 521–524 (1973). Zbl.286.35043

Novruzov, A.A. [1979] On the Dirichlet problem for second order elliptic equations. Dokl. Akad. Nauk SSSR *246*, 11–14. English transl.: Sov. Math., Dokl. 20, 437–440 (1979). Zbl.422.35040

Novruzov, A.A. [1983a] On an approach to the study of qualitative properties of second order elliptic equations not in divergence form, Mat. Sb., Nov. Ser. *122*, No. 3, 360–387. English transl.: Math. USSR, Sb. 50, 343–367 (1985). Zbl.541.35022

Novruzov, A.A. [1983b] $(s, \varphi^{\pm})$-capacities and their applications to second order elliptic equations. In: General Theory of Boundary Value Problems, Collect. Sci. Works, Naukova Dumka: Kiev 1983, 160–168. Zbl.603.35024

Nutsubidze, D.V. [1983] On the behavior of solutions of higher order parabolic equations in a neighborhood of a boundary point and at infinity. Usp. Mat. Nauk *38*, No. 6, 119–120. English transl.: Russ. Math. Surv. 38, No. 6, 127–128 (1983). Zbl.538.35039

Olejnik, O.A. [1949] On the Dirichlet problem for equations of elliptic type. Mat. Sb., Nov. Ser. *24*, 3–14. Zbl.35,187

Olejnik, O.A. [1983] On examples of non-uniqueness of solution of a boundary value problem for a parabolic equation in an unbounded domain. Usp. Mat. Nauk *38*, No. 1, 183–184. English transl.: Russ. Math. Surv. 38, No. 1, 209–210 (1983). Zbl.554.35054

Olejnik, O.A., Iosif'yan, G.A. [1977] On removable singularities at the boundary, and on the uniqueness of solutions of boundary value problems for second order elliptic and parabolic equations. Funkts. Anal. Prilozh. *11*, No. 3, 54–67. English transl.: Funct. Anal. Appl. 11, 206–217 (1978). Zbl.377.35010

Olejnik, O.A., Iosif'yan, G.A. [1980] On the behavior at infinity of solutions of the Neumann problem for second order elliptic equations in unbounded domains. Usp. Mat. Nauk *35*, No. 4, 197–198. English transl.: Russ. Math. Surv. 35, No. 4, 178–179 (1980). Zbl.439.35008.

Olejnik, O.A., Radkevich, E.B. [1978] The method of introducing a parameter in the study of evolution equations. Usp. Mat. Nauk *33*, No. 5, 7–76. English transl.: Russ. Math. Surv. 33, 7–84 (1978). Zbl.397.35033

Petrovskij, I.G. [1946] On some problems in the theory of partial differential equations. Usp. Mat. Nauk V. Ser. *1*(3–4), 44–70. Zbl.61,204

Petrowsky, I.G. (-Petrovskij, I.G.) [1935] Zur ersten Randwertaufgabe der Wärmeleitungsgleichung. Compos. Math. *1*(3), 383–419. Zbl.10,299

Pini, B. [1954] Sulla soluzione generalizzate di Wiener per il primo problema de valore al contoro nel caso parabolico. Rend. Semin. Mat. Univ. Padova *23*, 422–434. Zbl.57,328

Plís, A. [1963] On non-uniqueness in Cauchy problem for elliptic second order equation. Bull. Acad. Pol. Sci. Ser. Math. astron. phys. *11*, 95–100. Zbl.107,79

Porper, F.O., Ejdel'man, S.D. [1978] Theorems concerning asymptotic convergence and stabilization of solutions of multidimensional second order parabolic equations. In: Methods of Functional Analysis for Problems of Mathematical Physics, Kiev, 81–114. Zbl.414.35034

Porper, F.O., Ejdel'man, S.D. [1984] Two-sided estimates of fundamental solutions of second order parabolic equations and some of their applications. Usp. Mat. Nauk *39*, No. 3, 107–156. English transl.: Russ. Math. Surv. 39, No. 3, 119–178 (1984). Zbl.582.35052

Repnikov, V.D. [1963] Some stabilization theorems for solutions of the Cauchy problem for parabolic equations. Dokl. Akad. Nauk SSSR *148*, 527–530. English transl.: Sov. Math., Dokl. 4, 137–140 (1963). Zbl.178,456

Repnikov, V.D., Ejdel'man, S.D. [1967] A new proof of the stabilization theorem for the solution of the Cauchy problem for the heat equation. Mat. Sb. Nov. Ser. *73*, 155–159. English transl.: Math. USSR, Sb. 2, 135–139 (1967). Zbl.152,105

Safonov, M.V. [1980] Harnack's inequality for elliptic equations and the Hölder property of their solutions. Zap. Nauchn Semin. Leningr. Odt. Mat. Inst. Steklova *96*, 272–287. English transl: J. Sov. Math. 21, 851–863 (1983). Zbl.458.35028

Safonov, M.V. [1987] Optimality of the estimates of the Hölder exponents for solutions of linear elliptic equations with measurable coefficients. Mat. Sb. Nov. Ser. *132*, No. 2, 275–288. English transl.: Math. USSR, Sb. 60, No. 1, 269–281 (1988). Zbl.656.35027

Shifrin, M.A. [1972] On the possible decay rate of the solution of an elliptic equation. Mat. Sb., Nov. Ser. *89*, 616–629. English transl.: Math. USSR, Sb. 18, 621–634 (1974). Zbl.288.35002

Sobolev, S.L. [1950] Some Applications of Functional Analysis in Mathematical Physics. Sib. Otd. Akad. Nauk SSSR: Novosibirsk. English transl.: Ann. Math. Soc. (1963). Zbl.123.90

Stampacchia, G. [1958] Contributi alla regolarizzazione delli soluzioni del problemi al contoro per equazioni del secondo ordine ellitici. Ann. Sc. Norm. Super. Pisa, III. Ser. *12*, 223–245. Zbl.82,97

Stampacchia, G. [1965] Le probléme de Dirichlet pour les équations elliptiques du second ordre á coefficients discontinus. Ann. Inst. Fourier *15*, 189–258. Zbl.151,154

Tikhonov, A.N. [1935] Uniqueness theorems for the heat equation. Mat. Sb., Nov. Ser. *42*(2), 199–215. Zbl.12,355

Tikhonov, A.N. [1938] On the heat equation in several variables. Bull. Univ. Etat. Mosk., Ser. Int., Sect. A. Mat. Mech 1, No. 9, 1–44. Zbl.24,112

Tikhonov, A.N. [1950] On boundary value problems with derivatives of order higher than the order of the equation. Mat. Sb., Nov. Ser. *26*, 35–51. Zbl.41,66

Valitskij, Yu.N., Ejdel'man S.D. [1976] Necessary and sufficient conditions for the stabilization of positive solutions of the heat equation. Sib. Mat. Zh. *17*, 744–756. English transl.: Sib. Math. J. 17, 564–572 (1977). Zbl.362.35036

Verzhbinskij, G.M., Maz'ya, V.G. [1971, 1972] Asymptotic behaviour of second order elliptic equations near the boundary. I-Sib. Mat. Zh. *12*, 1217–1249, II idem, *13*, 1239–1271. English transl.: Sib. Math. J. 12, 874–899 (1972) and idem. 13, 858–885 (1973). Zbl.243.35025 and Zbl.308.35009.

Watson, N.A. [1978] Thermal capacity. Proc. Lond. Math. Soc., III. Ser. *37*, 342–362. Zbl.395.35034

Weinberger, H.F. [1962] Symmetrization in uniformly elliptic problems. In: Collection of Papers Dedicated to G. Pólya. pp. 424–428. Stanford. Zbl.123,72

Wiener, N. [1932] Tauberian theorems. Ann. Math. II. Ser. *33*, 1–100. Zbl.4,59

Zaremba, S. [1910] Sur un probléme mixte relatif á l'equation de Laplace. Bull. Acad. Sci. Math. Natur. *A*, 313–314

Zhikov, V.V. [1977] On the stabilization of solutions of parabolic equations. Mat. Sb., Nov. Ser. *104*, 597–616. English transl.: Math. USSR, Sb. 33, 519–537 (1977). Zbl.374.35025

Zhikov, V.V., Kozlov, S.M., Olejnik, O.A. [1981] On *G*-convergence of parabolic operators. Usp. Mat. Nauk *36*, No. 1, 11–59. English transl.: Russ. Math. Surv. 36, No. 1, 9–60 (1981). Zbl.467.35056.

Author Index

Subject Index

Encyclopaedia of Mathematical Sciences
Editor-in-chief: R. V. Gamkrelidze

Dynamical Systems

Volume 1: **D. V. Anosov, V. I. Arnol'd** (Eds.)
Dynamical Systems I
Ordinary Differential Equations and Smooth Dynamical Systems
1988. IX, 233 pp. 25 figs. ISBN 3-540-17000-6

Volume 2: **Ya. G. Sinai** (Ed.)
Dynamical Systems II
Ergodic Theory with Applications to Dynamical Systems and Statistical Mechanics
1989. IX, 281 pp. 25 figs. ISBN 3-540-17001-4

Volume 3: **V. I. Arnol'd** (Ed.)
Dynamical Systems III
1988. XIV, 291 pp. 81 figs.
ISBN 3-540-17002-2

Volume 4: **V. I. Arnol'd, S. P. Novikov** (Eds.)
Dynamical Systems IV
Symplectic Geometry and its Applications
1989. VII, 283 pp. 62 figs.
ISBN 3-540-17003-0

Volume 5: **V. I. Arnol'd** (Ed.)
Dynamical Systems V
Bifurcation Theory and Catastrophe Theory
1991. Approx. 280 pp. 130 figs.
ISBN 3-540-18173-3

Volume 6: **V. I. Arnol'd** (Ed.)
Dynamical Systems VI
Singularity Theory I
1992. Approx. 250 pp. ISBN 3-540-50583-0

Volume 16: **V. I. Arnol'd, S. P. Novikov** (Eds.)
Dynamical Systems VII
Nonholonomic Dynamical Systems. Integrable Hamiltonian Systems
1992. Approx. 290 pp. ISBN 3-540-18176-8

Partial Differential Equations

Volume 30: **Yu. V. Egorov, M. A. Shubin** (Eds.)
Partial Differential Equations I
Foundations of the Classical Theory
1991. Approx. 260 pp. 4 figs. ISBN 3-540-52002-3

Volume 31: **Yu. V. Egorov, M. A. Shubin** (Eds.)
Partial Differential Equations II
1992. Approx. 260 pp. ISBN 3-540-52001-5

Commutative Harmonic Analysis

Volume 15: **V. P. Khavin, N. K. Nikol'skij** (Eds.)
Commutative Harmonic Analysis I
General Survey. Classical Aspects
1991. IX, 268 pp. 1 fig. ISBN 3-540-18180-6

Volume 25: **N. K. Nikol'skij** (Ed.)
Commutative Harmonic Analysis II
Group-Theoretic Methods in Commutative Harmonic Analysis
1992. Approx. 300 pp. ISBN 3-540-51998-X

Volume 42: **V. P. Khavin, N. K. Nikol'skij** (Eds.)
Commutative Harmonic Analysis IV
Harmonic Analysis in $\mathbb{R}^n$
1991. Approx. 225 pp. ISBN 3-540-53379-6

Springer-Verlag
Berlin
Heidelberg
New York
London
Paris
Tokyo
Hong Kong
Barcelona
Budapest

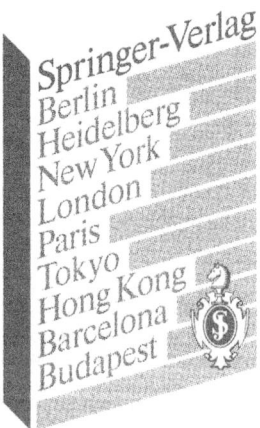

The manufacturer's authorised representative in the EU is Springer
Nature Customer Service Centre GmbH, Europaplatz 3, 69115 Heidelberg,
Germany. If you have any concerns regarding our products, please
contact ProductSafety@springernature.com

Printed and bound by CPI Group (UK) Ltd, Croydon, CR0 4YY
28/04/2026
02098503-0001